WORKSHEETS
FOR CLASSROOM OR LAB PRACTICE

with contributions from

BEVERLY FUSFIELD

STEVE OUELLETTE

JAMES J. BALL
Indiana State University

INTRODUCTORY & INTERMEDIATE ALGEBRA
FOURTH EDITION

Margaret L. Lial
American River College

John Hornsby
University of New Orleans

Terry McGinnis

Addison-Wesley
is an imprint of

Reproduced by Pearson Addison-Wesley from electronic files supplied by the author.

Publishing as Pearson Addison-Wesley, 75 Arlington Street, Boston, MA 02116.

ISBN-13: 978-0-321-57618-7
ISBN-10: 0-321-57618-7

1 2 3 4 5 6 CRS 12 11 10 09 08

Addison-Wesley
is an imprint of

www.pearsonhighered.com

CONTENTS

CHAPTER R PREALGEBRA REVIEW 1

CHAPTER 1 THE REAL NUMBER SYSTEM 11

CHAPTER 2 EQUATIONS, INEQUALITIES, AND APPLICATIONS 43

CHAPTER 3 GRAPHS OF LINEAR EQUATIONS AND INEQUALITIES; FUNCTIONS 79

CHAPTER 4 SYSTEMS OF LINEAR EQUATIONS AND INEQUALITIES 135

CHAPTER 5 EXPONENTS AND POLYNOMIALS 177

CHAPTER 6 FACTORING AND APPLICATIONS 217

CHAPTER 7 RATIONAL EXPRESSIONS AND FUNCTIONS 261

CHAPTER 8 EQUATIONS, INEQUALITIES AND SYSTEMS REVISITED 309

CHAPTER 9 ROOTS, RADICALS, AND ROOT FUNCTIONS 355

CHAPTER 10 QUADRATIC EQUATIONS, INEQUALITIES, AND FUNCTIONS 399

CHAPTER 11 INVERSE, EXPONENTIAL, AND LOGARITHMIC FUNCTIONS 461

CHAPTER 12 NONLINEAR FUNCTIONS, CONIC SECTIONS, AND NONLINEAR SYSTEMS 497

ANSWERS 543

Name: Date:
Instructor: Section:

Chapter R PREALGEBRA REVIEW

R.1 Fractions

Learning Objectives

1. Identify prime numbers.
2. Write numbers in prime factored form.
3. Write fractions in lowest terms.
4. Convert between improper fractions and mixed numbers.
5. Multiply and divide fractions.
6. Add and subtract fractions.
7. Solve applied problems that involve fractions.

Key Terms

Use the vocabulary terms listed below to complete each statement in exercises 1–9.

numerator	**denominator**	**proper fraction**
improper fraction	**equivalent fractions**	**lowest terms**
prime number	**composite number**	**prime factorization**

1. Two fractions are ______________________ when they represent the same portion of a whole.

2. A fraction whose numerator is larger than its denominator is called an ______________.

3. In the fraction $\frac{2}{9}$, the 2 is the ______________________.

4. A fraction whose denominator is larger than its numerator is called a ______________.

5. The ______________________ of a fraction shows the number of equal parts in a whole.

6. A ______________________ has at least one factor other than itself and 1.

7. In a ______________________ every factor is a prime number.

8. The factors of a ______________________ are itself and 1.

9. A fraction is written in ______________________ when its numerator and denominator have no common factor other than 1.

Name: Date:
Instructor: Section:

Objective 1 Identify prime numbers.

Tell whether each number is prime, composite, *or* neither.

1. 29 **1.** ________________

2. 35 **2.** ________________

3. 1 **3.** ________________

4. 2 **4.** ________________

Objective 2 Write numbers in prime factored form.

Write each number in prime factored form.

5. 98 **5.** ________________

6. 24 **6.** ________________

7. 256 **7.** ________________

8. 546 **8.** ________________

Objective 3 Write fractions in lowest terms.

Write each fraction in lowest terms.

9. $\frac{42}{150}$ **9.** ________________

10. $\frac{180}{216}$ **10.** ________________

Name: Date:
Instructor: Section:

11. $\frac{42}{12}$ **11.** ________________

12. $\frac{292}{132}$ **12.** ________________

Objective 4 Convert between improper fractions and mixed numbers.

Write each improper fraction as a mixed number.

13. $\frac{43}{9}$ **13.** ________________

14. $\frac{321}{15}$ **14.** ________________

Write each mixed number as an improper fraction.

15. $13\frac{5}{9}$ **15.** ________________

16. $22\frac{2}{11}$ **16.** ________________

Objective 5 Multiply and divide fractions.

Find each product or quotient, and write it in lowest terms.

17. $\frac{25}{11} \cdot \frac{33}{10}$ **17.** ________________

18. $\frac{5}{4} \div \frac{25}{28}$ **18.** ________________

Name: Date:
Instructor: Section:

19. $4\frac{3}{8}\cdot 2\frac{4}{7}$ **19.** ______________

20. $\frac{4}{11}\div 32$ **20.** ______________

Objective 6 Add and subtract fractions.

Find each sum or difference, and write it in lowest terms.

21. $\frac{23}{45}+\frac{47}{75}$ **21.** ______________

22. $\frac{17}{18}-\frac{1}{4}$ **22.** ______________

23. $2\frac{3}{4}+7\frac{2}{3}$ **23.** ______________

24. $12\frac{5}{6}-7\frac{7}{8}$ **24.** ______________

Objective 7 Solve applied problems that involve fractions.

Solve each applied problem. Write each answer in lowest terms.

25. Arnette worked $24\frac{1}{2}$ hours and earned \$9 per hour. How much did she earn? **25.** ______________

Name: Date:
Instructor: Section:

26. A dental office plays taped music constantly. Each tape takes $1\frac{1}{4}$ hours. How many tapes are played during $7\frac{1}{2}$ hours?

26. ______________

27. Debbie bought 15 yards of material at a sale. She made a shirt with $3\frac{1}{8}$ yards of the material, a dress with $4\frac{7}{8}$ yards, and a jacket with $3\frac{3}{4}$ yards. How many yards of material were left over?

27. ______________

28. Three sides of a parking lot are $35\frac{1}{4}$ yards, $42\frac{7}{8}$ yards, and $32\frac{3}{4}$ yards. If the total distance around the lot is $145\frac{1}{2}$ yards, find the length of the fourth side.

28. ______________

29. The formula for the area of a triangle is $A = \frac{1}{2}bh$, where b is the length of the base of the triangle, and h is the height of the triangle. Find the area of the triangle.

$5\frac{1}{2}$ in.

$3\frac{1}{4}$ in.

$4\frac{1}{2}$ in.

29. ______________

30. Thomas worked $25\frac{1}{2}$ hours over the last four days. If he worked the same amount each day, how long was he at work each day?

30. ______________

Name: Date:
Instructor: Section:

Chapter R PREALGEBRA REVIEW

R.2 Decimals and Percents

Learning Objectives

1	Write decimals as fractions.
2	Add and subtract decimals.
3	Multiply and divide decimals.
4	Write fractions as decimals.
5	Convert percents to decimals and decimals to percents.
6	Use fraction, decimal, and percent equivalents.

Key Terms

Use the vocabulary terms listed below to complete each statement in exercises 1–3.

decimals **place value** **percent**

1. We use ______________________ to show parts of a whole.

2. A ____________________________ is assigned to each place to the left or right of the decimal point.

3. ____________________ means per one hundred.

Objective 1 Write decimals as fractions.

Write each decimal as a fraction or mixed number. Do not write in lowest terms.

1. .42 **1.** ________________

2. .007 **2.** ________________

3. 2.0054 **3.** ________________

4. 18.03 **4.** ________________

5. 30.0005 **5.** ________________

Objective 2 Add and subtract decimals.

Add or subtract as indicated.

6. 45.83 + 20.923 + 5.7 **6.** ________________

Name: Date:
Instructor: Section:

7. $7.26 + 3.7 + 4.09$ 7. ______

8. $768.5 - 13.402$ 8. ______

9. $689 - 79.832$ 9. ______

10. $405.2 - 314.48 + 29.029$ 10. ______

Objective 3 Multiply and divide decimals.

Multiply or divide as indicated.

11. $14.64 \times .16$ 11. ______

12. $498.624 \div 21.2$ 12. ______

13. $516.096 \div 12.8$ 13. ______

14. 42.789×100 14. ______

15. $429.2 \div 1000$ 15. ______

Name: Date:
Instructor: Section:

Objective 4 Write fractions as decimals.

Write each fraction as a decimal. For repeating decimals, write the answer two ways: using the bar notation and rounding to the nearest thousandth.

16. $\frac{3}{7}$ 16. ______________

17. $\frac{9}{16}$ 17. ______________

18. $\frac{4}{9}$ 18. ______________

19. $\frac{151}{200}$ 19. ______________

20. $\frac{41}{40}$ 20. ______________

Objective 5 Convert percents to decimals and decimals to percents.

Convert each percent to a decimal and each decimal to a percent.

21. 27.5% 21. ______________

22. 362% 22. ______________

23. .4% 23. ______________

24. .0409 24. ______________

25. .0084 25. ______________

Name: Date:
Instructor: Section:

Objective 6 Use fraction, decimal, and percent equivalents.

Solve each problem.

26. A television set sells for $750 plus 8% sales tax. Find the price of the set including sales tax. **26.** ______________

27. Jane eats 1500 calories a day. If she eats 350 calories for breakfast, what percent of her daily calories is her breakfast? **27.** ______________

28. The total in sales at Hill's Market last month was \$87,428. If the profit was $1\frac{1}{2}$% of the sales, how much was the profit? **28.** ______________

29. Geishe's Shoes sells shoes at $33\frac{1}{3}$% off the regular price. Find the price of a pair of shoes normally priced at \$54, after the discount is given. **29.** ______________

30. A house costs $225,000. The Lees paid $45,000 as a down payment. What percent of the cost of the house is their down payment? **30.** ______________

Name: Date:
Instructor: Section:

Chapter 1 THE REAL NUMBER SYSTEM

1.1 Exponents, Order of Operations, and Inequality

Learning Objectives

1. Use exponents.
2. Use the rules for order of operations.
3. Use more than one grouping symbol.
4. Know the meanings of $\neq$, $<$, $>$, $\leq$, and $\geq$.
5. Translate word statements to symbols.
6. Write statements that change the direction of inequality symbols.

Key Terms

Use the vocabulary terms listed below to complete each statement in exercises 1–3.

exponent **base** **exponential expression**

1. A number written with an exponent is an ______________________.

2. The ________________ is the number that is a repeated factor when written with an exponent.

3. An ________________ is a number that indicates how many times a factor is repeated.

Objective 1 Use exponents.

Find the value of each exponential expression.

1. 3^3 **1.** ________________

2. $\left(\frac{1}{2}\right)^6$ **2.** ________________

3. $\left(\frac{2}{3}\right)^4$ **3.** ________________

4. $(.4)^2$ **4.** ________________

5. $(2.4)^3$ **5.** ________________

Name: Date:
Instructor: Section:

Objective 2 Use the rules for order of operations.

Find the value of each expression.

6. $20 \div 5 - 3 \cdot 1$ **6.** ______

7. $3 \cdot 5^2 - 3 \cdot 7 - 9$ **7.** ______

8. $6^2 \div 3^2 - 4 \cdot 3 - 2 \cdot 5$ **8.** ______

9. $\dfrac{3 \cdot 15 + 10^2}{12^2 - 8^2}$ **9.** ______

10. $\dfrac{10(5-3) - 9(6-2)}{2(4-1) - 2^2}$ **10.** ______

Objective 3 Use more than one grouping symbol.

Find the value of each expression.

11. $8[14 - 3(9-4)]$ **11.** ______

12. $19 - 3[8(5-2) + 6]$ **12.** ______

13. $4[5 + 2(8-6)] + 12$ **13.** ______

Name: Date:
Instructor: Section:

14. $3^3\left[(6+5)-2^2\right]$ **14.** ______________

15. $2\left[4+2\left(5^2-3\right)\right]$ **15.** ______________

Objective 4 Know the meanings of ≠, <, >, ≤, and ≥.

Tell whether each statement is true *or* false.

16. $3 \cdot 4 \div 2^2 \neq 3$ **16.** ______________

17. $3.25 > 3.52$ **17.** ______________

18. $2\left[7(4)-3(5)\right] \leq 45$ **18.** ______________

19. $4\frac{1}{2}+2\frac{3}{4}<7$ **19.** ______________

20. $4 \geq \frac{2(3+1)-3(2+1)}{3 \cdot 2-1}$ **20.** ______________

Objective 5 Translate word statements to symbols.

Write each word statement in symbols.

21. Seven equals thirteen minus six. **21.** ______________

Name: Date:
Instructor: Section:

22. Seven is greater than the quotient of fifteen and five. **22.** ________________

23. The difference between thirty and seven is greater than twenty. **23.** ________________

24. Five times the sum of two and nine is less than one hundred six. **24.** ________________

25. Twenty is greater than or equal to the product of two and seven. **25.** ________________

Objective 6 Write statements that change the direction of inequality symbols.

Write each statement with the inequality symbol reversed.

26. $\frac{2}{7} \le \frac{4}{5}$ **26.** ________________

27. $\frac{3}{4} > \frac{2}{3}$ **27.** ________________

28. $12 \ge 8$ **28.** ________________

29. $.002 > .0002$ **29.** ________________

30. $\frac{3}{8} \le \frac{3}{7}$ **30.** ________________

Name: Date:
Instructor: Section:

Chapter 1 THE REAL NUMBER SYSTEM

1.2 Variables, Expressions, and Equations

Learning Objectives

1. Evaluate algebraic expressions, given values for the variables.
2. Translate phrases from words to algebraic expressions.
3. Identify solutions of equations.
4. Translate sentences to equations.
5. Distinguish between *expressions* and *equations*.

Key Terms

Use the vocabulary terms listed below to complete each statement in exercises 1–4.

variable **algebraic expression** **equation** **solution**

1. An ______________________ is a statement that says two expressions are equal.

2. A ___________________________ is a symbol, usually a letter, used to represent an unknown number.

3. A collection of numbers, variables, operation symbols, and grouping symbols is an____________________________________.

4. Any value of a variable that makes an equation true is a __________________ of the equation.

Objective 1 Evaluate algebraic expressions, given values for the variables.

Find the value of each expression if $x = 2$ *and* $y = 4$.

1. $9x - 3y + 2$ **1.** ________________

2. $.3(8x + 2y)$ **2.** ________________

3. $\frac{3x}{4} - \frac{3y}{2}$ **3.** ________________

Name: Date:
Instructor: Section:

4. $\dfrac{2x+3y}{3x-y+2}$ **4.** ______________

5. $\dfrac{3y^2+2x^2}{5x+y^2}$ **5.** ______________

6. $\dfrac{3x+y^2+2}{2x+3y}$ **6.** ______________

Objective 2 Translate phrases from words to algebraic expressions.

Write each word phrase as an algebraic expression. Use x as the variable.

7. One more than three times a number **7.** ______________

8. 17 less than nine times a number **8.** ______________

9. Ten times a number, added to 21 **9.** ______________

10. The difference between twice a number and 7 **10.** ______________

11. 11 fewer than eight times a number **11.** ______________

12. Half a number subtracted from two-thirds of the number **12.** ______________

Objective 3 Identify solutions of equations.

Decide whether the given number is a solution of the equation.

13. $6b+2(b+3)=14;\ 2$ **13.** ______________

Name: Date:
Instructor: Section:

14. $5+3x^2=19;\ 2$

14. ______________

15. $\dfrac{m+2}{3m-10}=1;\ 8$

15. ______________

16. $\dfrac{x^2-7}{x}=6;\ 2$

16. ______________

17. $3y+5(y-5)=7;\ 4$

17. ______________

18. $\dfrac{4-x}{x+2}=\dfrac{7}{5};\ \dfrac{1}{2}$

18. ______________

Objective 4 Translate sentences to equations.

Write each word sentence as an equation. Use x as the variable.

19. The sum of five times a number and two is 23.

19. ______________

20. Ten divided by a number is two more than the number.

20. ______________

21. The product of six and five more than a number is nineteen.

21. ______________

22. The quotient of twenty-four and a number is the difference between the number and two.

22. ______________

Name: Date:
Instructor: Section:

23. Seven times a number subtracted from 61 is 13 plus the number. **23.** ________________

24. Four times a number is equal to two more than three times the number. **24.** ________________

Objective 5 Distinguish between *expressions* and *equations*.

Identify each as an **expression** *or an* **equation**.

25. $4x+2y+7$ **25.** ________________

26. y^2-4y-3 **26.** ________________

27. $\frac{x+3}{15}=2x$ **27.** ________________

28. $y^2-7y+4=0$ **28.** ________________

29. $\frac{x+4}{5}$ **29.** ________________

30. $8x=2y$ **30.** ________________

Name: Date:
Instructor: Section:

Chapter 1 THE REAL NUMBER SYSTEM

1.3 Real Numbers and the Number Line

Learning Objectives

1. Classify numbers and graph them on number lines.
2. Tell which of two real numbers is less than the other.
3. Find the opposite of a real number.
4. Find the absolute value of a real number.

Key Terms

Use the vocabulary terms listed below to complete each statement in exercises 1–13.

natural numbers	**whole numbers**	**number line**	**opposite**
integers	**negative number**	**positive number**	
rational number	**set-builder notation**		**coordinate**
irrational number	**real numbers**	**absolute value**	

1. The set {0, 1, 2, 3, ...} is called the set of ______________________.

2. The ______________ of a number is the same distance from 0 on the number line as the original number, but located on the other side of 0.

3. The whole numbers together with their opposites and 0 are called ______________.

4. The set { 1, 2, 3, ...} is called the set of ______________________.

5. The ______________________ of a number is its distance from 0 on the number line.

6. A __________________ shows the ordering of the real numbers on an infinite line.

7. A real number that is not a rational number is called a(n) ______________.

8. The number that corresponds to a point on the number line is the ______________ of that point.

9. A number located to the left of 0 on a number line is a ______________.

10. A number located to the right of 0 on a number line is a ______________.

11. Numbers that can be represented by point on the number line are ______________.

Name: Date:
Instructor: Section:

12. ______________________ uses a variable and a description to describe a set.

13. A number that can be written as the quotient of two integers is a ______________________.

Objective 1 Classify numbers and graph them on number lines.

Use a real number to express each number in the following applications.

1. Last year Nina lost 75 pounds. **1.** ______________

2. Mt. Whitney, one of the highest mountains in the United States, has an altitude of 14,495 feet. **2.** ______________

3. The Dead Sea, the saltiest body of water in the world, lies 396 meters below the level of the Mediterranean Sea. **3.** ______________

4. Between 1970 and 1982, the population of Norway increased by 279,867. **4.** ______________

Graph each group of rational numbers on a number line.

5. $-2, 3, -4, 1$ **5.**

$-5\ -4\ -3\ -2\ -1\ 0\ 1\ 2\ 3\ 4\ 5$

6. $\frac{1}{2}, 0, -3, -\frac{5}{2}$ **6.**

$-5\ -4\ -3\ -2\ -1\ 0\ 1\ 2\ 3\ 4\ 5$

7. $-3\frac{1}{2}, -\frac{3}{2}, 0, \frac{7}{2}, 1$ **7.**

$-5\ -4\ -3\ -2\ -1\ 0\ 1\ 2\ 3\ 4\ 5$

8. $-4.5, -2.3, 1.7, 4.2$ **8.**

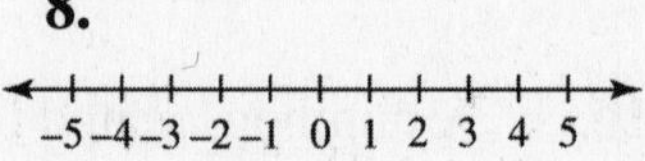

Name: Date:
Instructor: Section:

Objective 2 Tell which of two real numbers is less than the other.

Select the smaller number in each pair.

9. $-5.99, -6.01$ **9.** ______________

10. $\frac{2}{3}, -\frac{1}{2}$ **10.** ______________

11. $-(-4), -4$ **11.** ______________

12. $-\frac{2}{5}, -\frac{1}{4}$ **12.** ______________

Decide whether each statement is **true** *or* **false**.

13. $-76 < 45$ **13.** ______________

14. $-5 > -5$ **14.** ______________

15. $-12 > -10$ **15.** ______________

16. $3 < -4$ **16.** ______________

Objective 3 Find the opposite of a real number.

Find the opposite of each number.

17. -25 **17.** ______________

18. $\frac{3}{8}$ **18.** ______________

19. $-(-22)$ **19.** ______________

20. $2\frac{3}{7}$ **20.** ______________

21. 0 **21.** ______________

22. 4.5 **22.** ______________

Name: Date:
Instructor: Section:

23. $-\frac{5}{7}$ 23. ________________

Objective 4 Find the absolute value of a real number.

Simplify.

24. $-|95|$ 24. ________________

25. $-|49-39|$ 25. ________________

26. $\left|1\frac{1}{2}-2\frac{1}{4}\right|$ 26. ________________

27. $\left|\frac{1}{2}+\frac{1}{3}\right|$ 27. ________________

28. $|-7.52+6.3|$ 28. ________________

29. $|16-14|$ 29. ________________

30. $|0|$ 30. ________________

Name: Date:
Instructor: Section:

Chapter 1 THE REAL NUMBER SYSTEM

1.4 Adding Real Numbers

Learning Objectives

1. Add two numbers with the same sign.
2. Add numbers with different signs.
3. Add mentally.
4. Use the rules for order of operations with real numbers.
5. Translate words and phrases that indicate addition.

Key Terms

Use the vocabulary terms listed below to complete each statement in exercises 1–2.

sum **addends**

1. The answer to an addition problem is called the ____________________.

2. In an addition problem, the numbers being added are the ____________________.

Objective 1 Add two numbers with the same sign.

Find each sum.

1. $-20+(-20)$ 1. ____________

2. $9+12$ 2. ____________

3. $-7+(-11)$ 3. ____________

4. $-9+(-9)$ 4. ____________

5. $\frac{3}{5}+\frac{4}{5}$ 5. ____________

6. $-2\frac{3}{8}+\left(-3\frac{1}{4}\right)$ 6. ____________

Name: Date:
Instructor: Section:

Objective 2 Add numbers with different signs.

Use a number line to find each sum.

7. $7+(-12)$ **7.** ______________

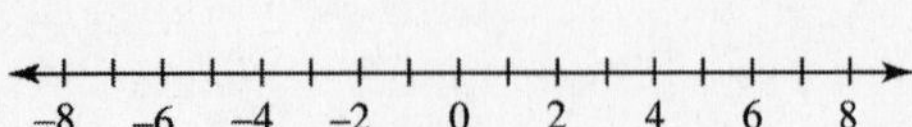

8. $-8+5$ **8.** ______________

–8 –6 –4 –2 0 2 4 6 8

9. $-4+4$ **9.** ______________

–8 –6 –4 –2 0 2 4 6 8

Find each sum.

10. $\frac{7}{12}+\left(-\frac{3}{4}\right)$ **10.** ______________

11. $-\frac{4}{7}+\frac{3}{5}$ **11.** ______________

12. $-10.475+6.325$ **12.** ______________

Objective 3 Add mentally.

Perform each operation and then determine whether the statement is true *or* false. *Try to do all work in your head.*

13. $-14+11=-3$ **13.** ______________

14. $(-14)+15+(-2)=-3$ **14.** ______________

15. $\frac{3}{5}+\left(-\frac{3}{10}\right)=-\frac{3}{10}$ **15.** ______________

Name: Date:
Instructor: Section:

16. $-\frac{3}{8}+\frac{11}{12}=-\frac{7}{24}$ 16. ______________

17. $5\frac{3}{8}+\left(-4\frac{1}{2}\right)=2\frac{1}{8}$ 17. ______________

18. $\left|-5+(-4)\right|=5+4$ 18. ______________

Objective 4 Use the rules for order of operations with real numbers.

Find each sum.

19. $-14+3+\left[8+(-13)\right]$ 19. ______________

20. $-2+\left[4+(-18+13)\right]$ 20. ______________

21. $\left[(-7)+14\right]+\left[(-16)+3\right]$ 21. ______________

22. $-8.9+\left[6.8+(-4.7)\right]$ 22. ______________

23. $\frac{3}{8}+\left[-\frac{2}{3}+\left(-\frac{7}{12}\right)\right]$ 23. ______________

24. $\left[-2\frac{3}{8}+\left(-3\frac{1}{4}\right)\right]+5\frac{3}{4}$ 24. ______________

Name: Date:
Instructor: Section:

Objective 5 Translate words and phrases that indicate addition.

Write a numerical expression for each phrase, and then simplify the expression.

25. The sum of –8 and –4 and –11 **25.** ________________

____

26. The sum of –14 and –29, increased by 27 **26.** ________________

____

27. –10 added to the sum of 20 and –4 **27.** ________________

____

Solve each problem.

28. A football team gained 4 yards from scrimmage on the first play, lost 21 yards on the second play, and gained 9 yards on the third play. How many yards did the team gain or lose altogether? Write the answer as a signed number. **28.** ________________

29. Pablo has $723 in his checking account. He write two checks, one for $358 and the other for $75. Finally, he deposits $205 in the account. How much does he now have in his account? **29.** ________________

30. The temperature at dawn in Blackwood was 24°F. During the day the temperature decreased 30°. Then it increased 11° by sunset. What was the temperature at sunset? **30.** ________________

Name: Date:
Instructor: Section:

Chapter 1 THE REAL NUMBER SYSTEM

1.5 Subtracting Real Numbers

Learning Objectives
1. Find a difference.
2. Use the definition of subtraction.
3. Work subtraction problems that involve brackets.
4. Translate words and phrases that indicate subtraction.

Key Terms

Use the vocabulary terms listed below to complete each statement in exercises 1–3.

minuend **subtrahend** **difference**

1. The number from which another number is being subtracted is called the ____________________.

2. The ____________________ is the number being subtracted.

3. The answer to a subtraction problem is called the ____________________.

Objective 1 Find a difference.

Use a number line to find the difference.

1. $8-5$ 1. ____________________

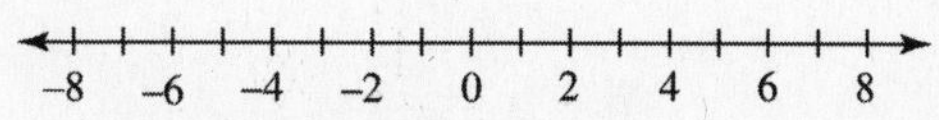

2. $7-10$ 2. ____________________

–8 –6 –4 –2 0 2 4 6 8

3. $4-4$ 3. ____________________

–8 –6 –4 –2 0 2 4 6 8

4. $3-9$ **4.** ________________

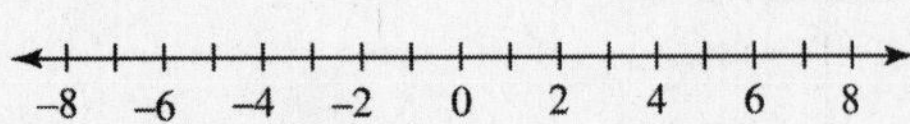

5. $-5-2$ **5.** ________________

–8 –6 –4 –2 0 2 4 6 8

6. $-3-5$ **6.** ________________

–8 –6 –4 –2 0 2 4 6 8

Objective 2 Use the definition of subtraction.

Find each difference.

7. $-7-(-14)$ **7.** ________________

8. $22-(-24)$ **8.** ________________

9. $-5.6-(-5.6)$ **9.** ________________

10. $-7.2-8.9$ **10.** ________________

11. $-3.2-(-7.6)$ **11.** ________________

12. $\frac{1}{10}-\frac{1}{2}$ **12.** ________________

Name: Date:
Instructor: Section:

13. $-\frac{3}{10}-\left(-\frac{4}{15}\right)$ 13. ______

14. $3\frac{3}{4}-\left(-2\frac{1}{8}\right)$ 14. ______

Objective 3 Work subtraction problems that involve brackets.

Perform each operation.

15. $[8-(-12)]-2$ 15. ______

16. $-.2-[.6+(-.9)]$ 16. ______

17. $[3-(-9)]-(-6)$ 17. ______

18. $3-[-4+(11-19)]$ 18. ______

19. $-2+[(-12+10)-(-4+2)]$ 19. ______

20. $\left(\frac{1}{2}-\frac{1}{3}\right)-\frac{5}{6}$ 20. ______

21. $\frac{2}{9}-\left[\frac{5}{6}-\left(-\frac{2}{3}\right)\right]$ 21. ______

Name: Date:
Instructor: Section:

22. $\left[\frac{5}{8}-\left(-\frac{1}{16}\right)\right]-\left(-\frac{3}{8}\right)$ 22. ________________

Objective 4 Translate words and phrases that indicate subtraction.

Write a numerical expression for each phrase, and then simplify the expression.

23. 4 less than –4 23. ________________

24. –12 subtracted from the sum of –4 and –2 24. ________________

25. The sum of –4 and 12, decreased by 9 25. ________________

26. 2 less than the difference between 10 and –4 26. ________________

Solve each problem.

27. Dr. Somers runs an experiment at –43.3°C. He then lowers the temperature by 7.9°C. What is the new temperature for the experiment? 27. ________________

28. David has a checking account balance of $439.42. He overdraws his account by writing a check for $702.58. Write his new balance as a negative number. 28. ________________

29. At 1:00 A.M., the temperature on the top of Mt. Washington in New Hampshire was –12°F. At 11:00 A.M., the temperature was 25°F. What was the rise in temperature? 29. ________________

30. The highest point in a country has an elevation of 1408 meters. The lowest point is 396 meters below sea level. Using zero as sea level, find the difference between the two elevations. 30. ________________

Name: Date:
Instructor: Section:

Chapter 1 THE REAL NUMBER SYSTEM

1.6 Multiplying and Dividing Real Numbers

Learning Objectives
1. Find the product of a positive number and a negative number.
2. Find the product of two negative numbers.
3. Use the reciprocal of a number to apply the definition of division.
4. Use the rules for order of operations when multiplying and dividing signed numbers.
5. Evaluate expressions involving variables.
6. Translate words and phrases involving multiplication and division.
7. Translate simple sentences into equations.

Key Terms

Use the vocabulary terms listed below to complete each statement in exercises 1–3.

product **quotient** **reciprocals**

1. The answer to a division problem is called the ________________.
2. Pairs of numbers whose product is 1 are called ________________.
3. The answer to a multiplication problem is called the ________________.

Objective 1 Find the product of a positive number and a negative number.

Find each product.

1. $7(-4)$ 1. ________________

2. $\left(\frac{1}{5}\right)\left(-\frac{2}{3}\right)$ 2. ________________

3. $\left(-\frac{3}{8}\right)\left(\frac{14}{9}\right)$ 3. ________________

4. $(-3.2)(4.1)$ 4. ________________

Name: Date:
Instructor: Section:

Objective 2 Find the product of two negative numbers.

Find each product.

5. $(-4)(-10)$ **5.** ____________

6. $\left(-\frac{2}{7}\right)\left(-\frac{14}{5}\right)$ **6.** ____________

7. $(-1.3)(-2.1)$ **7.** ____________

8. $(-.4)(-3.4)$ **8.** ____________

Objective 3 Use the reciprocal of a number to apply the definition of division.

Find each quotient.

9. $-\frac{3}{16} \div \frac{9}{8}$ **9.** ____________

10. $\frac{-120}{-20}$ **10.** ____________

11. $\frac{0}{-2}$ **11.** ____________

12. $\frac{10}{0}$ **12.** ____________

Name: Date:
Instructor: Section:

13. $5.5 \div (-2.2)$ 13. ________

Objective 4 Use the rules for order of operations when multiplying and dividing signed numbers.

Perform the indicated operations.

14. $-4[(-2)(7)-2]$ 14. ________

15. $-7[-4-(-2)(-3)]$ 15. ________

16. $\dfrac{-7(2)-(-3)}{5+(-3)}$ 16. ________

17. $\dfrac{-4[8-(-3+7)]}{-6[3-(-2)]-3(-3)}$ 17. ________

18. $\dfrac{(-9+1)^2-(-6)(-2)}{5(-5)+3(4)}$ 18. ________

Objective 5 Evaluate expressions involving variables.

Evaluate the following expressions if $x = -3$, $y = 2$, and $a = 4$.

19. $-x+[(-a+y)-2x]$ 19. ________

20. $(-4+x)(-a)-|x|$ 20. ________

Name: Date:
Instructor: Section:

21. $-x^2 + 2a^2 - 3y$ **21.** ______________

22. $\dfrac{4a - x}{y^2}$ **22.** ______________

Objective 6 Translate words and phrases involving multiplication and division.

Write a numerical expression for each phrase and simplify.

23. The product of –7 and 3, added to –7 **23.** ______________

24. Three-tenths of the difference between 50 and –10, subtracted from 85 **24.** ______________

25. The sum of –12 and the quotient of 49 and –7 **25.** ______________

26. The product of 40 and –3, divided by the difference between 5 and –10 **26.** ______________

Objective 7 Translate simple sentences into equations.

Write each statement in symbols, using x as the variable.

27. Two-thirds of a number is –7. **27.** ______________

28. –8 times a number is 72. **28.** ______________

29. When a number is divided by –4, the result is 1. **29.** ______________

30. The quotient of –2 and a number is –9. **30.** ______________

Name: Date:
Instructor: Section:

Chapter 1 THE REAL NUMBER SYSTEM

1.7 Properties of Real Numbers

Learning Objectives

1	Use the commutative properties.
2	Use the associative properties.
3	Use the identity properties.
4	Use the inverse properties.
5	Use the distributive property.

Key Terms

Use the vocabulary terms listed below to complete each statement in exercises 1–2.

identity element for addition

identity element for multiplication

1. When the ______________________________, 0, is added to a number, the number is unchanged.

2. When a number is multiplied by the ______________________________, 1, the number is unchanged.

Objective 1 Use the commutative properties.

Complete each statement. Use a commutative property.

1. $y+4=$ _____ $+\,y$ 1. ______________

2. $5(2)=$ _____ (5) 2. ______________

3. $2+[10+(-9)]=$ _____ $+\,2$ 3. ______________

4. $-4(4+z)=$ _____ (-4) 4. ______________

5. $10\left(\frac{1}{4}\cdot 2\right)=$ _____ (10) 5. ______________

Name: Date:
Instructor: Section:

6. $3\cdot(-2)+12=12+$_____ 6. __________

Objective 2 Use the associative properties.

Complete each statement. Use an associative property.

7. $(4\cdot5)(-7)=$_____$[5(-7)]$ 7. __________

8. $[-4+(-2)]+y=$_____$+(-2+y)$ 8. __________

9. $4(ab)=$_____$\cdot b$ 9. __________

10. $[x+(-4)]+3y=x+$_____ 10. __________

11. $(-r)[(-p)(-1)]=$_____(-1) 11. __________

12. $4r+(3s+14t)=$_____$+14t$ 12. __________

Objective 3 Use the identity properties.

Use an identity property to complete each statement.

13. $4+0=$_____ 13. __________

14. $1(-4)=$_____ 14. __________

15. _____$\cdot1=12$ 15. __________

Name: Date:
Instructor: Section:

Use an identity property to simplify each expression.

16. $\frac{30}{35}$ 16. ____________

17. $\frac{7}{10}+\frac{9}{30}$ 17. ____________

18. $\frac{27}{25}-\frac{8}{5}$ 18. ____________

Objective 4 Use the inverse properties.

Complete the statements so that they are examples of either an identity property or an inverse property. Identify which property is used.

19. $-4+$ _____ $=0$ 19. ____________

20. $\frac{2}{7}\cdot$ _____ $=1$ 20. ____________

21. $-9+$ _____ $=-9$ 21. ____________

22. $-\frac{3}{5}\cdot$ _____ $=1$ 22. ____________

23. _____ $\cdot -2\frac{5}{6}=1$ 23. ____________

24. _____ $\cdot -2\frac{5}{6}=-2\frac{5}{6}$ 24. ____________

Name: Date:
Instructor: Section:

Objective 5 Use the distributive property.

Use the distributive property to rewrite each expression. Simplify if possible.

25. $-(-4b-8)$ **25.** ____________

26. $n(2a-4b+6c)$ **26.** ____________

27. $-2(5y-9z)$ **27.** ____________

28. $-(-2k+7)$ **28.** ____________

29. $-14x+(-14y)$ **29.** ____________

30. $2(7x)+2(8z)$ **30.** ____________

Name: Date:
Instructor: Section:

Chapter 1 THE REAL NUMBER SYSTEM

1.8 Simplifying Expressions

Learning Objectives

1. Simplify expressions.
2. Identify terms and numerical coefficients.
3. Identify like terms.
4. Combine like terms.
5. Simplify expressions from word phrases.

Key Terms

Use the vocabulary terms listed below to complete each statement in exercises 1–3.

term **numerical coefficient** **like terms**

1. In the term $4x^2$, "4" is the________________________________.

2. A number, a variable, or a product or quotient of a number and one or more variables raised to powers is called a ________________________________.

3. Terms with exactly the same variables, including the same exponents, are called ________________________________.

Objective 1 Simplify expressions.

Simplify each expression.

1. $4(2x+5)+7$ **1.** ________________

2. $11-(d-2)+(-6)$ **2.** ________________

3. $-4+s-(12-21)$ **3.** ________________

4. $-2(-5x+2)+7$ **4.** ________________

5. $7(5n-2)-(6-11)$ **5.** ________________

Name: Date:
Instructor: Section:

6. $4(-6p-2)+2-4$ 6. ________

Objective 2 Identify terms and numerical coefficients.

Give the numerical coefficient of each term.

7. $-2y^2$ 7. ________

8. 125 8. ________

9. z^5 9. ________

10. $-\frac{3}{5}a^2b$ 10. ________

11. $\frac{7x}{9}$ 11. ________

12. $5.6r^5$ 12. ________

Objective 3 Identify like terms.

Identify each group of terms as **like** *or* **unlike**.

13. $4x^2, -7x^2$ 13. ________

14. $-8m, -8m^2$ 14. ________

15. $7xy, -6xy^2$ 15. ________

16. $2w, 4w, -w$ 16. ________

17. $5z^3, 5z^2, 5z^2$ 17. ________

18. $\frac{1}{3}, -\frac{3}{4}, 4$ 18. ________

Name: Date:
Instructor: Section:

Objective 4 Combine like terms.

Simplify.

19. $7r-(2r+4)$ **19.** ______________

20. $12y-7y^2+4y-3y^2$ **20.** ______________

21. $.8y^2-.2xy-.3xy+.9y^2$ **21.** ______________

22. $-4(x+4)+2(3x+1)$ **22.** ______________

23. $2.5(3y+1)-4.5(2y-3)$ **23.** ______________

24. $\frac{1}{2}(2x-4)-\frac{3}{4}(8x+12)$ **24.** ______________

Objective 5 Simplify expressions from word phrases.

Write each phrase as a mathematical expression and simplify by combining like terms. Use x as the variable.

25. The sum of six times a number and 12, added to four times the number. **25.** ______________

Name: Date:
Instructor: Section:

26. The sum of seven times a number and 2, subtracted from three times the number.

26. ______________

27. Three times the sum of 9 and twice a number, added to four times the number.

27. ______________

28. The sum of ten times a number and 7, subtracted from the difference between 2 and nine times the number.

28. ______________

29. Four times the difference between twice a number and six times the number, added to six times the sum of the number and 9.

29. ______________

30. Four times the difference between twice a number and –10, subtracted from three times the sum of –7 and five times the number.

30. ______________

Name: Date:
Instructor: Section:

Chapter 2 EQUATIONS, INEQUALITIES, AND APPLICATIONS

2.1 The Addition Property of Equality

Learning Objectives
1. Identify linear equations.
2. Use the addition property of equality.
3. Simplify, and then use the addition property of equality.

Key Terms

Use the vocabulary terms listed below to complete each statement in exercises 1–3.

linear equation **solution set** **equivalent equations**

1. Equations that have exactly the same solutions sets are called
______________________.

2. An equation that can be written in the form $Ax + B = C$, where A, B, and C are real numbers and $A \neq 0$, is called a ______________________.

3. The set of all numbers that satisfy an equation is called its
______________________.

Objective 1 Identify linear equations.

Tell whether each of the following is a linear equation.

1. $9x + 2 = 0$ 1. ______________

2. $3x^2 + 4x + 3 = 0$ 2. ______________

3. $7x^2 = 10$ 3. ______________

4. $3x^3 = 2x^2 + 5x$ 4. ______________

5. $\frac{5}{x} - \frac{3}{2} = 0$ 5. ______________

6. $4x - 2 = 12x + 9$ 6. ______________

Name: Date:
Instructor: Section:

Objective 2 Use the addition property of equality.

Solve each equation by using the addition property of equality. Check each solution.

7. $y-4=16$ **7.** ________

8. $r+9=8$ **8.** ________

9. $3x+2=5x+12$ **9.** ________

10. $3y=7y-4$ **10.** ________

11. $p-\frac{2}{3}=\frac{5}{6}$ **11.** ________

12. $y+4\frac{1}{2}=3\frac{3}{4}$ **12.** ________

13. $\frac{2}{3}t-5=\frac{5}{3}t$ **13.** ________

14. $\frac{9}{8}p-\frac{1}{2}=\frac{1}{8}p$ **14.** ________

15. $5.7x+12.8=4.7x$ **15.** ________

Name: Date:
Instructor: Section:

16. $9.5y - 2.4 = 10.5y$ **16.** ______________

17. $2z + 8 = -12$ **17.** ______________

18. $-7t + 12 = -4t$ **18.** ______________

Objective 3 Simplify, and then use the addition property of equality.

Solve each equation. First simplify each side of the equation as much as possible. Check each solution.

19. $3(t+3) - (2t+7) = 9$ **19.** ______________

20. $5x + 4(2x+1) - (5x - 1 - 2) = 9$ **20.** ______________

21. $-4(5g - 7) + 3(8g - 3) = 15 - 4 + 3g$ **21.** ______________

22. $10x + 4x - 11x + 4 - 7 = 2 - 4x - 3 + 8x$ **22.** ______________

23. $4(3a - 2) - 6(2 + a) = 5(2a - 5)$ **23.** ______________

Name: Date:
Instructor: Section:

24. $2(4t+6)-3(2t-3)=-3(3t-4)+5-t$

24. ________________

25. $-7(1+2b)-6(3-5b)=5(4+3b)-45$

25. ________________

26. $8(2-4b)+3(5-b)=4(1-9b)+22$

26. ________________

27. $\frac{8}{5}t+\frac{1}{3}=\frac{5}{6}+\frac{3}{5}t-\frac{1}{6}$

27. ________________

28. $\frac{5}{12}+\frac{7}{6}s-\frac{1}{6}=\frac{5}{6}s+\frac{1}{4}-\frac{2}{3}s$

28. ________________

29. $3.6p+4.8+4.0p=8.6p-3.1+.7$

29. ________________

30. $.03x+0.6+.09x-.9=2.1$

30. ________________

Name: Date:
Instructor: Section:

Chapter 2 EQUATIONS, INEQUALITIES, AND APPLICATIONS

2.2 The Multiplication Property of Equality

Learning Objectives
1 Use the multiplication property of equality.
2 Simplify, and then use the multiplication property of equality.

Key Terms

Use the vocabulary terms listed below to complete each statement in exercises 1–2.

multiplication property of equality **addition property of equality**

1. The ____________________________ states that multiplying both sides of an equation by the same nonzero number will not change the soltuion.

2. When the same quantity is added to both sides of an equation, the ________________________________ is being applied.

Objective 1 Use the multiplication property of equality.

Solve each equation and check your solution.

1. $8x = 24$ 1. ________________

2. $-3w = 51$ 2. ________________

3. $-16a = -48$ 3. ________________

4. $\frac{b}{5} = 4$ 4. ________________

5. $\frac{3p}{7} = -6$ 5. ________________

Name: Date:
Instructor: Section:

6. $\frac{b}{-2} = 21$ 6. ______________

7. $-\frac{7}{2}t = -4$ 7. ______________

8. $\frac{y}{4} = \frac{1}{3}$ 8. ______________

9. $\frac{6}{7}y = \frac{2}{3}$ 9. ______________

10. $\frac{3}{4}r = -27$ 10. ______________

11. $.81m = 2.916$ 11. ______________

12. $2.1a = 9.03$ 12. ______________

13. $7.5p = -61.5$ 13. ______________

Name: Date:
Instructor: Section:

14. $-2.7v = -17.28$ **14.** ______________

15. $4.3r = -11.61$ **15.** ______________

Objective 2 Simplify, and then use the multiplication property of equality.

Solve each equation and check your solution.

16. $12r + 3r = -90$ **16.** ______________

17. $-7b + 12b = 125$ **17.** ______________

18. $3b - 4b = 8$ **18.** ______________

19. $7q - 10q = -24$ **19.** ______________

20. $3w - 7w = 20$ **20.** ______________

21. $10a - 7a = -24$ **21.** ______________

22. $6m - 14m = -56$ **22.** ______________

Name: Date:
Instructor: Section:

23. $.8c + .6c = -2.1$ **23.** ______________

24. $8f + 4f - 3f = 108$ **24.** ______________

25. $8s - 3s + 4s = 90$ **25.** ______________

26. $4x - 8x + 2x = 18$ **26.** ______________

27. $2f + 3f - 7f = 48$ **27.** ______________

28. $-11h - 6h + 14h = -21$ **28.** ______________

29. $18r - 6r + 3r = -105$ **29.** ______________

30. $17x + 9x - 11x = -9$ **30.** ______________

Name: Date:
Instructor: Section:

Chapter 2 EQUATIONS, INEQUALITIES, AND APPLICATIONS

2.3 More on Solving Linear Equations

Learning Objectives

1. Learn and use the four steps for solving a linear equation.
2. Solve equations with fractions or decimals as coefficients.
3. Solve equations that have no solution or infinitely many solutions.
4. Write expressions for two related unknown quantities.

Key Terms

Use the vocabulary terms listed below to complete each statement in exercises 1–3.

conditional equation **identity** **contradiction**

1. An equation with no solution is called a(n) ______________________________.

2. A(n) ______________________________ is an equation that is true for some values of the variable and false for other values.

3. An equation that is true for all values of the variable is called a(n) ______________________.

Objective 1 Learn and use the four steps for solving a linear equation.

Solve each equation and check your solution.

1. $7t+6=11t-4$ 1. ________________

2. $4(z-2)-(3z-1)=2z-6$ 2. ________________

3. $3(x+4)=6-2(x-8)$ 3. ________________

4. $-(v+2)=3+v$ 4. ________________

Name: Date:
Instructor: Section:

5. $3-(1-y)=3+5y$ 5. ______________

6. $3a-6a+4(a-4)=-2(a+2)$ 6. ______________

7. $3(t+5)=6-2(t-4)$ 7. ______________

8. $4r-3(3r-2)=8-3(r-4)$ 8. ______________

Objective 2 Solve equations with fractions or decimals as coefficients.

Solve each equation and check your solution.

9. $\frac{3}{8}x-\frac{1}{3}x=\frac{1}{12}$ 9. ______________

10. $\frac{1}{3}(2m-1)-\frac{3}{4}m=\frac{5}{6}$ 10. ______________

11. $\frac{5}{6}(r-2)-\frac{2}{9}(r+4)=\frac{7}{18}$ 11. ______________

12. $\frac{3}{8}x-\left(x-\frac{3}{4}\right)=\frac{5}{8}(x+3)$ 12. ______________

Name: Date:
Instructor: Section:

13. $.90x = .40(30) + .15(100)$ **13.** ______________

14. $.35(20) + .45y = .125(200)$ **14.** ______________

15. $.24x - .38(x+2) = -.34(x+4)$ **15.** ______________

16. $.45a - .35(20-a) = .02(50)$ **16.** ______________

Objective 3 Solve equations that have no solution or infinitely many solutions.

Solve each equation and check your solution.

17. $3(6x-7) = 2(9x-6)$ **17.** ______________

18. $6y - 3(y+2) = 3(y-2)$ **18.** ______________

19. $-1-(2+y) = -(-4+y)$ **19.** ______________

20. $6(6t+1) = 9(4t-3) + 33$ **20.** ______________

Name: Date:
Instructor: Section:

21. $3(r-2)-r+4=2r+6$ **21.** ______________

22. $8(2d-4)-3(7d+8)=-5(d+2)$ **22.** ______________

23. $2(5w-3)-6=3(3w+1)+5(w-3)-4w$ **23.** ______________

Objective 4 Write expressions for two related unknown quantities.

Write an expression for the two related unknown quantities.

24. Two numbers have a sum of 36. One is m. Find the other number. **24.** ______________

25. The product of two numbers is 17. One number is p. What is the other number? **25.** ______________

26. A cashier has q dimes. Find the value of the dimes in cents. **26.** ______________

27. The length of a rectangle is x inches. Its width is four times the length. Find the width of the rectangle. **27.** ______________

28. Temperature in degrees Kelvin is always 273 more than temperature in degrees Celsius. If the Kelvin temperature is k°, what is the temperature in degree Celsius? **28.** ______________

29. Shirley is x years old. Her mother is 28 years older. How old is her mother? **29.** ______________

30. Admission to the circus costs x dollars for an adult and y dollars for a child. Find the total cost of 6 adults and 4 children. **30.** ______________

Name: Date:
Instructor: Section:

Chapter 2 EQUATIONS, INEQUALITIES, AND APPLICATIONS

2.4 An Introduction to Applications of Linear Equations

Learning Objectives

1. Learn the six steps for solving applied problems.
2. Solve problems involving unknown numbers.
3. Solve problems involving sums of quantities.
4. Solve problems involving supplementary and complementary angles.
5. Solve problems involving consecutive integers.

Key Terms

Use the vocabulary terms listed below to complete each statement in exercises 1–5.

complementary angles **right angle** **supplementary angles**

straight angle **consecutive integers**

1. Two angles whose measures sum to 180º are ________________________________.
2. Two angles whose measures sum to 90º are ________________________________.
3. An angle whose measure is exactly 90º is a ______________________________.
4. An angle whose measure is exactly 180º is a ______________________________.
5. Two integers that differ by 1 are ________________________________.

Objective 1 Learn the six steps for solving applied problems.

1. Write the six problem-solving steps. **1.** ______________

____

____

____

____

____

Name: Date:
Instructor: Section:

Objective 2 Solve problems involving unknown numbers.

Write an equation for each of the following and then solve the problem. Use x as the variable.

2. If 4 is added to 3 times a number, the result is 7. Find the number. **2.** ________________

____

3. If 2 is subtracted from four times a number, the result is 3 more than six times the number. What is the number? **3.** ________________

____

4. If –2 is multiplied by the difference between 4 and a number, the result is 24. Find the number. **4.** ________________

____

5. Six times the difference between a number and 4 equals the product of the number and –2. Find the number. **5.** ________________

____

6. When the difference between a number and 4 is multiplied by –3, the result is two more than –5 times the number. Find the number. **6.** ________________

____

7. If four times a number is added to 7, the result is five less than six times the number. Find the number. **7.** ________________

____

Name: Date:
Instructor: Section:

Objective 3 Solve problems involving sums of quantities.

Write an equation for each of the following and then solve the problem. Use x as the variable.

8. A rope 116 inches long is cut into three pieces. The middle-sized piece is 10 inches shorter than twice the shortest piece. The longest piece is $\frac{5}{3}$ as long as the shortest piece. What is the length of the shortest piece?

8. ____________

____

9. George and Al were opposing candidates in the school board election. George received 21 more votes than Al, with 439 votes cast. How many votes did Al receive?

9. ____________

____

10. On a psychology test, the highest grade was 38 points more than the lowest grade. The sum of the two grades was 142. Find the lowest grade.

10. ____________

____

11. Mount McKinley is Alaska is 5910 feet higher than Mount Rainier in Washington. Together, their heights total 34,730 feet. How high is each mountain?

11. ____________

Mt. Rainier ____________

Mt. McKinley ____________

12. Charles bought five general admission tickets and four student tickets for a movie. He paid $35.25. If each student ticket cost $3.50, how much did each general admission ticket cost?

12. ____________

____

Name: Date:
Instructor: Section:

13. Penny is making punch for a party. The recipe requires twice as much orange juice as cranberry juice and 8 times as much ginger ale as cranberry juice. If she plans to make 176 ounces of punch, how much of each ingredient should she use?

13. ______________

cranberry juice ______________

orange juice ______________

ginger ale ______________

14. Pablo, Faustino, and Mark swim at a public pool each day for exercise. One day Pablo swam five more than three times as many laps as Mark, and Faustino swam four times as many laps as Mark. If the men swam 29 laps altogether, how many laps did each one swim?

14. ______________

Mark ______________

Pablo ______________

Faustino ______________

15. Linda wishes to build a rectangular dog pen using 52 feet of fence and the back of her house, which is 36 feet long to enclose the pen. How wide will the dog pen be if the pen is 36 feet long?

15. ______________

Objective 4 Solve problems involving supplementary and complementary angles.

Solve each problem.

16. Find the measure of an angle if the measure of the angle is 8° less than three times the measure of its supplement.

16. ______________

17. Find the measure of an angle whose supplement measures 20° more than twice its complement.

17. ______________

Name: Date:
Instructor: Section:

18. Find the measure of an angle such that the sum of the measures of its complement and its supplement is 138°. **18.** ____________

19. Find the measure of an angle such that the difference between the measure of its supplement and twice the measure of its complement is 49°. **19.** ____________

20. Find the measure of an angle whose complement is 9° more than twice its measure. **20.** ____________

21. Find the measure of an angle such that the difference between the measures of an angle and its complement is 20°. **21.** ____________

22. Find the measure of an angle if its supplement measures 4° less than three times its complement. **22.** ____________

Objective 5 Solve problems involving consecutive integers.

Solve each problem.

23. Find two consecutive even integers whose sum is 154. **23.** ____________

Name: Date:
Instructor: Section:

24. Find two consecutive even integers such that the smaller, added to twice the larger, is 292. **24.** ____________

25. Find two consecutive integers such that the larger, added to three times the smaller, is 109. **25.** ____________

26. Find two consecutive odd integers such that if three times the smaller is added to twice the larger, the sum is 69. **26.** ____________

27. Find two consecutive odd integers such that the larger, added to eight times the smaller, equals 119. **27.** ____________

28. Find three consecutive odd integers whose sum is 363. **28.** ____________

29. Find three consecutive integers such that the sum of the first two is 74 more than the third. **29.** ____________

30. The sum of four consecutive even integers is 4. Find the integers. **30.** ____________

Name: Date:
Instructor: Section:

Chapter 2 EQUATIONS, INEQUALITIES, AND APPLICATIONS

2.5 Formulas and Additional Applications from Geometry

Learning Objectives

1 Solve a formula for one variable, given the values of the other variables.
2 Use a formula to solve an applied problem.
3 Solve problems involving vertical angles and straight angles.
4 Solve a formula for a specified variable.

Key Terms

Use the vocabulary terms listed below to complete each statement in exercises 1–4.

formula **area** **perimeter** **vertical angles**

1. The nonadjacent angles formed by two intersecting lines are called ______________________.

2. An equation in which variables are used to describe a relationship is called a(n) ______________________.

3. The distance around a figure is called its ______________________.

4. A measure of the surface covered by a figure is called its ______________________.

Objective 1 Solve a formula for one variable, given the values of the other variables.

In the following exercises, a formula is given, along with the values of all but one of the variables in the formula. Find the value of the variable that is not given.

1. $V = LWH$; $L = 2$, $W = 4$, $H = 3$ 1. ______________

2. $P = 2L + 2W$; $P = 42$; $W = 6$ 2. ______________

3. $S = \dfrac{a}{1-r}$; $S = 60$, $r = .4$ 3. ______________

Name: Date:
Instructor: Section:

4. $I = prt$; $I = 288$, $r = .04$, $t = 3$ **4.** __________

5. $C = \frac{5}{9}(F - 32)$; $F = 104$ **5.** __________

6. $A = \frac{1}{2}(b + B)h$; $b = 6$, $B = 16$, $A = 132$ **6.** __________

7. $V = \frac{1}{3}\pi r^2 h$; $r = 4$, $h = 6$, $\pi = 3.14$ **7.** __________

Objective 2 Use a formula to solve an applied problem.

Use a formula to write an equation for each of the following applications; then solve the application. (Use 3.14 as an approximation for π.)

8. Find the length of a rectangular garden if its perimeter is 96 feet and its width is 12 feet. **8.** __________

9. Find the height of a triangular banner whose area is 48 square inches and base is 12 inches. **9.** __________

Name: Date:
Instructor: Section:

10. Ruth has 42 feet of binding for a rectangular rug that she is weaving. If the rug is 9 feet wide, how long can she make the rug if she wishes to use all the binding on the perimeter of the rug?

10. ______________

11. Linda invests \$5000 at 6% simple interest and earns \$450. How long did Linda invest her money?

11. ______________

12. A tent has the shape of a right pyramid. The volume is 200 cubic feet and the height is 12 feet. Find the area of the floor of the tent.

12. ______________

13. A spherical balloon has a radius of 9 centimeters. Find the amount of air required to fill the balloon. (Round your answer to the nearest hundredth.)

13. ______________

14. Find the height of an ice cream cone if the diameter is 6 centimeters and the volume is 37.68 cubic centimeters. (Round your answer to the nearest hundredth.)

14. ______________

15. The circumference of a circular garden is 628 feet. Find the area of the garden. (Hint: First find the radius of the garden.)

15. ______________

Name: Date:
Instructor: Section:

Objective 3 Solve problems involving vertical angles and straight angles.

Find the measure of each marked angle.

16.

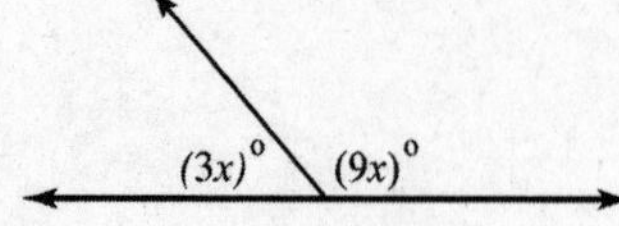

16. ________________

17.

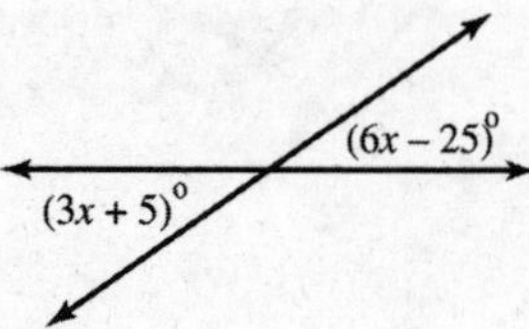

17. ________________

18.

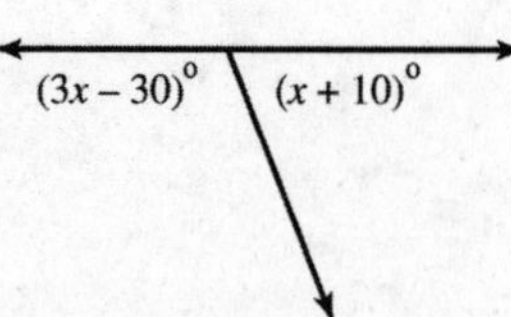

18. ________________

19.

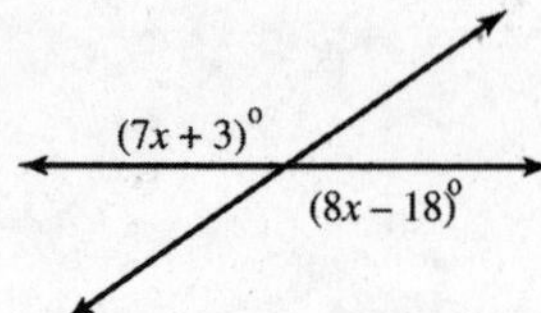

19. ________________

20.

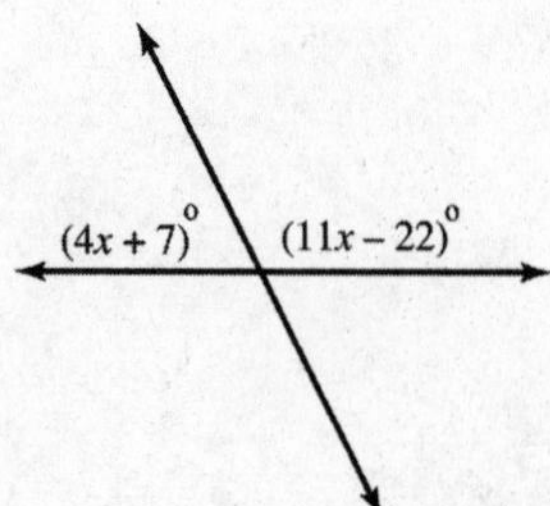

20. ________________

Name: Date:
Instructor: Section:

21. 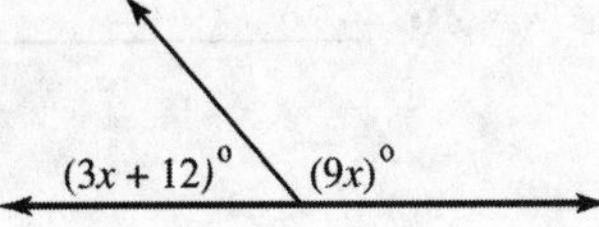

21. ______________

22. 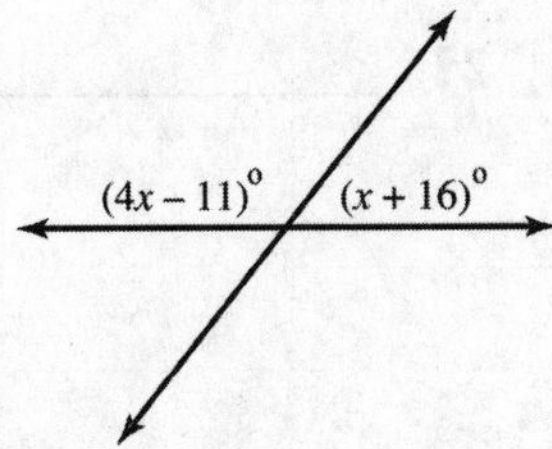

22. ______________

23. 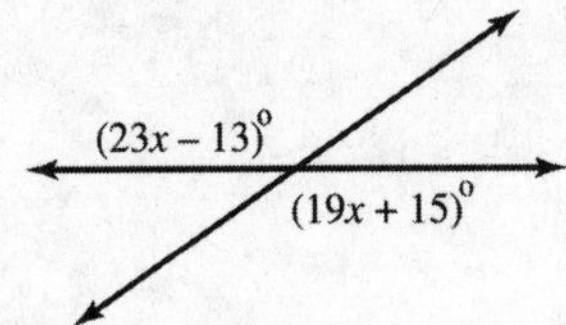

23. ______________

Objective 4 Solve a formula for a specified variable.

Solve each formula for the specified variable.

24. $V = LWH$ for H

24. ______________

25. $S = \dfrac{a}{1-r}$ for r

25. ______________

Name: Date:
Instructor: Section:

26. $a_n = a_1 + (n-1)d$ for n

26. ________________

27. $P = A - Art$ for A

27. ________________

28. $S_n = \frac{n}{2}(a_1 + a_n)$ for a_1

28. ________________

29. $S = (n-2)180$ for n

29. ________________

30. $V = \frac{1}{3}\pi r^2 h$ for h

30. ________________

Name: Date:
Instructor: Section:

Chapter 2 EQUATIONS, INEQUALITIES, AND APPLICATIONS

2.6 Ratio, Proportion, and Percent

Learning Objectives
1. Write ratios.
2. Solve proportions.
3. Solve applied problems using proportions.
4. Find percentages and percents.

Key Terms

Use the vocabulary terms listed below to complete each statement in exercises 1–4.

ratio **proportion** **cross products** **terms**

1. A ______________________ shows that two ratios are equal.

2. A ______________________ is a comparison of two quantities using a quotient.

3. In the proportion, $\frac{a}{b}=\frac{c}{d}$, a, b, c, and d are called the ______________.

4. To see whether a proportion is true, determine if the ______________ are equal.

Objective 1 Write ratios.

Write a ratio for each word phrase. Write fractions in lowest terms.

1. 8 men to 3 men — 1. ____________

2. 10 days to 2 weeks — 2. ____________

3. 9 dollars to 48 quarters — 3. ____________

A supermarket was surveyed and the following prices were charged for items in various sizes. Find the best buy (based on price per unit) for each of the following items.

4. Rice: — 4. ____________
   1-pound box: $1.99
   3-pound box: $4.99
   5-pound box: $6.79
   10-pound box: $9.99

Name: Date:
Instructor: Section:

5. Trash bags
10-count box: $2.89
20-count box: $5.29
45-count box: $6.69
85-count box: $13.99

5. ________________

6. Applesauce
16-ounce jar: $1.19
24-ounce jar: $1.29
48-ounce jar: $2.69
64-ounce jar: $3.49

6. ________________

7. Olive oil
16-ounce bottle: $6.99
25.5-ounce bottle: $9.99
32-ounce bottle: $12.99
44-ounce bottle: $14.99

7. ________________

Objective 2 Solve proportions.

Solve each equation.

8. $\frac{z}{20}=\frac{25}{125}$

8. ________________

Name: Date:
Instructor: Section:

9. $\frac{25}{3} = \frac{125}{x}$ 9. ________________

10. $\frac{m}{5} = \frac{m-2}{2}$ 10. ________________

11. $\frac{6y-4}{y} = \frac{11}{5}$ 11. ________________

12. $\frac{4}{z+1} = \frac{2}{z+7}$ 12. ________________

13. $\frac{3x+4}{x-2} = \frac{1}{3}$ 13. ________________

14. $\frac{s+2}{s+8} = \frac{9}{14}$ 14. ________________

Objective 3 Solve applied problems using proportions.

Solve each problem.

15. Ginny can type 8 pages of her term paper in 30 minutes. How long will it take her to type the paper if it has 20 pages? 15. ________________

Name: Date:
Instructor: Section:

16. Margie earns $168.48 in 26 hours. How much does she earn in 40 hours?

16. ______________

17. On a road map, 6 inches represents 50 miles. How many inches would represent 125 miles?

17. ______________

18. If 3 ounces of medicine must be mixed with 10 ounces of water, how many ounces of medicine must be mixed with 15 ounces of water?

18. ______________

19. A certain lawn mower uses 7 tanks of gas to cut 15 acres of lawn. How many tanks of gas are needed to cut 30 acres of lawn?

19. ______________

20. If 12 rolls of tape cost $4.60, how much will 15 rolls cost?

20. ______________

21. If four pounds of fertilizer will cover 50 square feet of garden, how many pounds would be needed for 125 square feet?

21. ______________

Name: Date:
Instructor: Section:

22. A garden service charges $30 to install 50 square feet of sod. Find the charge to install 225 square feet. **22.** ______________

Objective 4 Find percentages and percents.

Answer each question about percent.

23. What is 2.5% of 3500? **23.** ______________

24. What percent of 5200 is 104? **24.** ______________

25. 2.75% of what number is 20.625? **25.** ______________

Solve each problem.

26. The number of students enrolled in a calculus course is 145. If 40% of these students are female, how many are male? **26.** ______________

27. Paul recently bought a duplex for $144,000. He expects to earn $6120 per year on this investment. What percent of the purchase price will he earn? **27.** ______________

Name: Date:
Instructor: Section:

28. A pair of jeans with a regular price of $45 is on sale this week at 45% off. Find the amount of discount and the sale price of the jeans.

28. discount__________

sale price__________

29. An advertisement for a DVD player gives a sale price of $175.50. The regular price is $225. Find the percent discount on this DVD player.

29. __________

30. Jasmine solved 40 problems correctly on a test, giving her a score of $62\frac{1}{2}\%$. How many problems were on the test?

30. __________

Name: Date:
Instructor: Section:

Chapter 2 EQUATIONS, INEQUALITIES, AND APPLICATIONS

2.7 Solving Linear Inequalities

Learning Objectives

1. Graph intervals on a number line.
2. Use the addition property of inequality.
3. Use the multiplication property of inequality.
4. Solve inequalities using both properties of inequality.
5. Solve linear inequalities with three parts.
6. Use inequalities to solve applied problems.

Key Terms

Use the vocabulary terms listed below to complete each statement in exercises 1–5.

inequalities **interval** **interval notation** **linear inequality**

three-part inequality

1. An inequality that says that one number is between two other numbers is a(n)____________________.

2. A portion of a number line is called a(n) ____________________.

3. A(n) ____________________ can be written in the form $Ax + B > C$, $Ax + B \geq C$, $Ax + B < C$, or $Ax + B \leq C$, where A, B, and C are real numbers with $A \neq 0$.

4. Algebraic expressions related by $>$, $\geq$, $<$, or $\leq$ are called ____________________.

5. The ____________________ for $a \leq x < b$ is $[a, b)$.

Objective 1 Graph intervals on a number line.

Write each inequality in interval notation and graph the interval.

1. $3 < a$ 1. ____________________

2. $y \geq -2$ 2. ____________________

3. $-1 < x < 3$ 3. ____________________

Name: Date:
Instructor: Section:

4. $-3 \le y < 0$ 4. ______

5. $-3 < a \le 2$ 5. ______

Objective 2 Use the addition property of inequality.

Solve each inequality. Write the solution set in interval notation and then graph it.

6. $5a + 3 \le 6a$ 6. ______

7. $-2 + 8b \ge 7b - 1$ 7. ______

8. $6 + 3x < 4x + 4$ 8. ______

9. $3 + 5p \le 4p + 3$ 9. ______

10. $9 + 8b > 9b + 11$ 10. ______

Objective 3 Use the multiplication property of inequality.

Solve each inequality. Write the solution set in interval notation and then graph it.

11. $-2s < 4$ 11. ______

Name: Date:
Instructor: Section:

12. $4k \geq -16$

12. ______________

13. $\frac{3}{5}n \geq 0$

13. ______________

14. $-5t \leq -35$

14. ______________

15. $-9m > -36$

15. ______________

Objective 4 Solve inequalities using both properties of inequality.

Solve each inequality. Write the solution set in interval notation and then graph it.

16. $4(y-3)+2>3(y-2)$

16. ______________

17. $-3(m+4)+1 \leq -4(m-2)$

17. ______________

18. $7(2-x)-3 \leq -2(x-4)-x$

18. ______________

Name: Date:
Instructor: Section:

19. $3x - \frac{3}{4} \geq 2x + \frac{1}{3}$

19. ____________

20. $5(2z+2) - 2(z-3) > 3(2z+5) + z$

20. ____________

Objective 5 Solve linear inequalities with three parts.

Solve each inequality. Write the solution set in interval notation and then graph it.

21. $9 < 2x + 1 \leq 15$

21. ____________

22. $-17 \leq 3x - 2 < -11$

22. ____________

23. $6 < 2x - 4 < 8$

23. ____________

24. $-5 < -2 - x \leq 4$

24. ____________

Name: Date:
Instructor: Section:

25. $1 < 3z + 4 < 19$

25. ________________

Objective 6 Use inequalities to solve applied problems.

Solve each problem.

26. Lauren has grades of 98 and 86 on her first two chemistry quizzes. What must she score on her third quiz to have an average of at least 91 on the three quizzes?

26. ________________

27. Nina has a budget of $230 for gifts for this year. So far she has bought gifts costing $47.52, $38.98, and $26.98. If she has three more gifts to buy, find the average amount she can spend on each gift and still stay within her budget.

27. ________________

28. Ruth tutors mathematics in the evenings in an office for which she pays $600 per month rent. If rent is her only expense and she charges each student $40 per month, how many students must she teach to make a profit of at least $1600 per month?

28. ________________

Name: Date:
Instructor: Section:

29. Two sides of a triangle are equal in length, with the third side 8 feet longer than one of the equal sides. The perimeter of the triangle cannot be more than 38 feet. Find the largest possible value for the length of the equal sides.

29. ________________

30. If twice the sum of a number and 7 is subtracted from three times the number, the result is more than –9. Find all such numbers.

30. ________________

Name: Date:
Instructor: Section:

Chapter 3 GRAPHS OF LINEAR EQUATIONS AND INEQUALITIES; FUNCTIONS

3.1 Reading Graphs; Linear Equations in Two Variables

Learning Objectives
1 Interpret graphs.
2 Write a solution as an ordered pair.
3 Decide whether a given ordered pair is a solution of a given equation.
4 Complete ordered pairs for a given equation.
5 Complete a table of values.
6 Plot ordered pairs.

Key Terms

Use the vocabulary terms listed below to complete each statement in exercises 1–14.

bar graph **line graph** **linear equation in two variables**

ordered pair **table of values** ***x*-axis**

***y*-axis** **rectangular (Cartesian) coordinate system**

quadrants **origin** **plane** **coordinates**

plot **scatter diagram**

1. A ______________________________ uses dots connected by line to show trends.

2. An equation that can be written in the form $Ax + By = C$, where A, B, and C are real numbers and $A, B \neq 0$, is called a ______________________________.

3. A ______________________________ uses bars of various heights or lengths to show quantity or frequency.

4. ______________________________ are the numbers in the ordered pair that specify the location of a point on a rectangular coordinate system.

5. In a rectangular coordinate system, the horizontal axis is called the ________________________.

6. In a rectangular coordinate system, the vertical axis is called the ________________________.

7. A pair of numbers written between parentheses in which order is important is called a(n) ______________________________.

8. Together, the x-axis and the y-axis form a ____________________.

9. A coordinate system divides the plane into four regions called ________________.

Name: Date:
Instructor: Section:

10. The axis lines in a coordinate system intersect at the ________________.

11. To ________________ an ordered pair is to find the corresponding point on a coordinate system.

12. A graph of ordered pairs is called a ________________.

13. A table showing selected ordered pairs of numbers that satisfy an equation is called a ________________.

14. A flat surface determined by two intersecting lines is a ________________.

Objective 1 Interpret graphs.

The bar graph shows the enrollment for the fall semester at a small college for the past five years. Use this graph for problems 1–2.

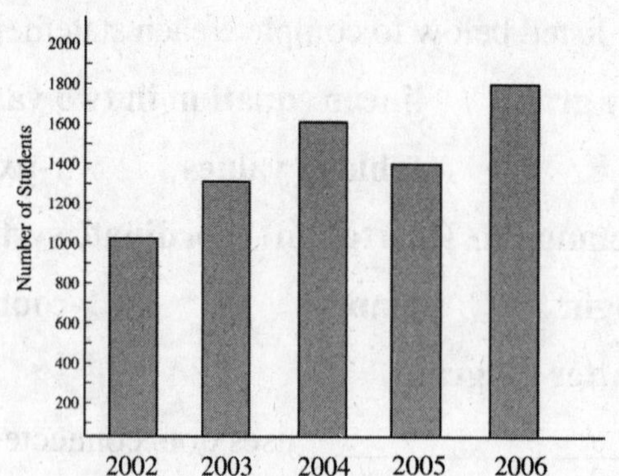

1. In what years was enrollment less than 1500 students? **1.** ________________

2. How many more students were enrolled in 2004 than in 2003? **2.** ________________

Name:
Instructor:

Date:
Section:

The line graph shows the number of degrees awarded by a university for the years 2000–2005. Use this graph to answer exercises 3–5.

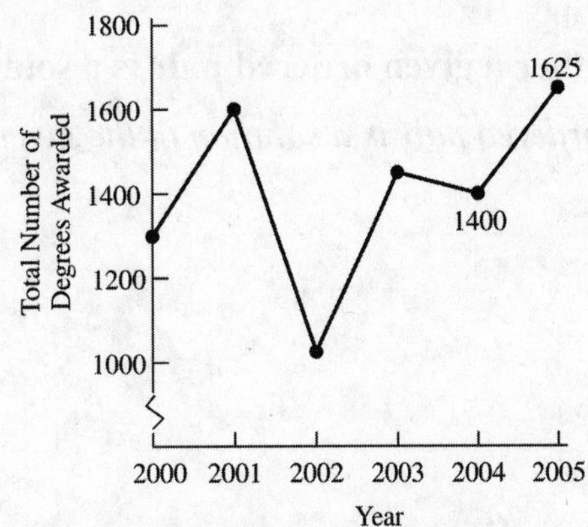

3. Between which two years did the total number of degrees awarded show the smallest change?

3. ______________

4. Between which two years did the total number of degrees awarded show the greatest decline?

4. ______________

5. If 20% of the degrees awarded in 2005 were M.B.A. degrees, how many M.B.A.'s were awarded in 2005?

5. ______________

Objective 2 Write a solution as an ordered pair.

Write each solution as an ordered pair.

6. $x = 4$ and $y = 7$

6. ______________

7. $x = \frac{1}{3}$ and $y = -9$

7. ______________

8. $y = \frac{1}{3}$ and $x = 0$

8. ______________

9. $x = .2$ and $y = .3$

9. ______________

Name: Date:
Instructor: Section:

10. $x = -4$ and $y = 0$ **10.** ______________

Objective 3 Decide whether a given ordered pair is a solution of a given equation.

Decide whether the given ordered pair is a solution of the given equation.

11. $4x - 3y = 10; (1, 2)$ **11.** ______________

12. $2x - 3y = 1; \left(0, \frac{1}{3}\right)$ **12.** ______________

13. $x = -7; (-7, 9)$ **13.** ______________

14. $x = 1 - 2y; \left(0, -\frac{1}{2}\right)$ **14.** ______________

15. $2x = 3y; (3, 2)$ **15.** ______________

Name: Date:
Instructor: Section:

Objective 4 Complete ordered pairs for a given equation.

For each of the given equations, complete the ordered pairs beneath it.

16. $y = 2x - 5$

(a) $(2, \ )$
(b) $(0, \ )$
(c) $(\ , 3)$
(d) $(\ , -7)$
(e) $(\ , 9)$

16.
(a) ______
(b) ______
(c) ______
(d) ______
(e) ______

17. $y = 3 + 2x$

(a) $(-4, \ )$
(b) $(2, \ )$
(c) $(\ , 0)$
(d) $(-2, \ )$
(e) $(\ , -7)$

17.
(a) ______
(b) ______
(c) ______
(d) ______
(e) ______

18. $x = -2$

(a) $(\ , -2)$
(b) $(\ , 0)$
(c) $(\ , 19)$
(d) $(\ , 3)$
(e) $\left(\ , -\frac{2}{3}\right)$

18.
(a) ______
(b) ______
(c) ______
(d) ______
(e) ______

19. $y = 4$

(a) $(2, \ )$
(b) $(0, \ )$
(c) $(4, \ )$
(d) $(-4, \ )$
(e) $(.75, \ )$

19.
(a) ______
(b) ______
(c) ______
(d) ______
(e) ______

Name: Date:
Instructor: Section:

20. $5x+4y=10$

(a) $(2, \ )$

(b) $(4, \ )$

(c) $(\ ,3)$

(d) $(0, \ )$

(e) $(\ ,2)$

20. (a) ____________
(b) ____________
(c) ____________
(d) ____________
(e) ____________

Objective 5 Complete a table of values.

Complete each table of values. Write the results as ordered pairs.

21. $3x-4y=-6$

x	y
0	
	0
2	

21. ____________

22. $-7x+2y=-14$

x	y
	0
0	
3	

22. ____________

23. $2x+5=7$

x	y
	−3
	0
	5

23. ____________

Name: Date:
Instructor: Section:

24. $y-4=0$ **24.** ________________

x	y
−4	
0	
6	

25. $4x+3y=12$ **25.** ________________

x	y
0	
	0
	−1

Objective 6 Plot ordered pairs.

Plot the each ordered pair on a coordinate system.

26. $(0,-2)$

26.–30.

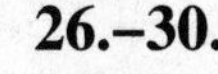

27. $(2,5)$

28. $(-2,-7)$

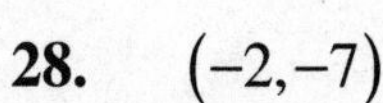

29. $(-3,4)$

30. $(4,-4)$

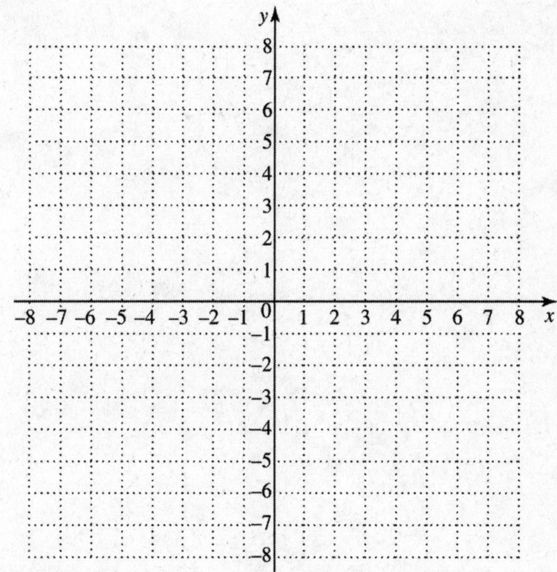

Name: Date:
Instructor: Section:

Chapter 3 GRAPHS OF LINEAR EQUATIONS AND INEQUALITIES; FUNCTIONS

3.2 Graphing Linear Equations in Two Variables

Learning Objectives

1 Graph linear equations by plotting ordered pairs.
2 Find intercepts.
3 Graph linear equations of the form $Ax + By = 0$.
4 Graph linear equations of the form $y = k$ or $x = k$.
5 Use a linear equation to model data.

Key Terms

Use the vocabulary terms listed below to complete each statement in exercises 1–4.

graph **graphing** ***y*-intercept** ***x*-intercept**

1. If a graph intersects the y-axis at k, then the ________________ is $(0, k)$.

2. If a graph intersects the x-axis at k, then the ________________ is $(k, 0)$.

3. The process of plotting the ordered pairs that satisfy a linear equation and drawing a line through them is called ____________________.

4. The set of all points that correspond to the ordered pairs that satisfy the equation is called the ____________________ of the equation.

Objective 1 Graph linear equations by plotting ordered pairs.

Complete the ordered pairs for each equation. Then graph the equation by plotting the points and drawing a line through them.

1. $x + y = 3$

   (0,)
   (,0)
   (2,)

1.

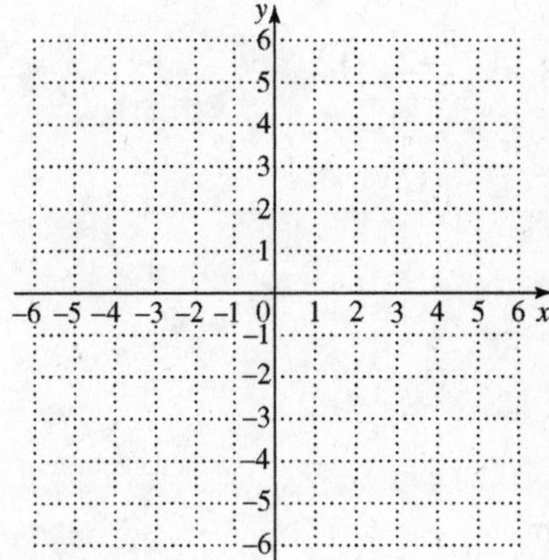

Name: Date:
Instructor: Section:

2. $y = 3x - 2$

(0,)

(,0)

(2,)

2.

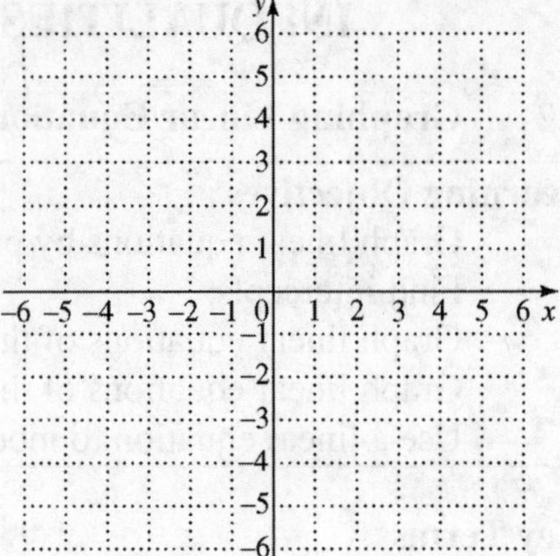

3. $x - y = 4$

(0,)

(,0)

(−2,)

3.

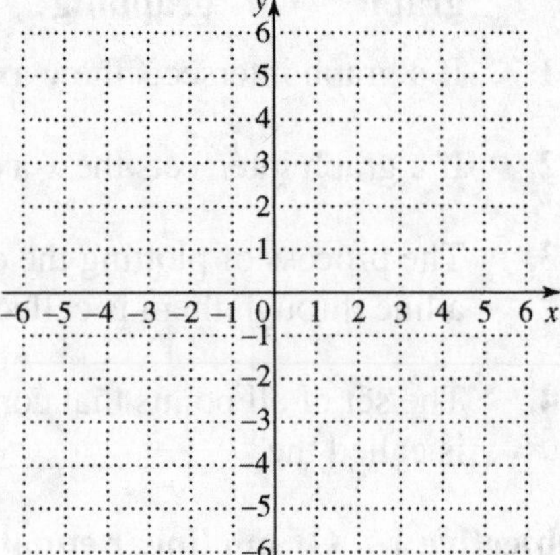

4. $2y - 4 = x$

(0,)

(,0)

(−2,)

4.

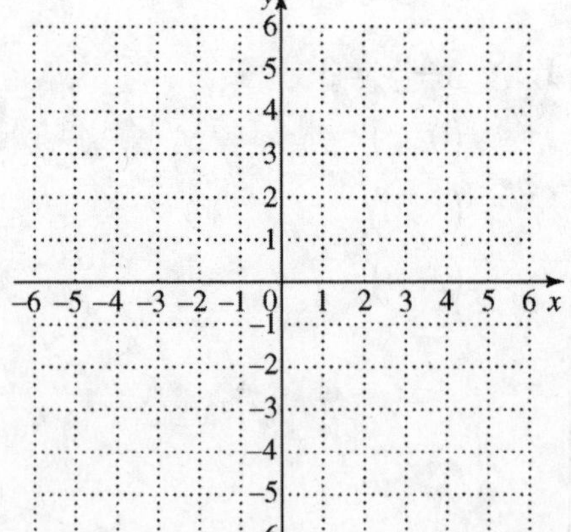

Name: Date:
Instructor: Section:

5. $2x + 3y = 6$

$(0, \quad)$

$(\quad, 0)$

$(-3, \quad)$

5.

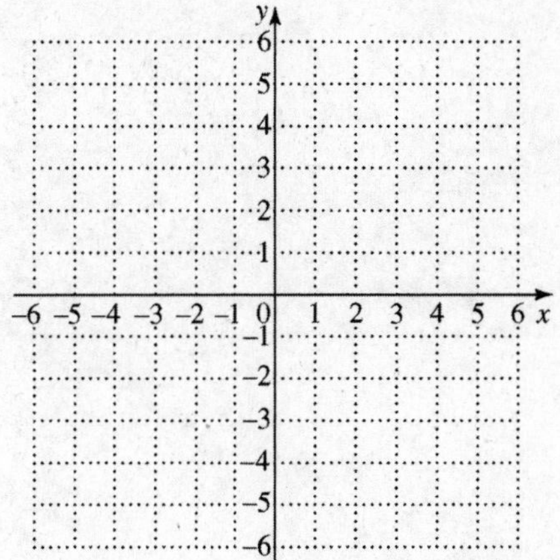

6. $x = 2y + 1$

$(0, \quad)$

$(\quad, 0)$

$(\quad, -2)$

6.

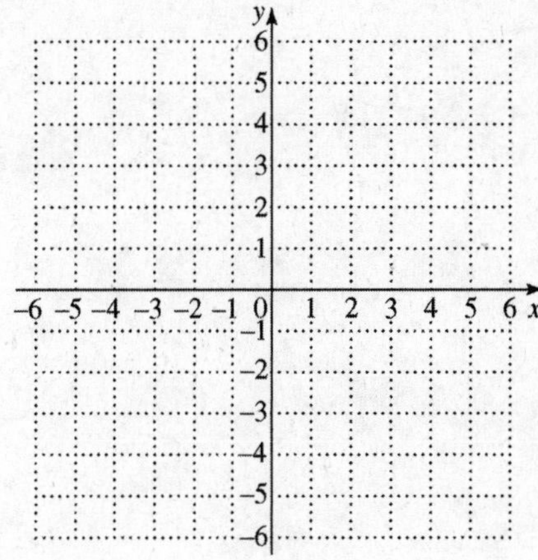

Objective 2 Find intercepts.

Find the intercepts for each equation. Then graph the equation.

7. $3x + y = 6$

7.

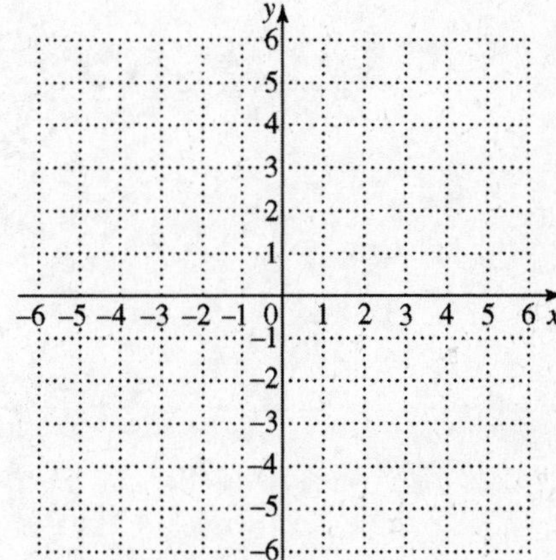

Name: Date:
Instructor: Section:

8. $4x - y = 4$

8.

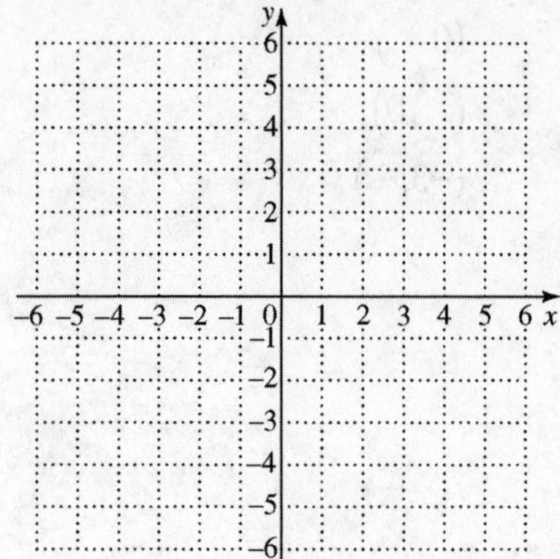

9. $5x - 2y = -10$

9.

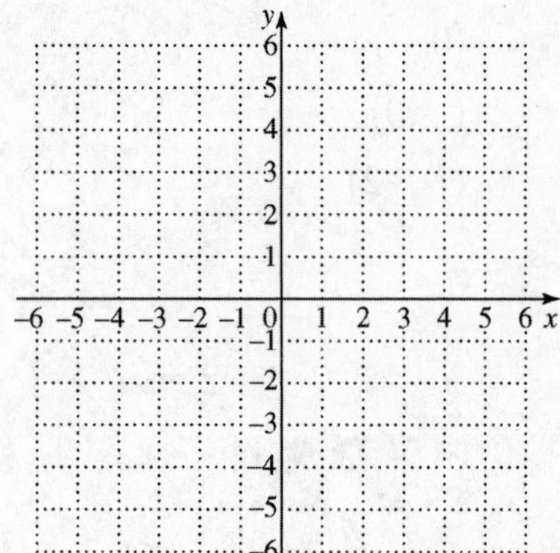

10. $2x - 3y = 6$

10.

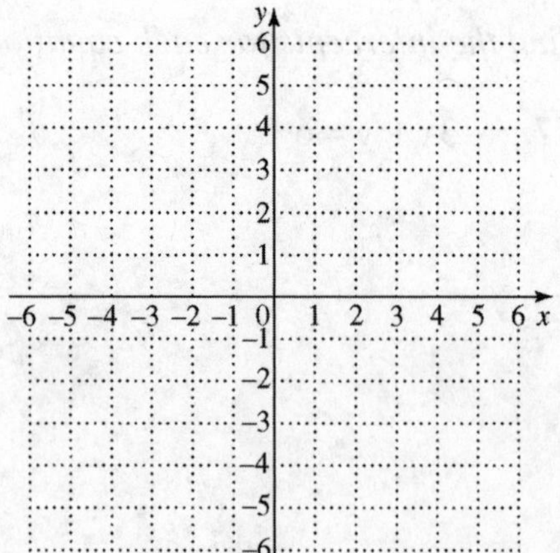

Name: Date:
Instructor: Section:

11. $4x - 7y = -8$

11.

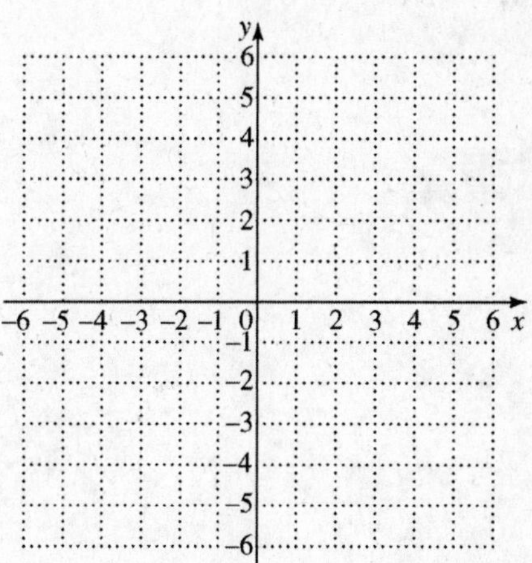

12. $3x - 2y = 8$

12.

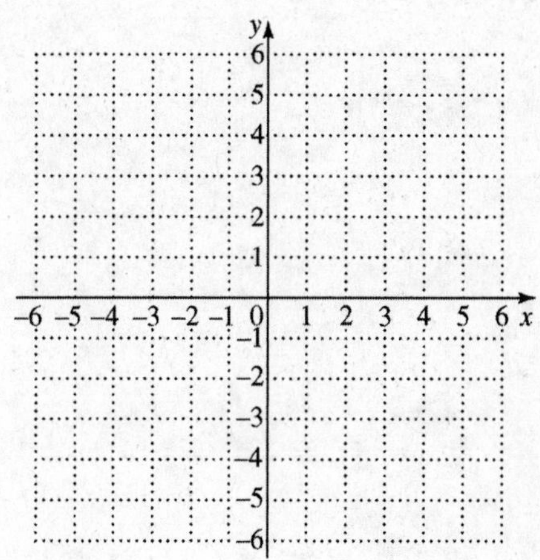

Objective 3 Graph linear equations of the form $Ax + By = 0$.

Graph each equation.

13. $-3x - 2y = 0$

13.

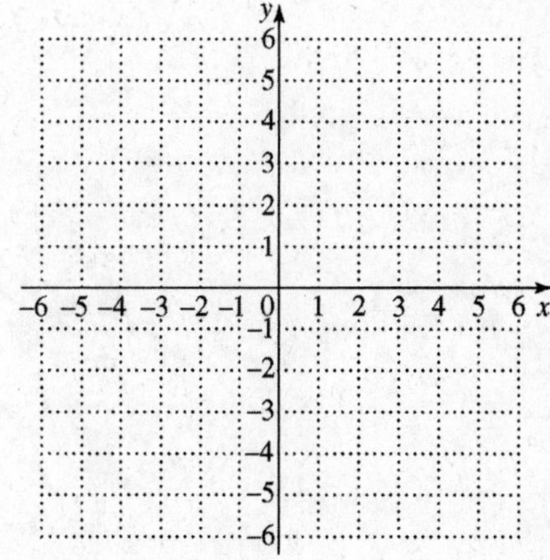

Name: Date:
Instructor: Section:

14. $3x - y = 0$

14.

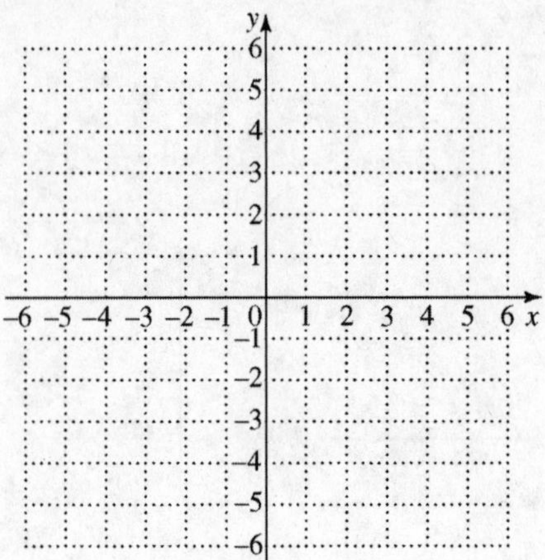

15. $x + 5y = 0$

15.

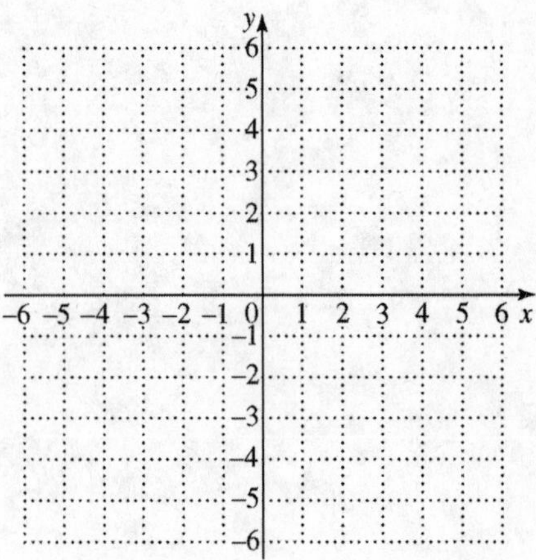

16. $x + y = 0$

16.

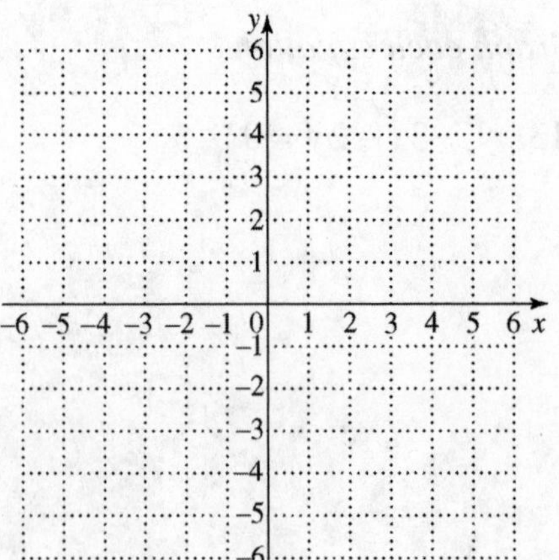

Name: Date:
Instructor: Section:

17. $4x = 3y$

17.

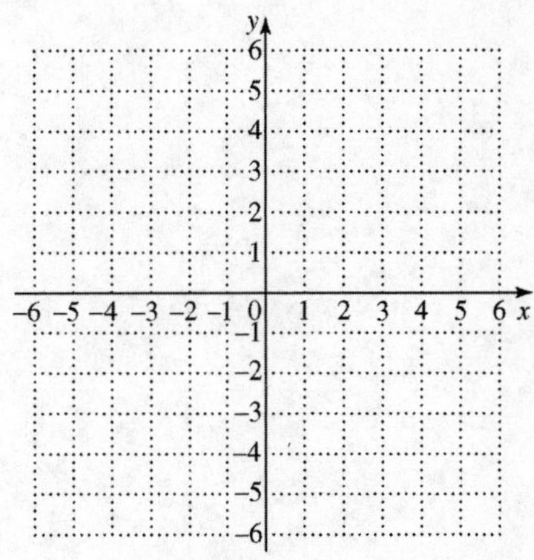

18. $y = 2x$

18.

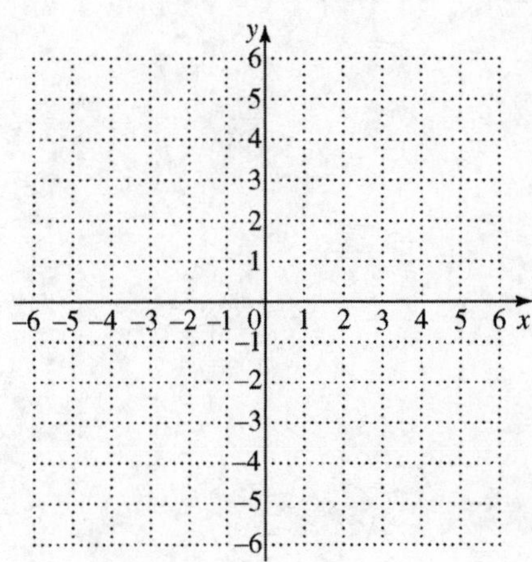

Objective 4 Graph linear equations of the form $y = k$ or $x = k$.

Graph each equation.

19. $y = -2$

19.

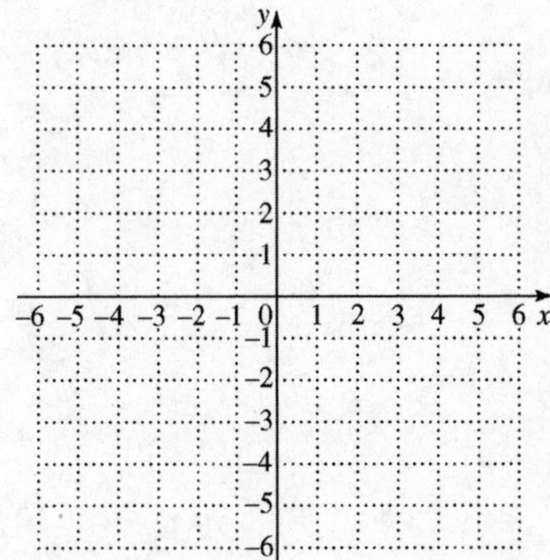

Name: Date:
Instructor: Section:

20. $x+4=0$

20.

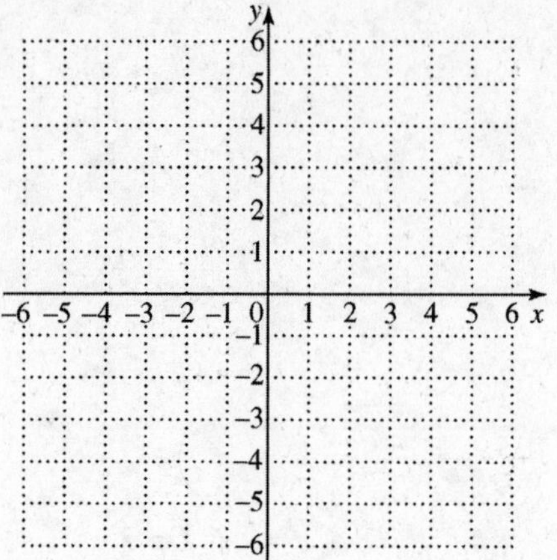

21. $x-1=0$

21.

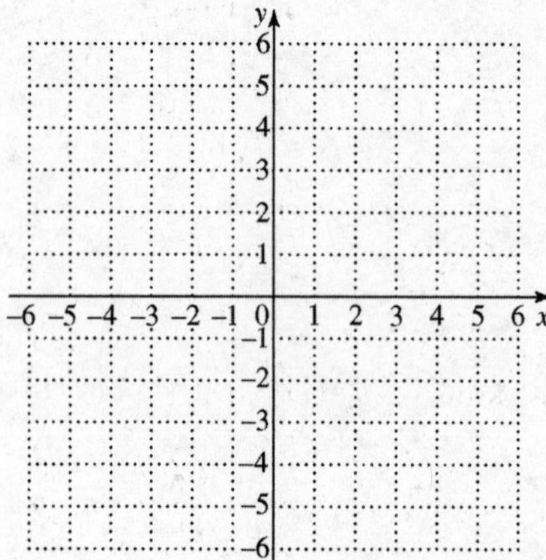

22. $y-4=0$

22.

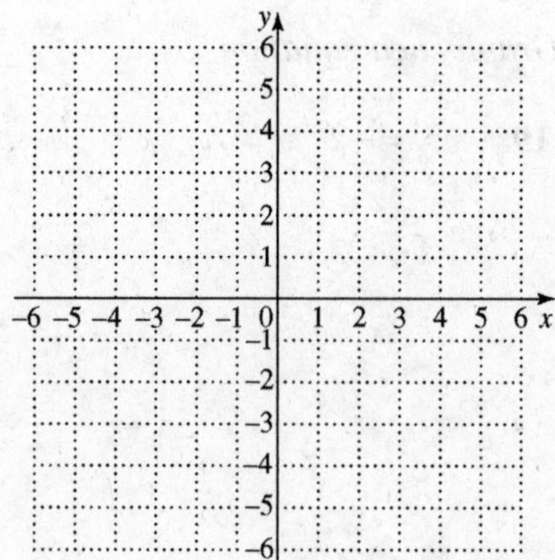

Name: Date:
Instructor: Section:

23. $y+3=0$

23.

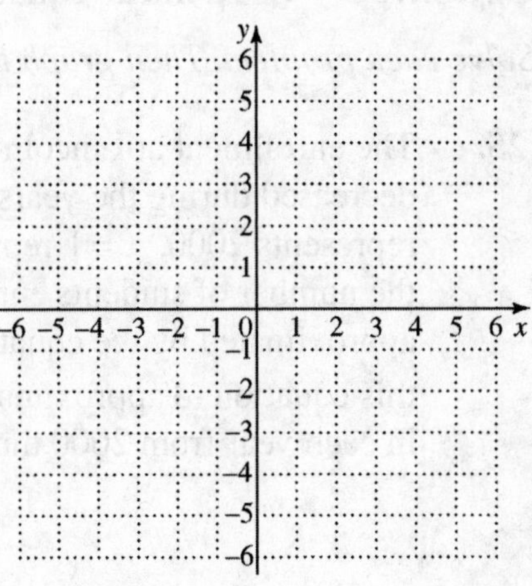

24. $x=0$

24.

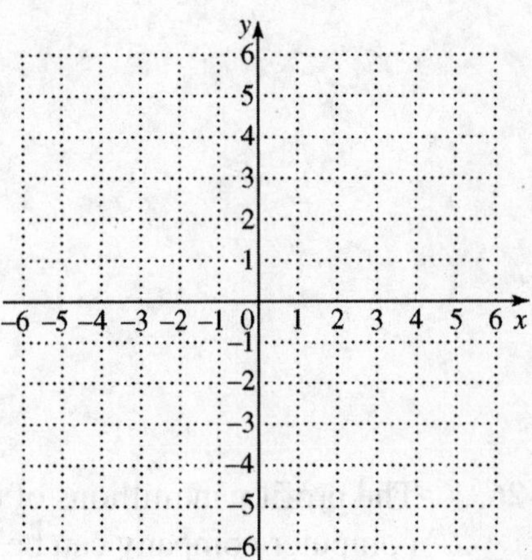

Name: Date:
Instructor: Section:

Objective 5 Use a linear equation to model data.

Solve each problem. Then graph the equation.

25. The enrollment at Lincolnwood High School decreased during the years 2000 to 2005. If $x = 0$ represents 2000, $x = 1$ represents 2001, and so on, the number of students enrolled in the school can be approximated by the equation $y = -85x + 2435$. Use this equation to approximate the number of students in each year from 2000 through 2005.

25. 2000 ________

2001 ________

2002 ________

2003 ________

2004 ________

2005 ________

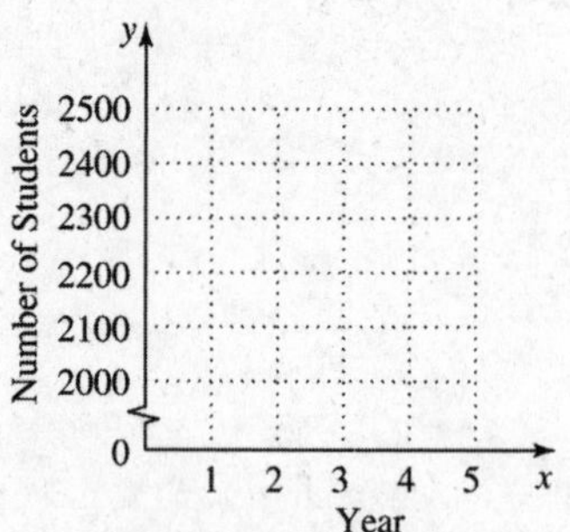

26. The profit y in millions of dollars earned by a small computer company can be approximated by the linear equation $y = .63x + 4.9$, where $x = 0$ corresponds to 2004, $x = 1$ corresponds to 2005, and so on. Use this equation to approximate the profit in each year from 2004 through 2007.

26. 2004 ________

2005 ________

2006 ________

2007 ________

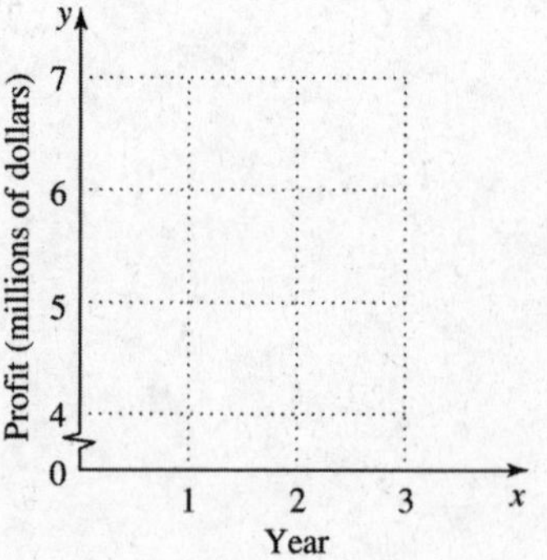

Name: Date:
Instructor: Section:

27. The number of band instruments sold by Elmer's Music Shop can be approximated by the equation $y = 325 + 42x$, where y is the number of instruments sold and x is the time in years, with $x = 0$ representing 2003. Use this equation to approximate the number of instruments sold in each year from 2003 through 2006.

27. 2003 ____________

2004 ____________

2005 ____________

2006 ____________

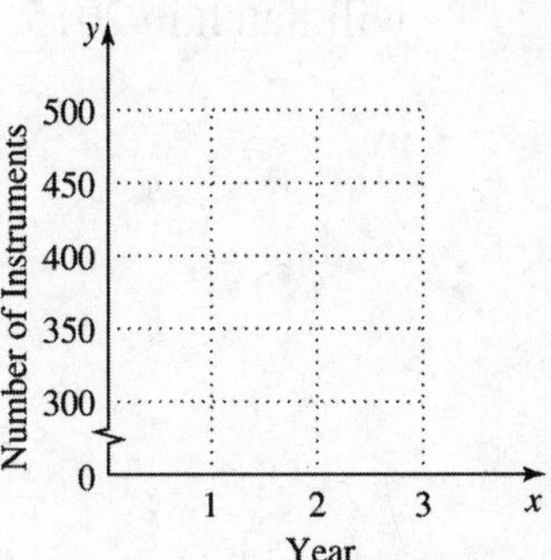

28. Suppose that the demand and price for a certain model of calculator are related by the equation $y = 45 - \frac{3}{5}x$, where y is the price (in dollars) and x is the demand (in thousands of calculators). Assuming that this model is valid for a demand up to 50,000 calculators, find the price at each of the following levels of demand.

(a) 0 calculators
(b) 5000 calculators
(c) 20,000 calculators
(d) 45,000 calculators

28. (a) ____________

(b) ____________

(c) ____________

(d) ____________

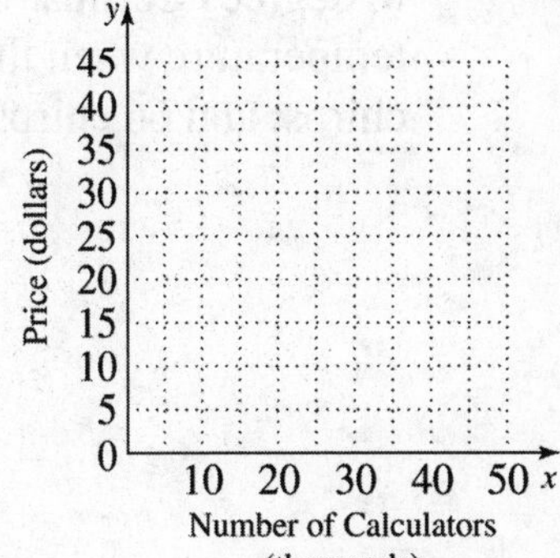

Name: Date:
Instructor: Section:

29. Every year sea turtles return to a certain group of islands to lay eggs. The number of turtle eggs that hatch can be approximated by the equation $y = -70x + 3260$, where y is the number of eggs that hatch and $x = 0$ representing 1990. Use this equation to find the number of eggs that hatched in 1995, 2000, and 2005. Estimate the number of eggs that will hatch in 2015.

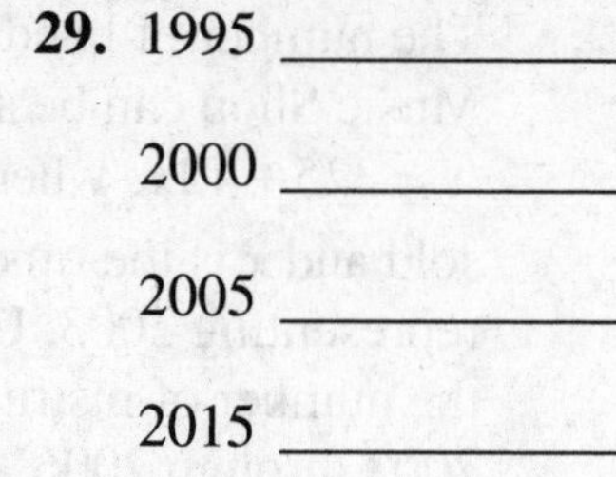

29. 1995 ________

2000 ________

2005 ________

2015 ________

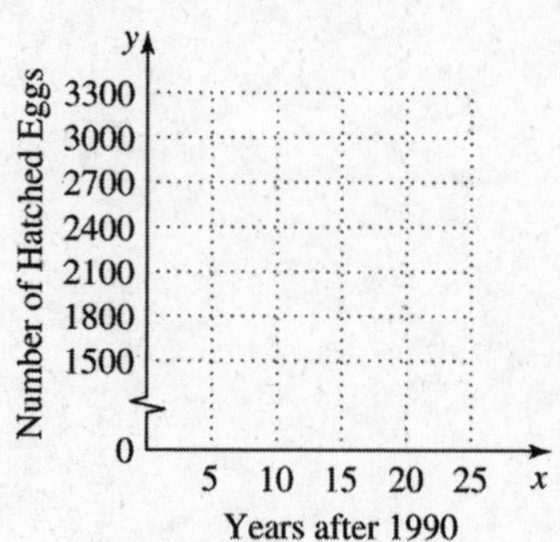

30. According to *The Old Farmer's Almanac*, the temperature in degrees Celsius can be determined by the equation $y = \frac{1}{3}x + 4$, where x is the number of cricket chirps in 25 seconds and y is the temperature in degrees Celsius. Use this equation to find the temperature when there are 48 chirps, 54 chirps, 60 chirps, and 66 chirps.

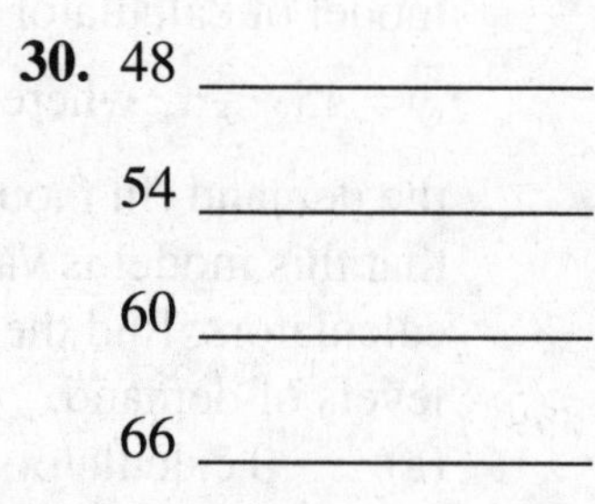

30. 48 ________

54 ________

60 ________

66 ________

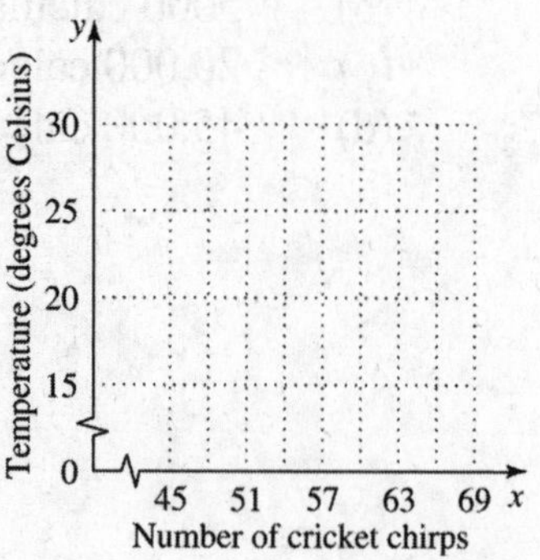

Name: Date:
Instructor: Section:

Chapter 3 GRAPHS OF LINEAR EQUATIONS AND INEQUALITIES; FUNCTIONS

3.3 Slope of a Line

Learning Objectives
1. Find the slope of a line given two points.
2. Find the slope from the equation of a line.
3. Use slope to determine whether two lines are parallel, perpendicular, or neither.
4. Solve problems involving average rate of change.

Key Terms

Use the vocabulary terms listed below to complete each statement in exercises 1–5.

rise **run** **slope** **parallel lines**

perpendicular lines

1. Two lines that intersect in a 90° angle are called ________________________.

2. The ________________ of a line is the ratio of the change in *y* compared to the change in *x* when moving along the line from one point to another.

3. The vertical change between two different points on a line is called the ______________________.

4. Two lines in a plane that never intersect are called ______________________.

5. The horizontal change between two different points on a line is called the ______________________.

Objective 1 Find the slope of a line given two points.

Find the slope of the line through the given points.

1. (4, 3) and (3, 5) **1.** ________________

2. (5, −2) and (2, 7) **2.** ________________

3. (7, 2) and (−7, 3) **3.** ________________

Name: Date:
Instructor: Section:

4. (–6, –6) and (2, –6) **4.** ____________

5. (0, 0) and (3, 5) **5.** ____________

6. (–4, 6) and (–4, –1) **6.** ____________

7. (–4, –7) and (–2, 1) **7.** ____________

8. (–3, 3) and (6, 3) **8.** ____________

Objective 2 Find the slope from the equation of a line.

Find the slope of each line.

8. $y = \frac{1}{2}x + 5$ **9.** ____________

10. $y = -5x$ **10.** ____________

11. $7y - 4x = 11$ **11.** ____________

Name: Date:
Instructor: Section:

12. $4x - 3y = 0$ **12.** ________________

13. $y = -4$ **13.** ________________

14. $3y = 2x - 1$ **14.** ________________

15. $y = -\frac{2}{5}x - 4$ **15.** ________________

16. $x = 0$ **16.** ________________

Objective 3 Use slope to determine whether two lines are parallel, perpendicular, or neither.

In each pair of equations, give the slope of each line, and then determine whether the two lines are **parallel**, **perpendicular**, *or* **neither**.

17. $y = -5x - 2$ **17.** ________________

$y = 5x + 11$

18. $y = 4x + 4$

$y = 3 - \frac{1}{4}x$

18. ____________

19. $4x + 2y = 8$

$x + 4y = -3$

19. ____________

20. $9x + 3y = 2$

$x - 3y = 5$

20. ____________

21. $4x + 2y = 7$

$5x + 3y = 11$

21. ____________

22. $y = 9$

$x = 0$

22. ____________

Name: Date:
Instructor: Section:

23. $8x + 2y = 7$

$x = 3 - y$

23. ________________

24. $y = 6$

$y + 2 = 9$

24. ________________

Objective 4 Solve problems involving average rate of change.

Solve each problem.

25. Suppose the sales of a company are given by the linear equation $y = 1250x + 10{,}000$, where x is the number of years after 2000, and y is the sales in dollars. What is the average rate of change in sales per year?

25. ________________

26. A small company had the following sales during their first three years of operation.

26. ________________

Year	Sales
2005	\$82,250
2006	\$89,790
2007	\$96,100

(a) What was the rate of change from 2005–2006?

(b) What was the rate of change from 2006–2007?

(c) What was the average rate of change from 2005–2007?

Name: Date:
Instructor: Section:

27. A plane had an altitude of 8500 feet at 4:02 P.M. and 12,700 feet at 4:39 P.M. What was the average rate of change in the altitude in feet per minute?

27. ______________

28. Enrollment in a college was 11,500 two years ago, 10,975 last year, and 10,800 this year.

(a) What is the average rate of change in enrollment per year for this 3–year period?

(b) Explain why the rate of change is negative.

28. ______________

29. A company had 41 employees during the first year of operation. During their eighth year, the company had 79 employees. What was the average rate of change in the number of employees per year?

29. ______________

30. A state had a population of 755,000 in 2000. The population is increasing at an average rate of 4500 people a year. At that rate, predict the population for the year 2012.

30. ______________

Name: Date:
Instructor: Section:

Chapter 3 GRAPHS OF LINEAR EQUATIONS AND INEQUALITIES; FUNCTIONS

3.4 Equations of Lines

Learning Objectives

1. Write an equation of a line given its slope and y-intercept.
2. Graph a line given its slope and a point on the line.
3. Write an equation of a line given its slope and any point on the line.
4. Write an equation of a line given two points on the line.
5. Write an equation of a line parallel or perpendicular to a given line.
6. Write an equation of a line that models real data.

Key Terms

Use the vocabulary terms listed below to complete each statement in exercises 1–3.

slope-intercept form **point-slope form** **standard form**

1. A linear equation in the form $y - y_1 = m(x - x_1)$ is written in

________________________.

2. A linear equation in the form $Ax + By = C$ is written in

________________________.

3. A linear equation in the form $y = mx + b$ is written in

________________________.

Objective 1 Write an equation of a line given its slope and y-intercept.

Write the slope-intercept form equation of the line with the given slope and y-intercept.

1. $m = \frac{3}{2}$; $b = -\frac{2}{3}$ **1.** ____________

2. slope -4; y-intercept $(0,0)$ **2.** ____________

3. slope 0; y-intercept $(0,-4)$ **3.** ____________

Name: Date:
Instructor: Section:

Use the geometric interpretation of slope to find the slope of each line. Then, by identifying the y-intercept from the graph, write the slope-intercept form of the equation of the line.

4.

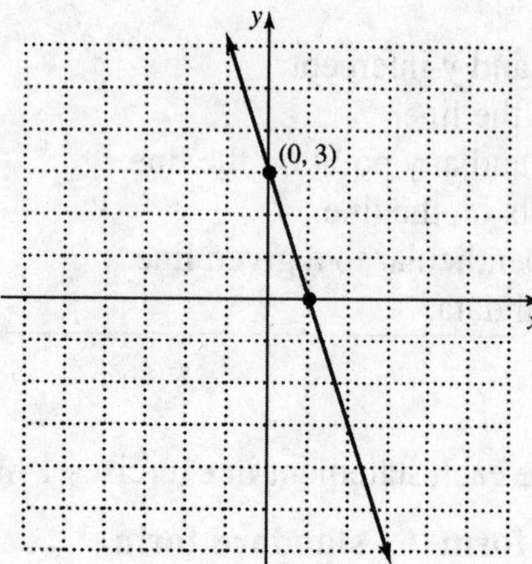

4. ______________

5.

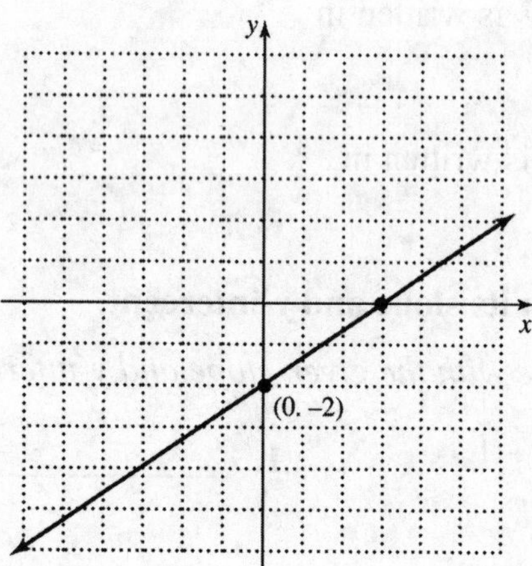

5. ______________

Name: Date:
Instructor: Section:

Objective 2 Graph a line given its slope and a point on the line.

Graph the line passing through the given point and having the given slope.

6. $(-3,-2)$; $m=\frac{2}{3}$

6.

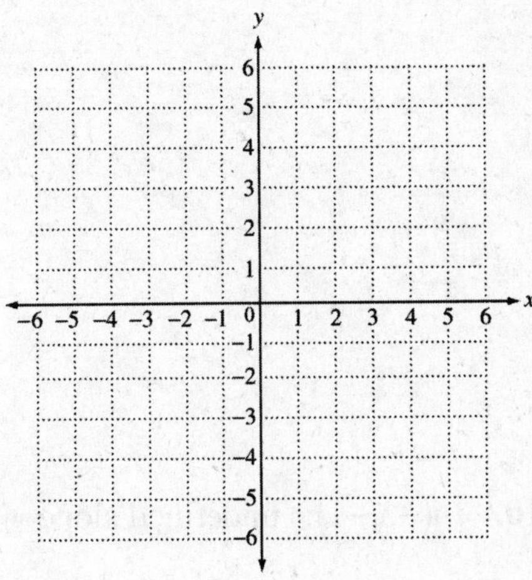

7. $(-2,-2)$; $m=0$

7.

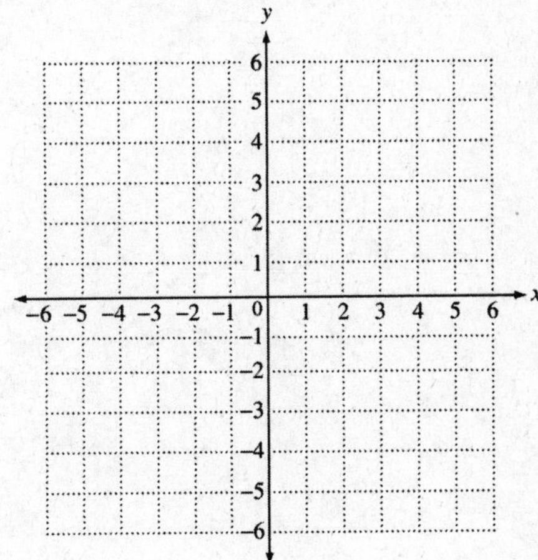

8. $(1,-3)$; $m=-\frac{5}{2}$

8.

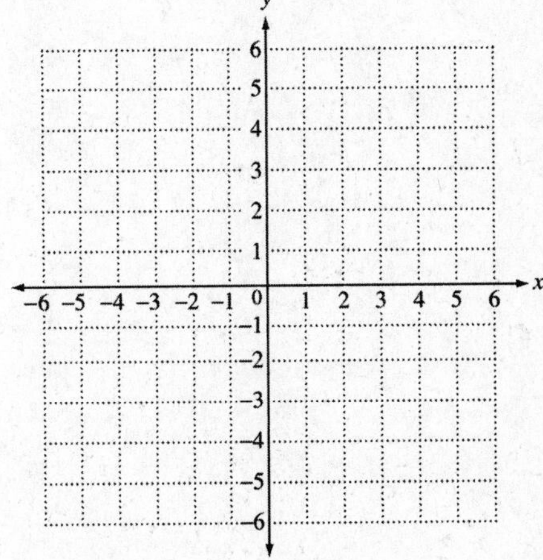

Name: Date:
Instructor: Section:

9. $(2,2)$; $m=\frac{1}{3}$ **9.**

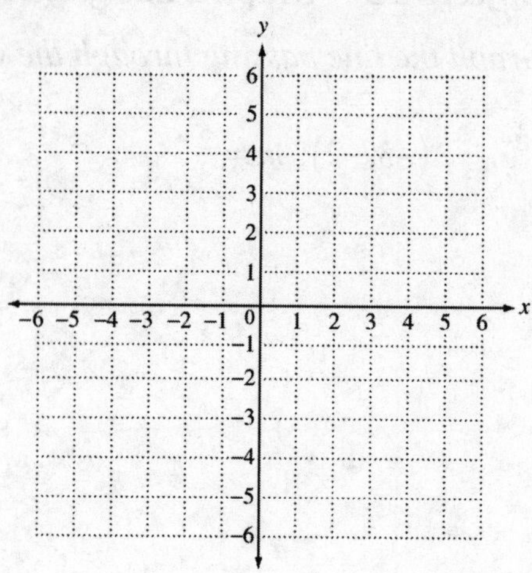

10. $(-3,-1)$; undefined slope **10.**

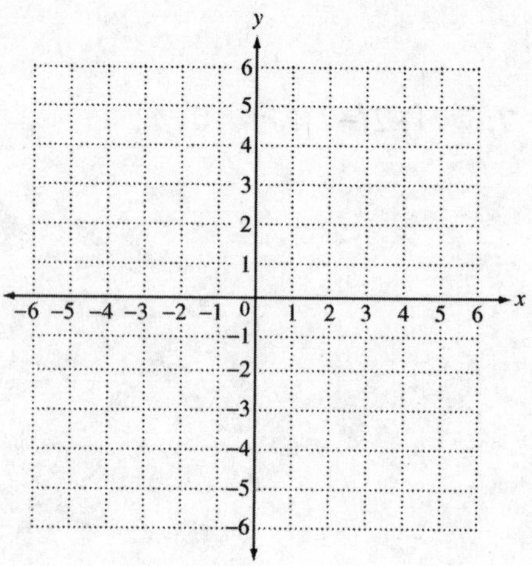

Name: Date:
Instructor: Section:

Objective 3 Write an equation of a line given its slope and any point on the line.

Write an equation for the line passing through the given point and having the given slope. Write the equations in slope-intercept form, if possible.

11. $(-4,-2)$; undefined slope **11.** ________

12. $(-3,4)$; $m=-\frac{3}{5}$ **12.** ________

13. $(-4,-7)$; $m=\frac{4}{3}$ **13.** ________

14. $(0,0)$; $m=0$ **14.** ________

15. $(2,2)$; $m=-\frac{3}{2}$ **15.** ________

Objective 4 Write an equation of a line given two points on the line.

Write an equation for the line passing through each pair of points. Write the equations in standard form.

16. $(-2,1)$ and $(3,11)$ **16.** ________

Name: Date:
Instructor: Section:

17. $\left(-\frac{4}{5},\frac{1}{8}\right)$ and $\left(-\frac{8}{5},-\frac{3}{8}\right)$ **17.** ____________

18. $(2,6)$ and $(-4,6)$ **18.** ____________

19. $(3,-4)$ and $(2,7)$. **19.** ____________

20. $(-2,-4)$ and $(-2,7)$. **20.** ____________

Objective 5 Write an equation of a line parallel or perpendicular to a given line.

Write the equation in standard form of the line satisfying the given conditions.

21. parallel to $2x+3y=-12$, through $(9,-3)$ **21.** ____________

22. perpendicular to $3x-2y=6$, through $(5,-3)$ **22.** ____________

Name: Date:
Instructor: Section:

23. perpendicular to $y = -1$, through (2, 5) **23.** ______________

24. parallel to $4x - 3y = 8$, through (–2, 3). **24.** ______________

25. perpendicular to $x - 3y = 0$, through (–10, 2) **25.** ______________

Objective 6 Write an equation of a line that models real data.

For each situation,

(a) *Write an equation in the form $y = mx + b$;*
(b) *Give the three ordered pairs associated with the equation for x-values 0, 5, and 10.*

26. To run a newspaper ad, there is a \$25 set up fee plus a charge of \$1.25 per line of type in the ad. Let x represent the number of lines of type in the ad so that y represents the total cost of the ad (in dollars).

26. (a) ______________

(b) ______________

27. A music teacher set up rows of chairs for a concert. There were 8 chairs in each row, plus 15 special reserved seats up front for faculty. Let x represent the number of rows of chairs so that y represents the total number of guests who can be seated.

27. (a) ______________

(b) ______________

Name: Date:
Instructor: Section:

Solve each problem.

28. A long distance phone call costs \$.35 plus \$.13 per minute for each minute of the call. Let x represent the number of minutes so that y represents the total cost of the call (in dollars).

(a) Write an equation in the form $y = mx + b$

(b) Find and interpret the ordered pair associated with the equation for $x = 5$

(c) If the call costs \$1.91, how long was the call in minutes?

28. (a) ____________

(b) ____________

(c) ____________

29. The table shows the U.S. municipal solid waste recycling percents since 1985, where year 0 represents 1985. (Source: www.epa.gov/epaoswer non-hw/muncpl/pubs/msw06.pdf)

Year	Recycling Percent
0	10.1
5	16.2
10	26.0
15	29.1
20	32.5

(a) Using the first and last points, find the equation of a line that models the data. Write the equation in slope-intercept form.

(b) Use the equation from part (a) to predict the percent of municipal solid waste recycling in the year 2015.

29. (a) ____________

(b) ____________

Name: Date:
Instructor: Section:

30. The table shows the average annual telephone expenditures for residential and pay telephones from 2001 to 2006, where year 0 represents 2001. (Source: http://www.bls.gov/cex/cellphones.htm)

30. (a) ______________

(b) ______________

Year	Annual Telephone Expenditures
0	$686
2	$620
3	$592
4	$570
5	$542

(a) Using the first and last points, find the equation of a line that models the data. Write the equation in slope-intercept form.

(d) Use the equation from part (c) to predict the annual telephone expenditures in 2010.

Name: Date:
Instructor: Section:

Chapter 3 GRAPHS OF LINEAR EQUATIONS AND INEQUALITIES; FUNCTIONS

3.5 Graphing Linear Inequalities in Two Variables

Learning Objectives
1 Graph linear inequalities.
2 Graph a linear inequality with boundary line through the origin.

Key Terms

Use the vocabulary terms listed below to complete each statement in exercises 1–2.

linear inequality in two variables **boundary line**

1. In the graph of a linear inequality, the ______________________________ separates the region that satisfies the inequality from the region that does not satisfy the inequality.

2. An inequality that can be written in the form $Ax + By < C$, $Ax + By > C$, $Ax + By \leq C$, or $Ax + By \geq C$ is called a________________________________.

Objective 1 Graph linear inequalities.

Graph each linear inequality.

1. $y \geq x - 1$

1.

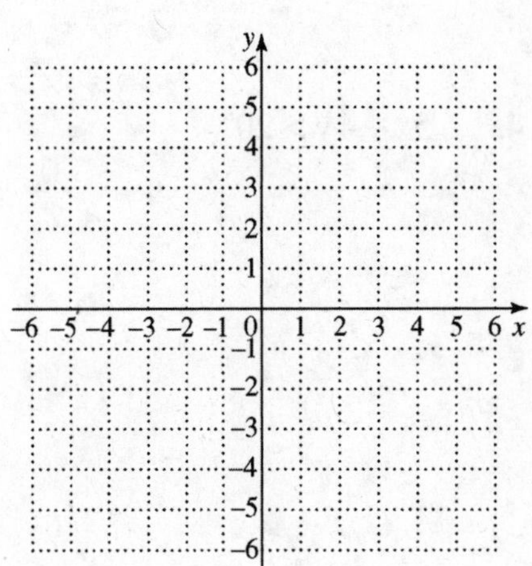

2. $y > -x + 2$

2.

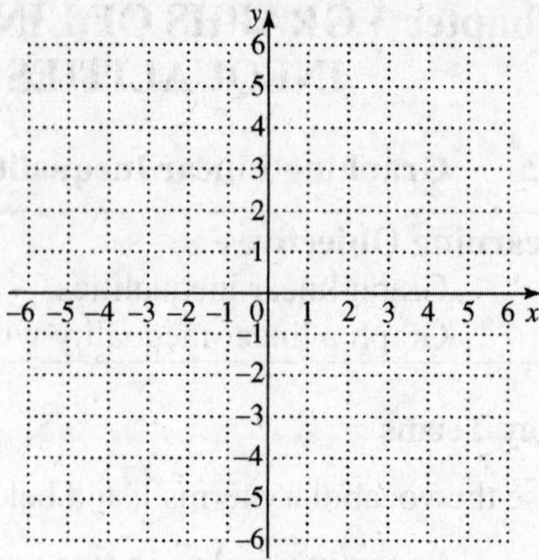

3. $3x - 2y \le 6$

3.

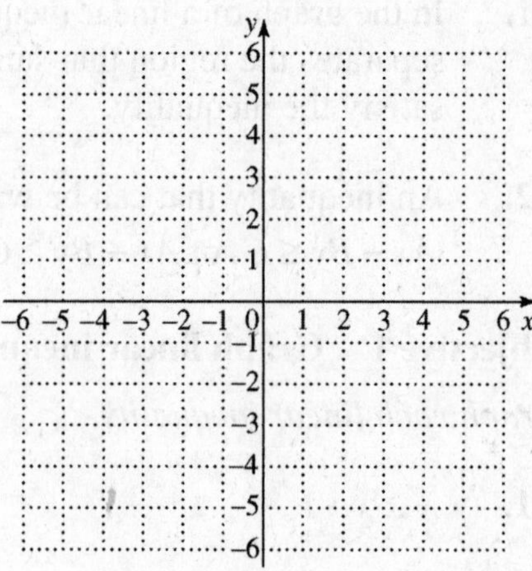

4. $5x + 4y > 20$

4.

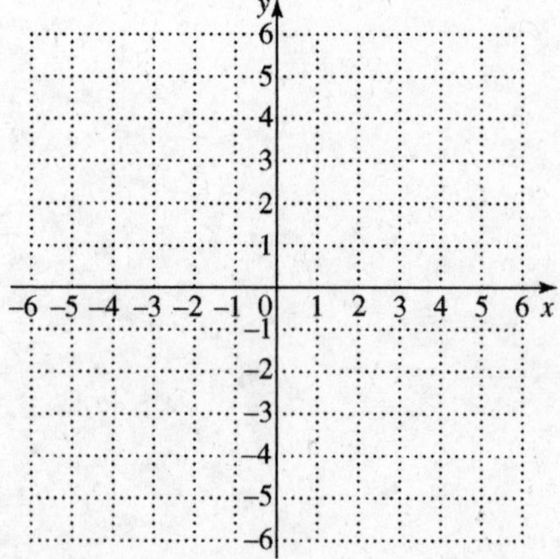

Name: Date:
Instructor: Section:

5. $x-4 \le -1$

5.

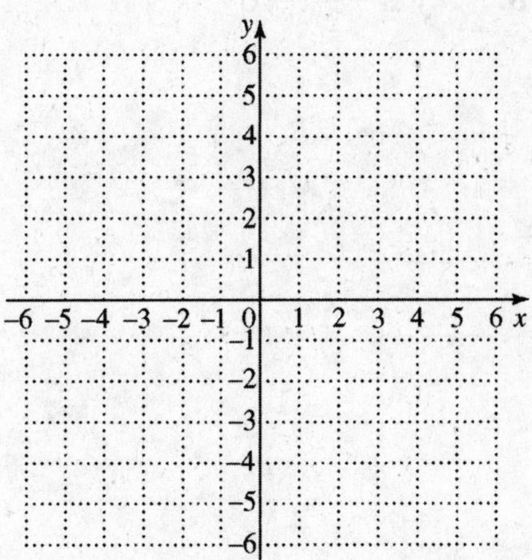

6. $2x+5y > -10$

6.

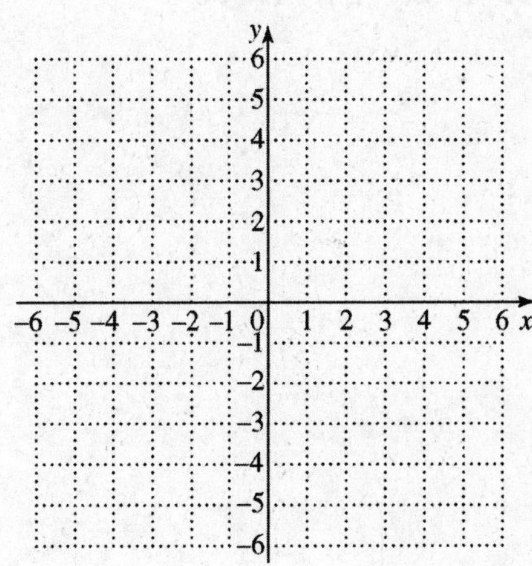

7. $2x+5y \le -8$

7.

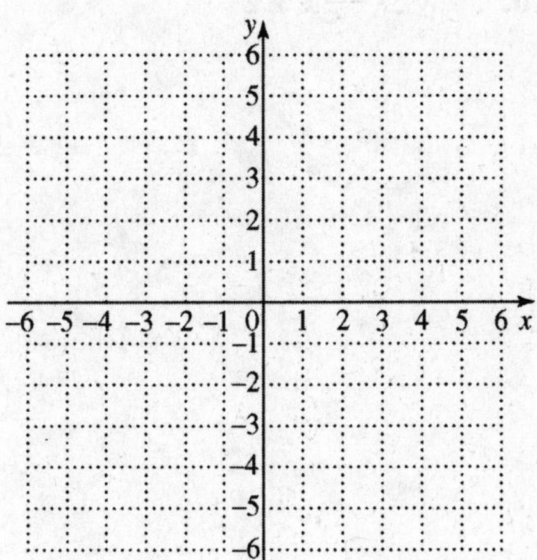

Name: Date:
Instructor: Section:

8. $y \le -\frac{1}{2}x + 6$

8.

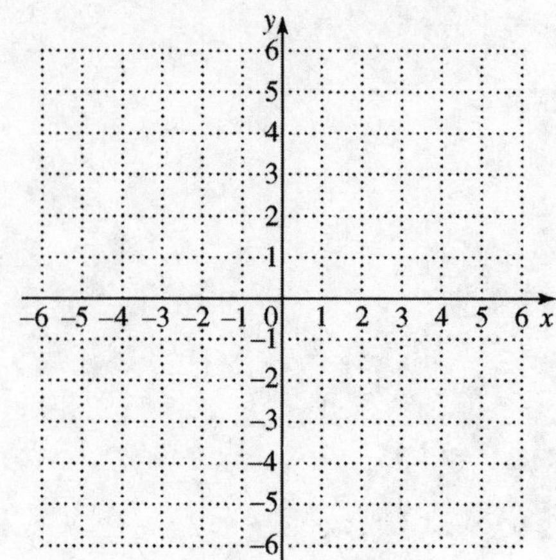

9. $5x - 2y + 10 < 0$

9.

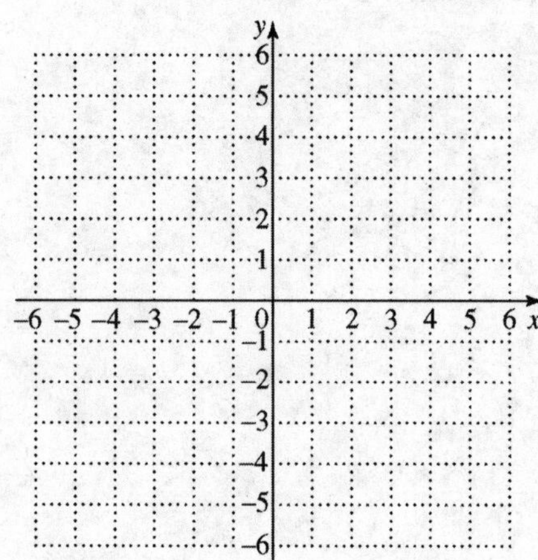

10. $y \le -\frac{2}{5}x + 2$

10.

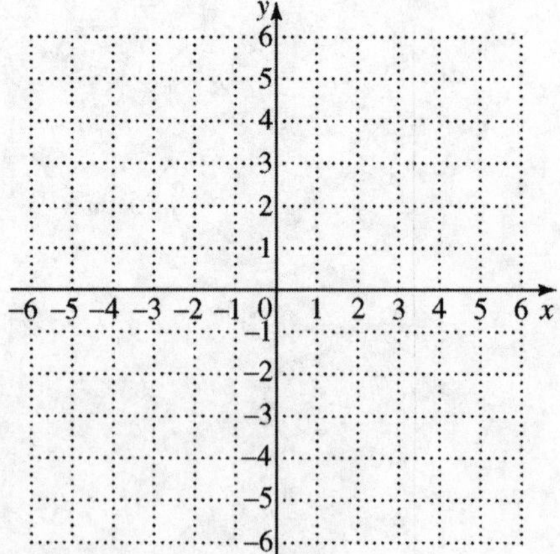

Name: Date:
Instructor: Section:

11. $3x - 5y > -15$

11.

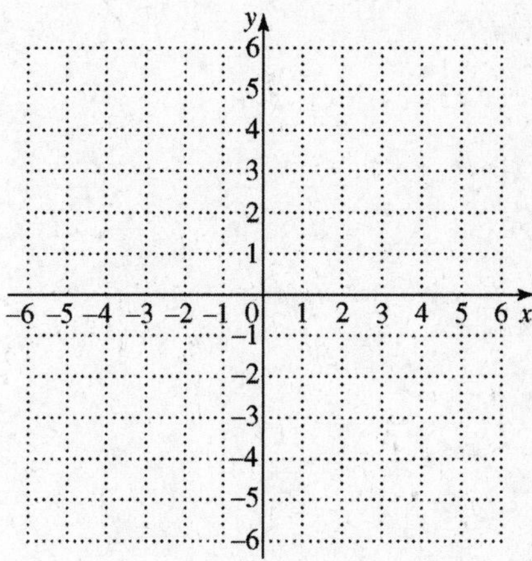

12. $y \geq -1$

12.

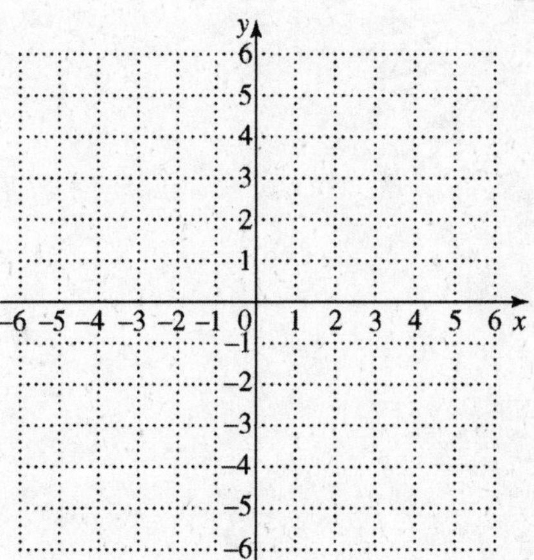

13. $3x + 2y \leq -6$

13.

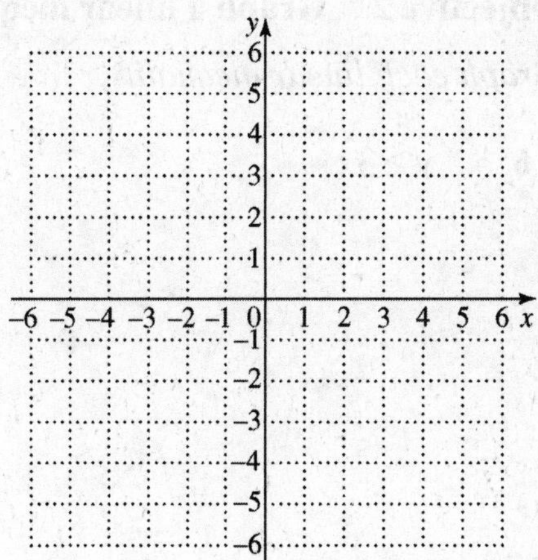

Name: Date:
Instructor: Section:

14. $2-3y>x$

14.

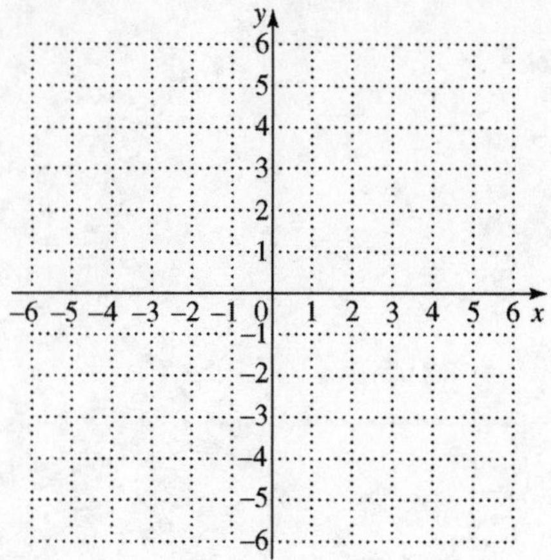

15. $3x-4y-12>0$

15.

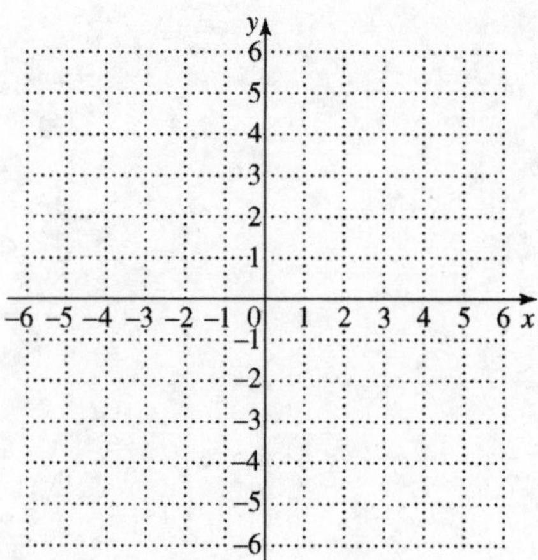

Objective 2 Graph a linear inequality with boundary through the origin.

Graph each linear inequality.

16. $y \geq 3x$

16.

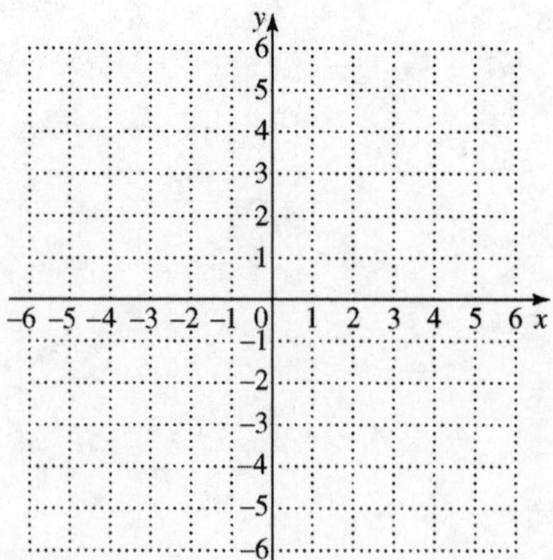

Name: Date:
Instructor: Section:

17. $y \le \frac{2}{5}x$

17.

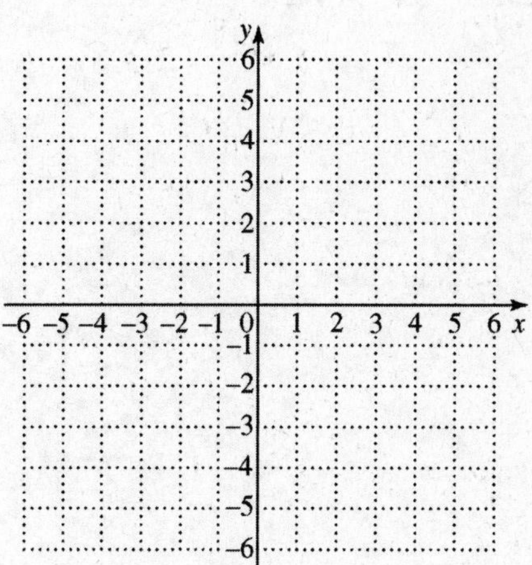

18. $y \ge \frac{1}{3}x$

18.

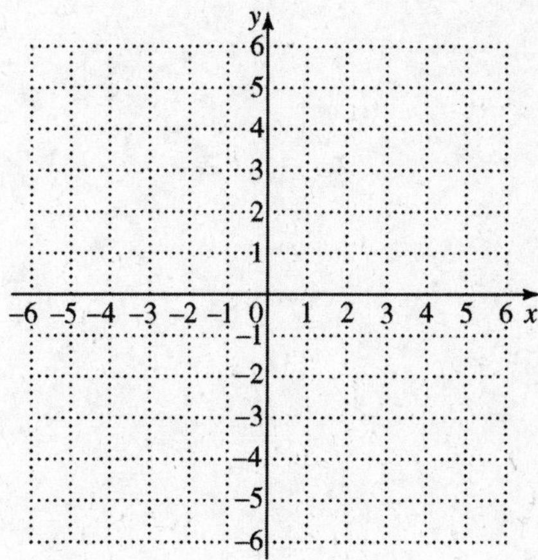

19. $y \ge x$

19.

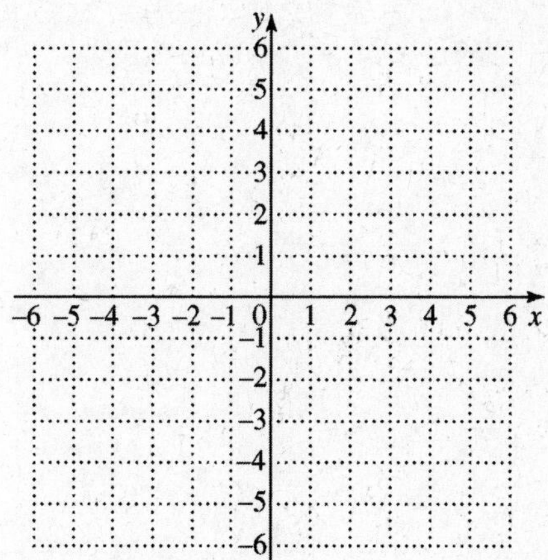

Name: Date:
Instructor: Section:

20. $3x - 4y \geq 0$

20.

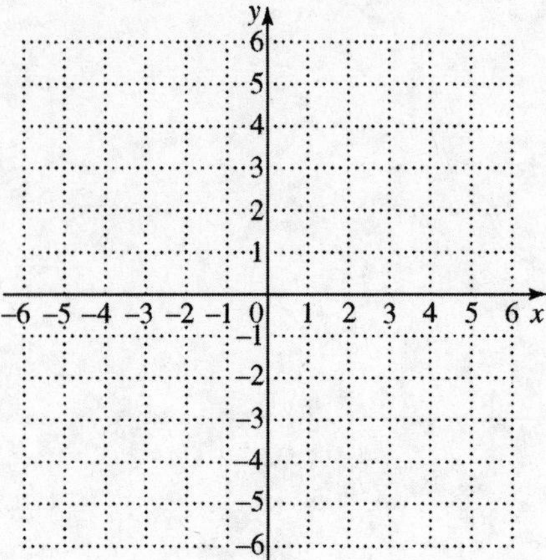

21. $x \geq -4y$

21.

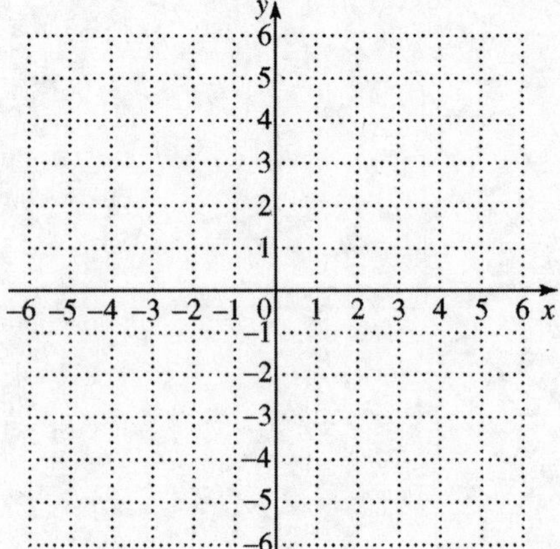

22. $x < 2y$

22.

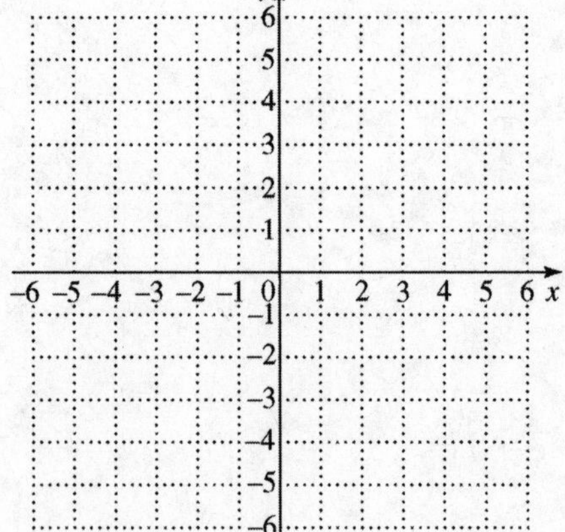

Name: Date:
Instructor: Section:

23. $x < -2y$

23.

24. $x > 4y$

24.

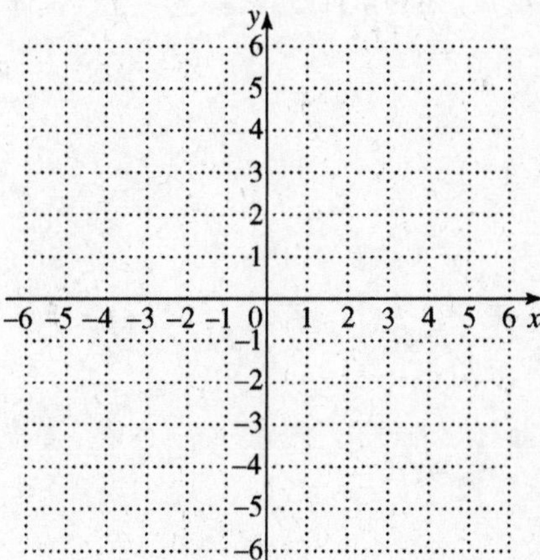

25. $3x - 2y < 0$

25.

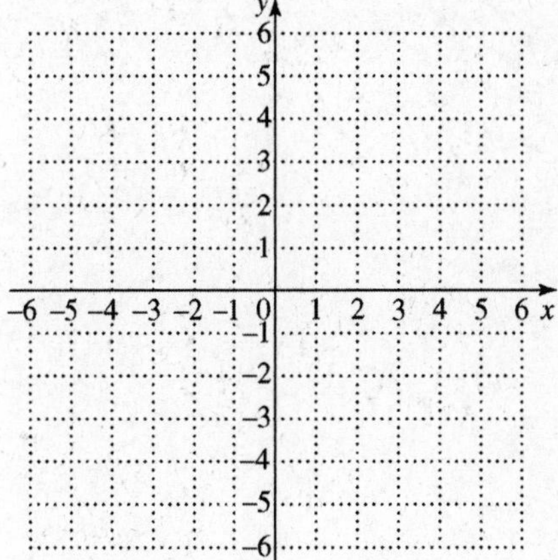

Name: Date:
Instructor: Section:

26. $y \geq -\frac{1}{2}x$

26.

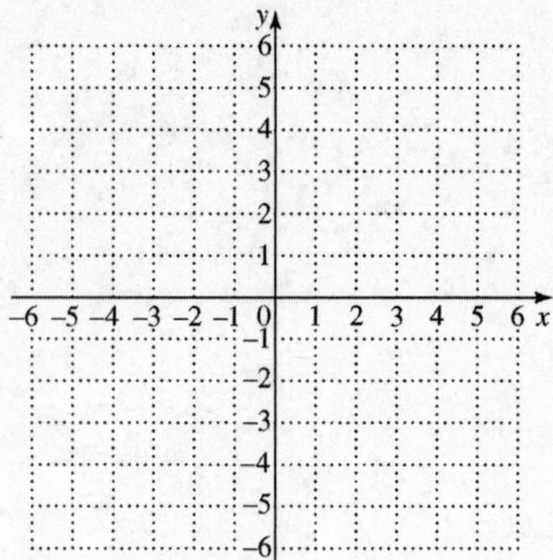

27. $x < \frac{1}{3}y$

27.

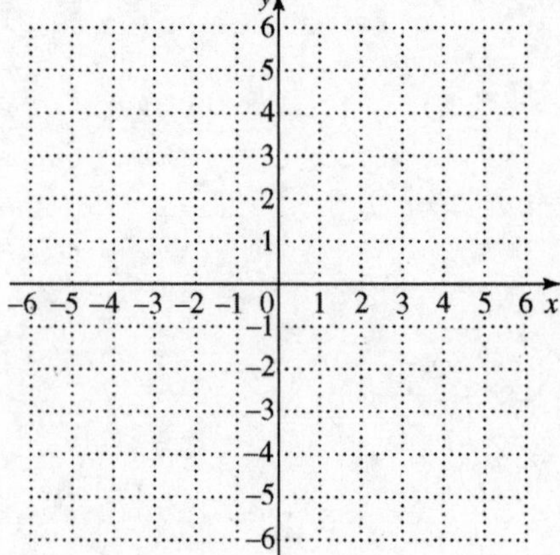

28. $3x \geq 5y$

28.

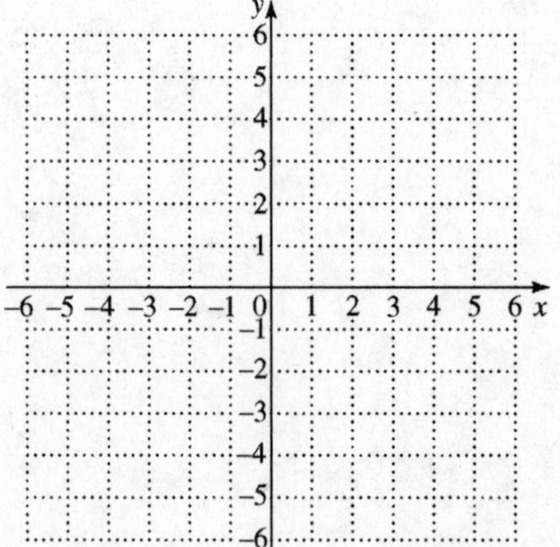

Name:
Instructor:

Date:
Section:

29. $2y - 3x \geq 0$

29.

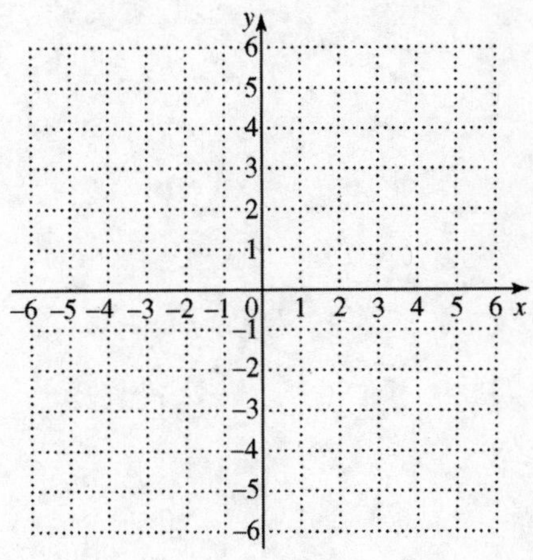

30. $x > -\frac{1}{3}y$

30.

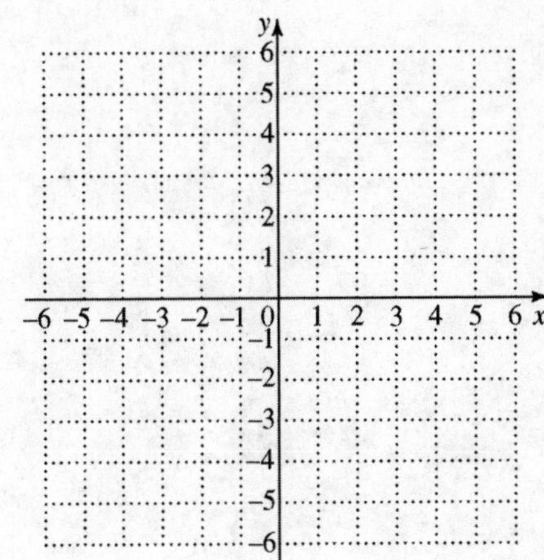

Name: Date:
Instructor: Section:

Chapter 3 GRAPHS OF LINEAR EQUATIONS AND INEQUALITIES; FUNCTIONS

3.6 Introduction to Functions

Learning Objectives

1. Define and identify relations and functions.
2. Find domain and range.
3. Identify functions defined by graphs and equations.
4. Use function notation.
5. Graph linear and constant functions.

Key Terms

Use the vocabulary terms listed below to complete each statement in exercises 1–9.

dependent variable	**independent variable**	**relation**
function	**domain**	**range**
function notation	**linear function**	**constant function**

1. The ____________________ of a relation is the set of second components (y-values) of the ordered pairs of the relation.

2. A ____________________ is a set of ordered pairs of real numbers.

3. If the quantity y depends on x, then y is called the ____________________ in a relation between x and y.

4. The ____________________ of a relation is the set of first components (x-values) of the ordered pairs of the relation.

5. The ____________________ $f(x)$ is another way to represent the dependent variable y for the function f.

6. A ____________________ is a linear function of the form $f(x) = b$, for a real number b.

7. A ____________________ is a set of ordered pairs in which each value of the first component, x, corresponds to exactly one value of the second component, y.

8. If the quantity y depends on x, then x is called the ____________________ in a relation between x and y.

9. A function that is defined by $y = mx + b$ is a ____________________.

Name: Date:
Instructor: Section:

Objective 1 Define and identify relations and functions.
Objective 2 Find domain and range.

Decide whether or not the relations graphed or defined are functions and give the domain and range of each.

1. $\{(2,7),(5,-4),(-3,-1),(0,-8),(5,2)\}$

1. ________

domain: ________

range: ________

2. $\{(1,3);(5,7),(11,9),(8,-2),(6,-7),(-4,-3)\}$

2. ________

domain: ________

range: ________

3. $\{(0,1),(2,6),(-3,7),(2,9),(-7,1),(-3,4)\}$

3. ________

domain: ________

range: ________

4. $\{(3,5),(3,8),(3,-4),(3,1),(3,0)\}$

4. ________

domain: ________

range: ________

5. $\{(1,4),(3,4),(7,4),(-2,4),(-5,4)\}$

5. ________

domain: ________

range: ________

6. $\{(-1.2,4),(1.8,-2.5),(3.7,-3.8),(3.7,3.8)\}$

6. ________

domain: ________

range: ________

Name: Date:
Instructor: Section:

7. $\{(-3,5),(-2,5),(-1,0),(0,-5),(1,5)\}$

7. ____________

domain: ____________

range: ____________

8. {(3, 4), (5, 2), (4, 3), (5, 3), (−2, 2)}

8. ____________

domain: ____________

range: ____________

9.

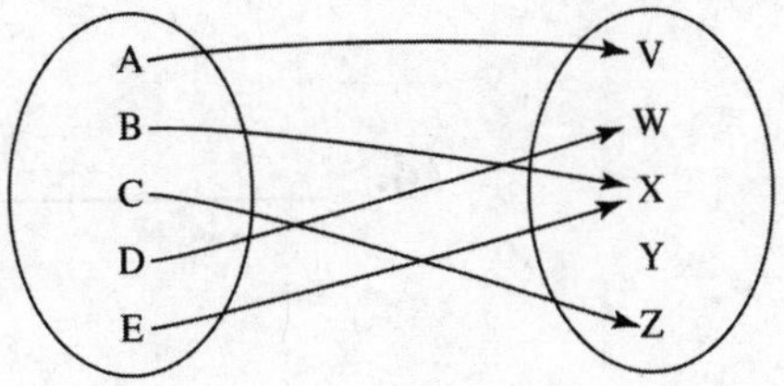

A → V
B → X
C → Z
D → W
E → X

Y

9. ____________

domain: ____________

range: ____________

10.

A → 50
B → 70
C → 90
D → 60
D → 80
E → 70

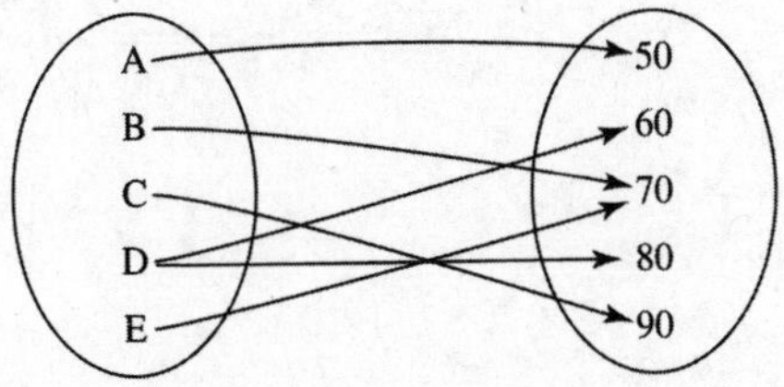

10. ____________

domain: ____________

range: ____________

11.

x	y
1	3
2	−1
−1	4
1	4

11. ____________

domain: ____________

range: ____________

12.

x	y
4	2
3	2
2	2
1	2

12. ____________

domain: ____________

range: ____________

Name: Date:
Instructor: Section:

Objective 3 Identify functions defined by graphs and equations.

Use the vertical line test to determine whether each relation graphed is a function.

13.

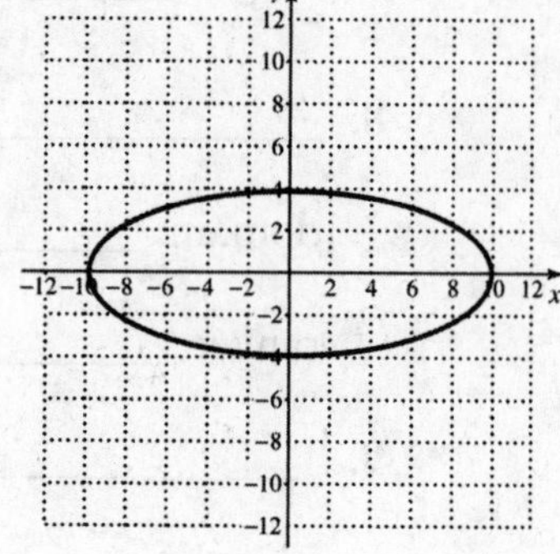

13. 

14.

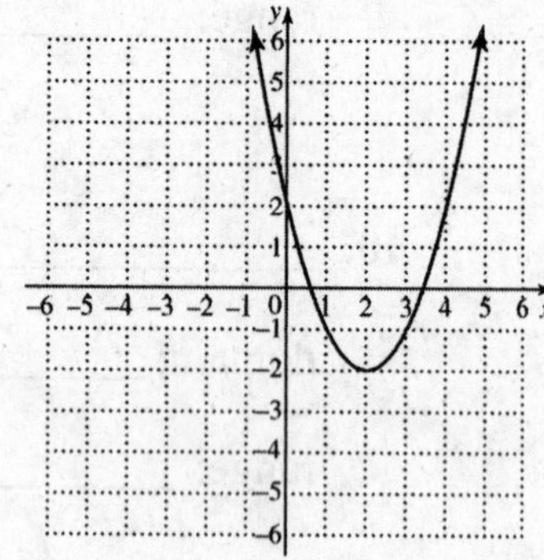

14. 

15.

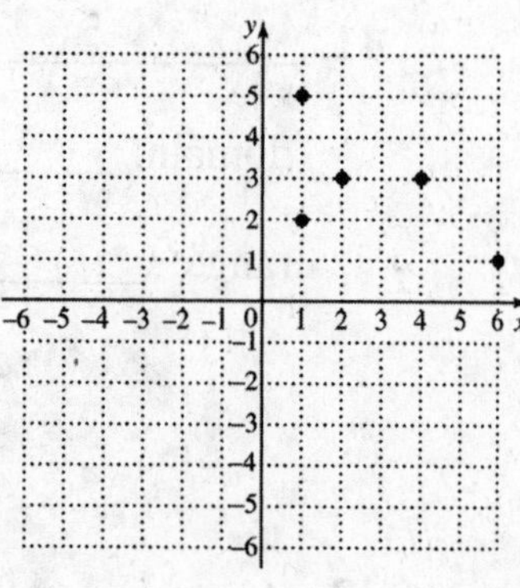

15. 

Decide whether each equation defines y as a function of x.

16. $y^2 = x + 1$

16. ______________

17. $y = \dfrac{1}{x}$

17. ______________

Name: Date:
Instructor: Section:

18. $y = 2x^2 + 3$ **18.** ______________

Objective 4 Use function notation.

For each function f, find (a) $f(-2)$, *(b)* $f(0)$, *and (c)* $f(-x)$.

19. $f(x) = 3x - 7$ **19. a.** ______________

b. ______________

c. ______________

20. $f(x) = x^2 - 3x + 2$ **20. a.** ______________

b. ______________

c. ______________

21. $f(x) = 2x^2 + x - 5$ **21. a.** ______________

b. ______________

c. ______________

22. $f(x) = |2x + 3|$ **22. a.** ______________

b. ______________

c. ______________

Name: Date:
Instructor: Section:

23. $f(x) = x^3 - 2x^2 + 4$

23. a.______________

b.______________

c.______________

24. $f(x) = 9$

24. a.______________

b.______________

c.______________

Objective 5 Graph linear and constant functions.

Graph each function. Give the domain and range.

25. $\frac{1}{3}y = x$

25. domain ____________

range____________

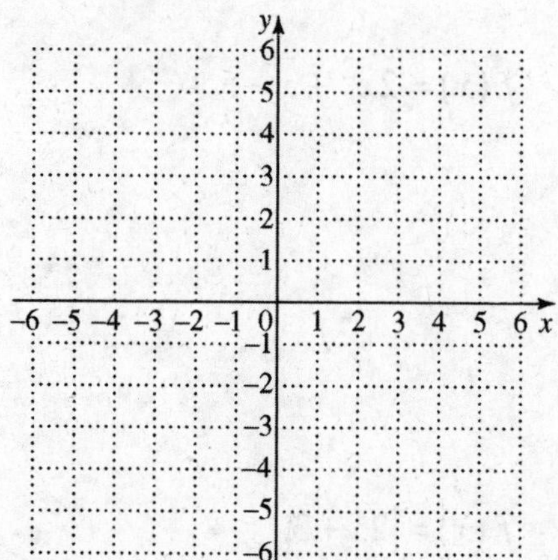

Name:
Instructor:

Date:
Section:

26. $2x - y = -2$

26. domain __________

range__________

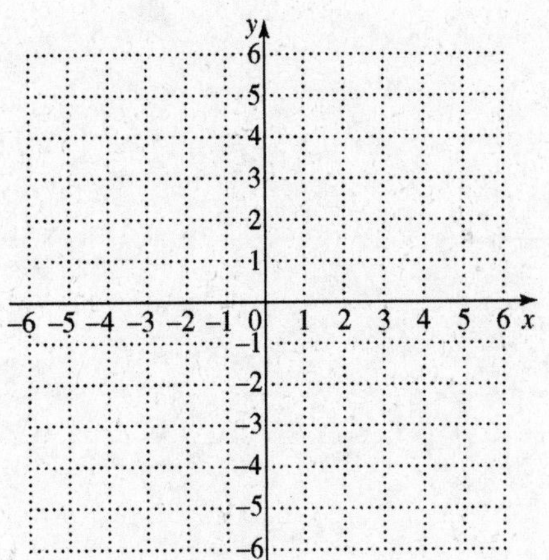

27. $y + \frac{1}{2}x = -2$

27. domain __________

range__________

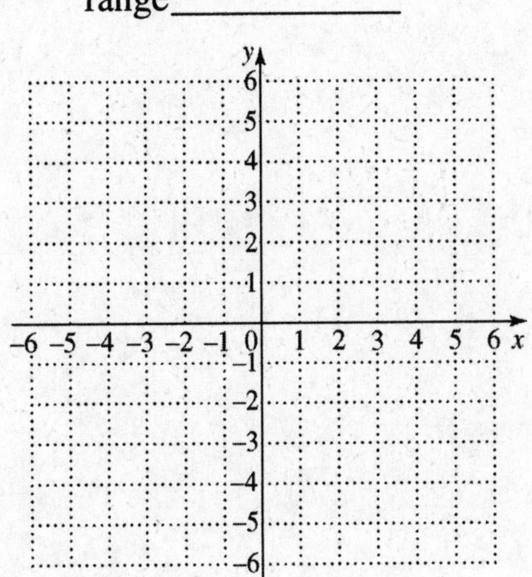

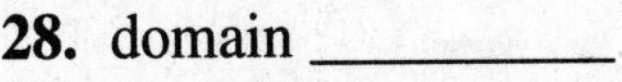

28. domain __________

range__________

28. $\frac{1}{2}x + \frac{1}{3}y = -1$

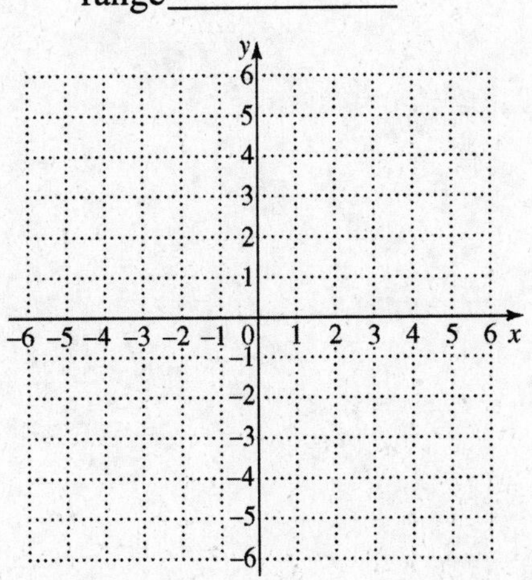

Name:
Instructor:

Date:
Section:

29. $y = 2$

29. domain__________

range__________

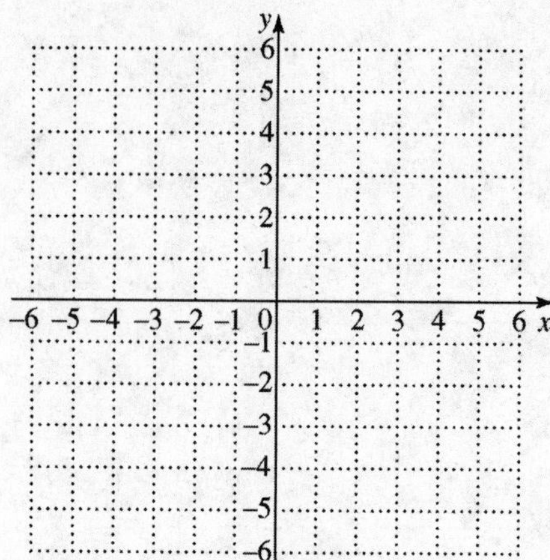

30. $x - 4y = 8$

30. domain__________

range__________

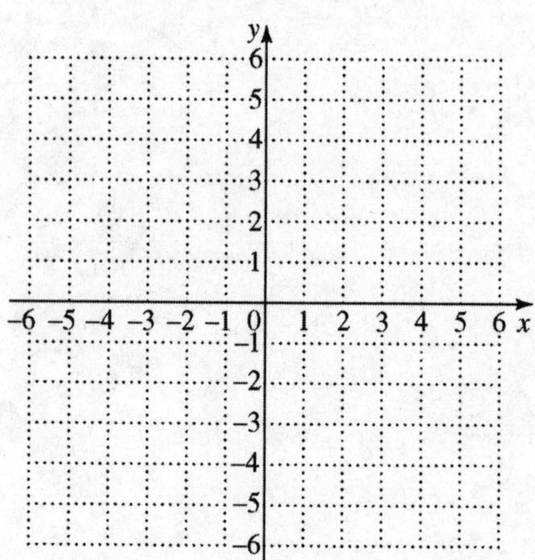

Name: Date:
Instructor: Section:

Chapter 4 SYSTEMS OF LINEAR EQUATIONS AND INEQUALITIES

4.1 Solving Systems of Linear Equations by Graphing

Learning Objectives

1. Decide whether a given ordered pair is a solution of a system.
2. Solve linear systems by graphing.
3. Solve special systems by graphing.
4. Identify special systems without graphing.

Key Terms

Use the vocabulary terms listed below to complete each statement in exercises 1–7.

system of linear equations **solution of the system**

solution set of the system **consistent system** **inconsistent system**

independent equations **dependent equations**

1. Equations of a system that have different graphs are called ____________________.

2. A system of equations with at least one solution is a ____________________.

3. The set of all ordered pairs that are solutions of a system is the ____________________.

4. The ____________________ of linear equations includes all the ordered pairs that make all the equations of the system true at the same time.

5. Equations of a system that have the same graph (because they are different forms of the same equation) are called ____________________.

6. A system with no solution is called a(n) ____________________.

7. A(n) ____________________ consists of two or more linear equations with the same variables.

Name: Date:
Instructor: Section:

Objective 1 Decide whether a given ordered pair is a solution of a system.

Decide whether the given ordered pair is a solution of the given system.

1. $(4,1)$

$2x+3y=11$

$3x-2y=9$

1. ____________

2. $(2,-4)$

$2x+3y=6$

$3x-2y=14$

2. ____________

3. $(-3,-1)$

$5x-3y=-12$

$2x+3y=-9$

3. ____________

4. $(4,0)$

$4x+3y=16$

$x-4y=-4$

4. ____________

5. $(-5,-4)$

$x-y=-1$

$4x+y=-24$

5. ____________

6. $(3,-7)$

$5x+y=8$

$2x-3y=26$

6. ____________

7. $(-1,-7)$

$x-y=6$

$-2x+3y=-19$

7. ____________

Name: Date:
Instructor: Section:

Objective 2 Solve linear systems by graphing.

Solve each system by graphing both equations on the same axes.

8. $x - 2y = 6$
$2x + y = 2$

8.

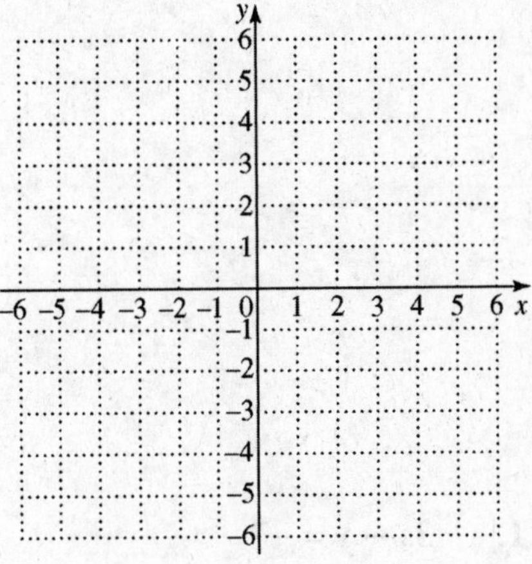

9. $2x + 3y = 5$
$3x - y = 13$

9.

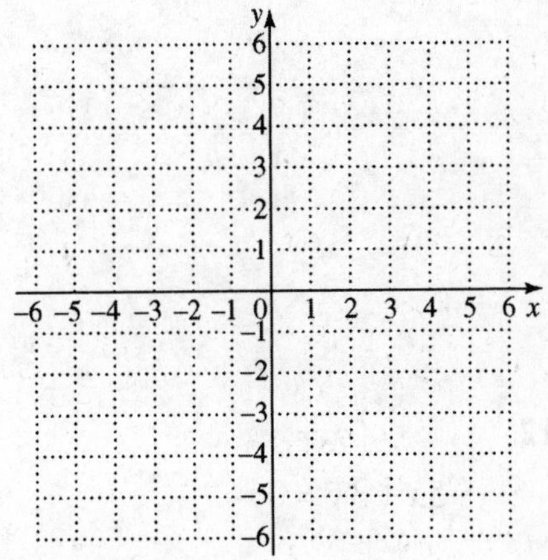

Name: Date:
Instructor: Section:

10. $6x - 5y = 4$
$2x - 5y = 8$

10.

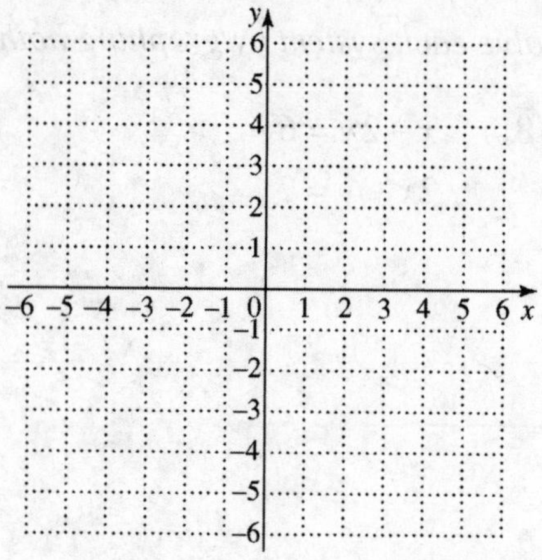

11. $3x - y = -7$
$2x + y = -3$

11.

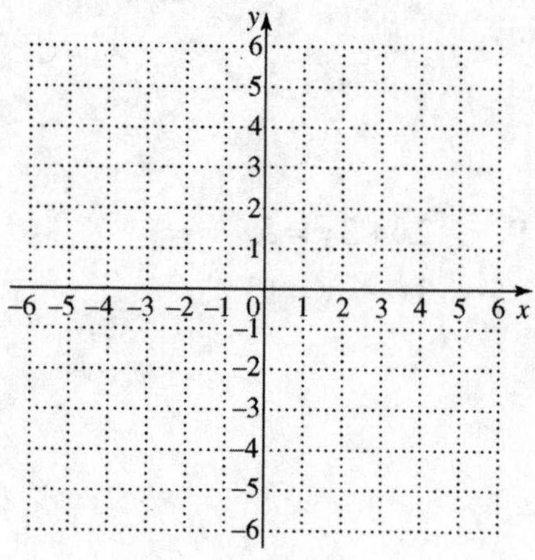

12. $2x = y$
$5x + 3y = 0$

12.

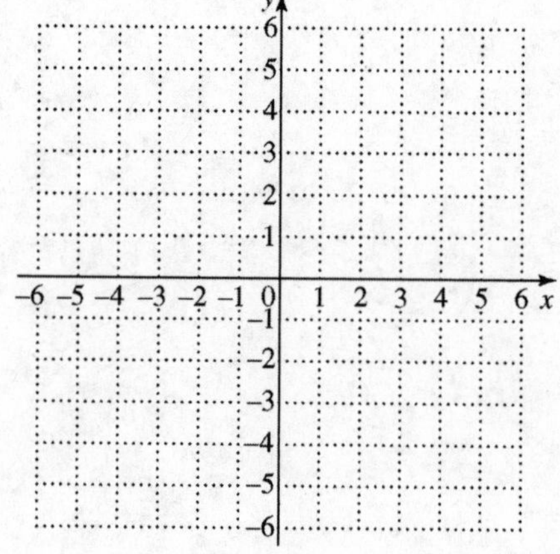

Name: Date:
Instructor: Section:

13. $y - 2 = 0$
$3x - 4y = -17$

13.

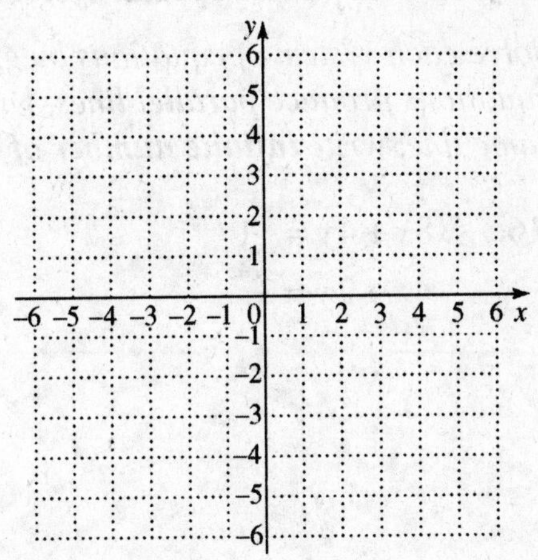

14. $3x + 2 = y$
$2x - y = 0$

14.

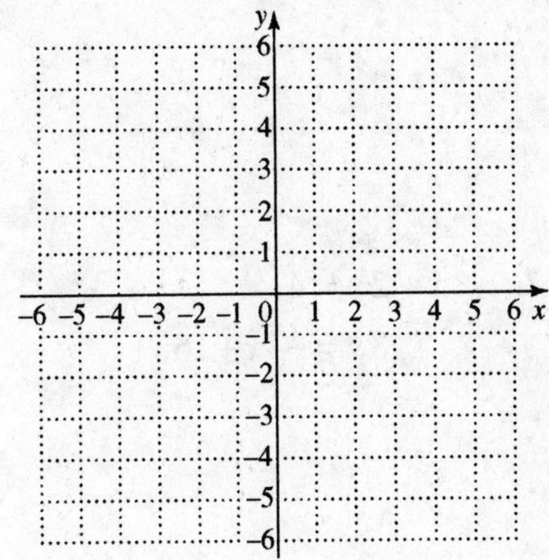

15. $x - y = -7$
$x + 11 = 2y$

15.

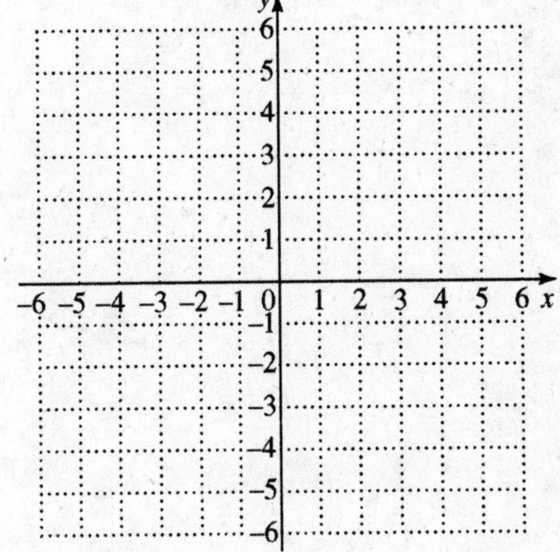

Name: Date:
Instructor: Section:

Objective 3 Solve special systems by graphing.

Solve each system of equations by graphing both equations on the same axes. If the two equations produce parallel lines, write ***no solution****. If the two equations produce the same line, write* ***infinite number of solutions****.*

16. $8x + 4y = -1$

$4x + 2y = 3$

16.

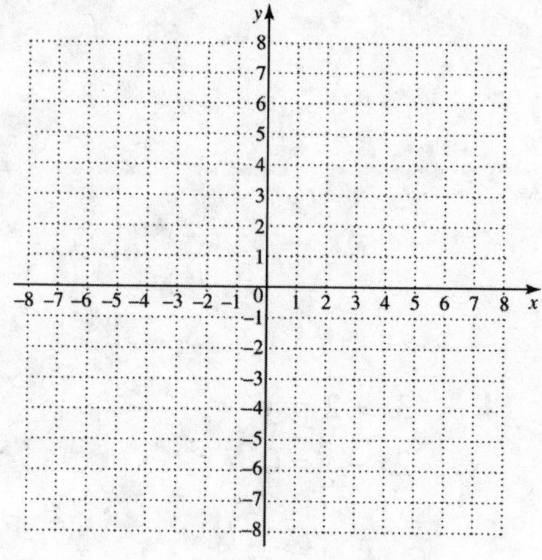

17. $x + 2y = 4$

$8y = -4x + 16$

17.

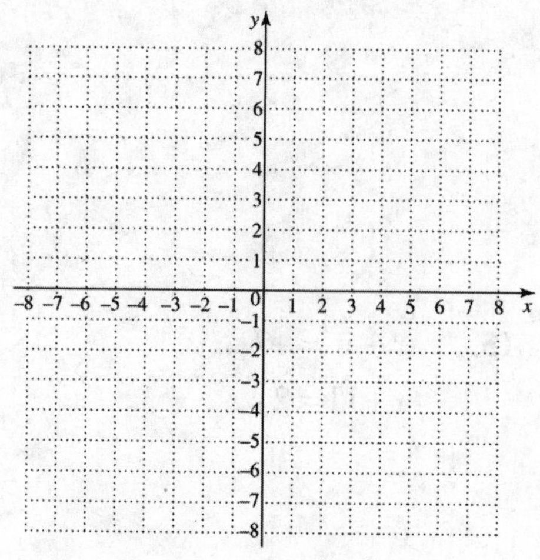

Name: Date:
Instructor: Section:

18. $4x+3y=12$

$6y+8x=-24$

18.

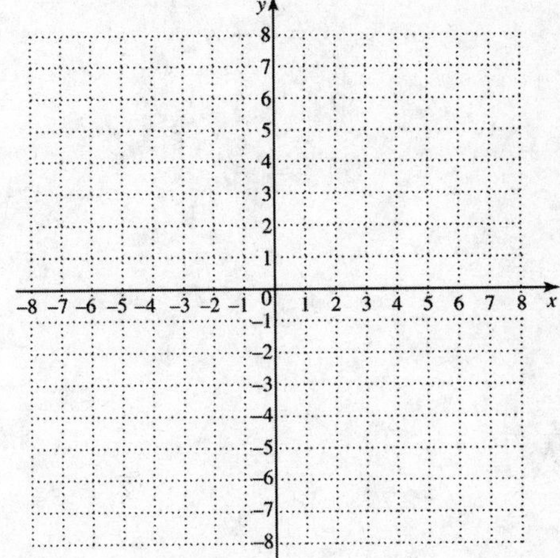

19. $2x+3y=0$

$6x=-9y$

19.

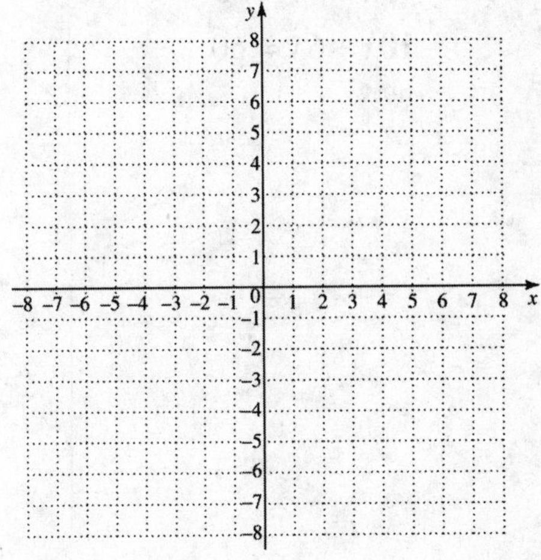

20. $-3x+2y=6$

$-6x+4y=12$

20.

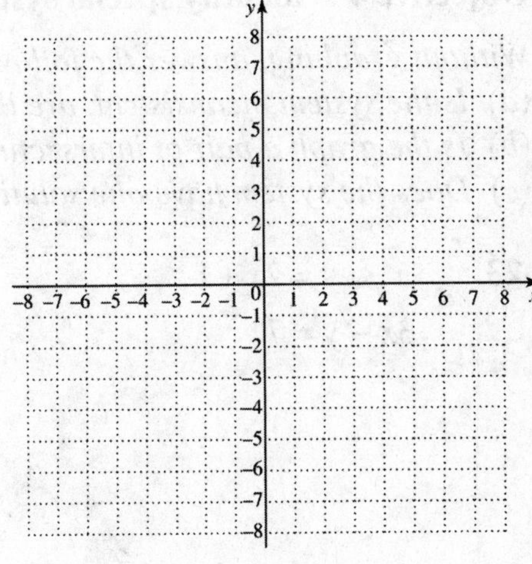

Name: Date:
Instructor: Section:

21. $3x+3y=8$

$x=4-y$

21.

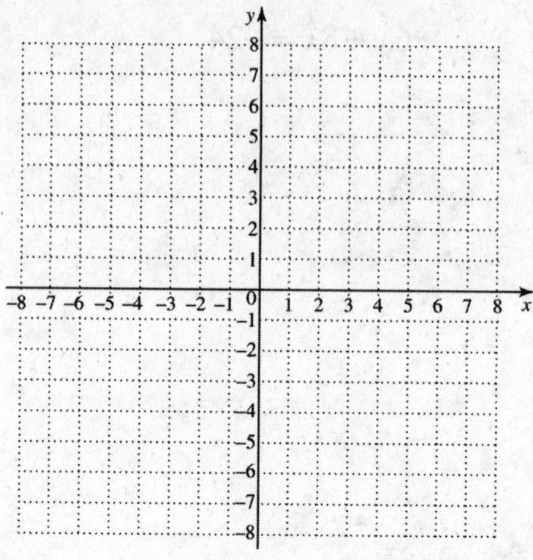

____

22. $5x+3y=30$

$10x+6y=60$

22.

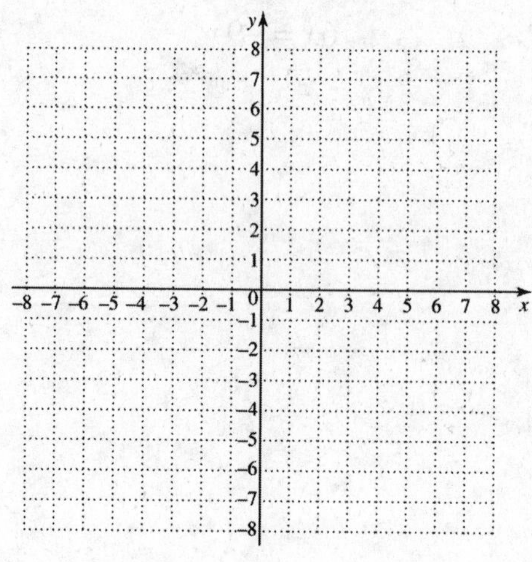

____

Objective 4 Identify special systems without graphing.

Without graphing, answer the following equations for each linear system.
(a) Is the system inconsistent, are the equations dependent, or neither?
(b) Is the graph a pair of intersecting lines, a pair of parallel lines, or one line?
(c) Does the system have one solution, no solution, or an infinite number of solutions?

23. $y=2x+1$

$3x-y=7$

23. (a)______________

(b)______________

(c)______________

Name: Date:
Instructor: Section:

24. $$\begin{aligned} -2x + y &= 4 \\ -4x + 2y &= -2 \end{aligned}$$

24. (a)____________

(b)____________

(c)____________

25. $$\begin{aligned} 4x + 3y &= 12 \\ -12x &= -36 + 9y \end{aligned}$$

25. (a)____________

(b)____________

(c)____________

26. $$\begin{aligned} x - y &= 1 \\ y &= x - 1 \end{aligned}$$

26. (a)____________

(b)____________

(c)____________

27. $$\begin{aligned} y &= -2x + 5 \\ 3x - 2y &= 4 \end{aligned}$$

27. (a)____________

(b)____________

(c)____________

28. $$\begin{aligned} y &= \frac{1}{2}x - 4 \\ 2y &= x - 8 \end{aligned}$$

28. (a)____________

(b)____________

(c)____________

Name: Date:
Instructor: Section:

29. $3x+5y=-5$

$2y+5=0$

29. (a)______________

(b)______________

(c)______________

30. $y=\frac{1}{3}x+2$

$3y=x-21$

30. (a)______________

(b)______________

(c)______________

Name: Date:
Instructor: Section:

Chapter 4 SYSTEMS OF LINEAR EQUATIONS AND INEQUALITIES

4.2 Solving Systems of Linear Equations by Substitution

Learning Objectives

1. Solve linear systems by substitution.
2. Solve special systems by substitution.
3. Solve linear systems with fractions and decimals by substitution.

Key Terms

Use the vocabulary terms listed below to complete each statement in exercises 1–4.

substitution **ordered pair** **inconsistent system**

dependent system

1. The solution of a linear system of equations is written as a(n) ________________________.

2. When one expression is replaced by another, ____________________ is being used.

3. A system of equations in which all solutions of the first equation are also solutions of the second equation is a(n) ______________________________.

4. A system of equations that has no common solution is called a(n) ________________________.

Objective 1 Solve linear systems by substitution.

Solve each system by the substitution method. Check each solution.

1. $x+y=7$
   $y=6x$

   1. __________________

2. $3x+2y=14$
   $y=x+2$

   2. __________________

Name: Date:
Instructor: Section:

3. $x + y = 9$
$5x - 2y = -4$

3. ______________

4. $x - 4y = 17$
$3x - 4y = 11$

4. ______________

5. $-8x + 5y = 11$
$x - y = -1$

5. ______________

6. $5x + y = 8$
$10x - 3y = -19$

6. ______________

7. $x - y = 6$
$-2x + 3y = -19$

7. ______________

8. $3x - 21 = y$
$y + 2x = -1$

8. ______________

9. $x + 6y = -1$
$-2x - 9y = 0$

9. ______________

Name: Date:
Instructor: Section:

10. $2x + 3y = -4$

$3x - 2y = 7$

10. ________________

11. $2x - 5y = 11$

$3x - 4y = 6$

11. ________________

12. $2x + 4y = -1$

$-4x - 6y = 1$

12. ________________

Objective 2 Solve special systems by substitution.

Solve each system by the substitution method. Use set-builder notation for dependent equations.

13. $y = -\frac{1}{3}x + 5$

$3y + x = -9$

13. ________________

14. $\frac{1}{2}x + 3 = y$

$6 = -x + 2y$

14. ________________

15. $3x + 5y = 7$

$6x + 10y = 3$

15. ________________

Name: Date:
Instructor: Section:

16. $4x + 3y = 2$
$8x + 6y = 4$

16. ______________

17. $3x + 5y = 3$
$6x + 10y = -2$

17. ______________

18. $4x - 2y = 3$
$-8x + 4y = -6$

18. ______________

Objective 3 Solve linear systems with fractions and decimals by substitution.

Solve each system by the substitution method. Check each solution

19. $\frac{5}{3}x + y = 12$
$x + \frac{1}{2}y = 7$

19. ______________

20. $\frac{5}{4}x - y = -\frac{1}{4}$
$-\frac{7}{8}x + \frac{5}{8}y = 1$

20. ______________

Name: Date:
Instructor: Section:

21. $x+\frac{5}{3}y=7$

$\frac{5}{6}x+\frac{2}{3}y=\frac{11}{3}$

21. ______________

22. $x-\frac{3}{4}y=3$

$\frac{1}{4}x-\frac{1}{2}y=\frac{1}{8}$

22. ______________

23. $\frac{1}{20}x-\frac{1}{15}y=\frac{1}{6}$

$\frac{1}{6}x+y=3$

23. ______________

24. $\frac{1}{4}x+\frac{3}{8}y=-3$

$\frac{5}{6}x-\frac{3}{7}y=-10$

24. ______________

Name: Date:
Instructor: Section:

25. $0.1x + 0.3y = 0.1$
$0.2x = -1 - 1.2y$

25. ______________

26. $0.2y + 0.8x = -0.2$
$0.1x + 0.3y = -1.4$

26. ______________

27. $0.3x - 2.1 = 0.1y$
$0.3y + 0.6x = -0.3$

27. ______________

28. $0.3x + 1.8y = -2.7$
$-0.2x - 0.6y = 0$

28. ______________

29. $0.6x + 0.8y = 1$
$0.4y = 0.5 - 0.3x$

29. ______________

30. $0.4x + 0.3y = -1$
$0.2x + 0.3y = -0.2$

30. ______________

Name: Date:
Instructor: Section:

Chapter 4 SYSTEMS OF LINEAR EQUATIONS AND INEQUALITIES

4.3 Solving Systems of Linear Equations by Elimination

Learning Objectives
1. Solve linear systems by elimination.
2. Multiply when using the elimination method.
3. Use an alternative method to find the second value in a solution.
4. Use the elimination method to solve special systems.

Key Terms

Use the vocabulary terms listed below to complete each statement in exercises 1–3.

addition property of equality **elimination method** **substitution**

1. Using the addition property to solve a system of equation is called the ______________________________.

2. The ______________________________ states that the same quantity to each side of an equation results in equal sums.

3. ______________________________ is being used when one expression is replaced by another.

Objective 1 Solve linear systems by elimination.

Solve each system by the elimination method. Check your answers.

1. $x + y = 5$
$x - y = -3$

1. ____________________

2. $3x - y = 5$
$2x + y = 0$

2. ____________________

Name: Date:
Instructor: Section:

3. $x - 4y = -4$
$-x + y = -5$

3. ______________

4. $2x - y = 10$
$3x + y = 10$

4. ______________

5. $4x + 3y = -4$
$2x - 3y = 16$

5. ______________

6. $8x + 2y = 14$
$3x - 2y = -14$

6. ______________

7. $x - 3y = 5$
$-x + 4y = -5$

7. ______________

Objective 2 Multiply when using the elimination method.

Solve each system by the elimination method. Check your answers.

8. $6x + 7y = 10$
$2x - 3y = 14$

8. ______________

Name: Date:
Instructor: Section:

9. $8x + 6y = 10$
$4x - y = 1$

9. ______________

10. $3x + 2y = 5$
$2x - 3y = 12$

10. ______________

11. $3x + 5y = 8$
$2x - y = -12$

11. ______________

12. $6x + y = 1$
$3x - 4y = 23$

12. ______________

13. $4x - 5y = -22$
$3x + 2y = -5$

13. ______________

14. $4x - 9y = 7$
$3x + 2y = 14$

14. ______________

Name: Date:
Instructor: Section:

15. $3x-7y=12$
$5x+3y=-2$

15. ______________

Objective 3 Use an alternative method to find the second value in a solution.

Solve each system by the elimination method. Check your answers.

16. $5x-3y=23$
$10+2y=2x$

16. ______________

17. $4y=2x-2$
$-9+3y=5x$

17. ______________

18. $4x-3y-20=0$
$6x+5y+8=0$

18. ______________

19. $5x+5y+15=0$
$3x+4y=-8$

19. ______________

20. $6x=16-7y$
$4x=3y+26$

20. ______________

Name: Date:
Instructor: Section:

21. $2x = 14 + 4y$

$6y = -5x + 3$

21. ______________

22. $7 - y = 2x$

$4x = 19 + 3y$

22. ______________

23. $5x + 3y + 4 = 0$

$4x + 5y - 2 = 0$

23. ______________

24. $2x = 21 - 3y$

$\frac{1}{3}x + \frac{2}{5}y = 3$

24. ______________

Objective 4 Use the elimination method to solve special systems.

Solve each system by the elimination method. Use set-builder notation for dependent equations. Check your answers.

25. $12x - 8y = 3$

$6x - 4y = 6$

25. ______________

Name: Date:
Instructor: Section:

26. $2x+4y=-6$
$-x-2y=3$

26. ______________

27. $6x-12y=3$
$2x-\ 4y=1$

27. ______________

28. $15x+6y=9$
$10x+4y=18$

28. ______________

29. $48x-56y=32$
$21y-18x=-12$

29. ______________

30. $15x-10y=6$
$-12x+8y=2$

30. ______________

Name: Date:
Instructor: Section:

Chapter 4 SYSTEMS OF LINEAR EQUATIONS AND INEQUALITIES

4.4 Applications of Linear Systems

Learning Objectives
1. Solve problems about unknown numbers.
2. Solve problems about quantities and their costs.
3. Solve problems about mixtures.
4. Solve problems about distance, rate (or speed), and time.

Key Terms

Use the vocabulary terms listed below to complete each statement in exercises 1–2.

system of linear equations $d = rt$

1. The formula that relates distance, rate, and time is ______________________.

2. A ______________________ consists of at least two linear equations with different variables.

Objective 1 Solve problems about unknown numbers.

Write a system of equations for each problem, then solve the problem.

1. The sum of two numbers is 20. Three times the smaller is equal to twice the larger. Find the numbers.

   1. larger number __________
   smaller number __________

2. The difference between two numbers is 14. If two times the smaller is added to one-half the larger, the result is 52. Find the numbers.

   2. larger number __________
   smaller number __________

3. Two towns have a combined population of 9045. There are 2249 more people living in one than in the other. Find the population in each town.

   3. larger town __________
   smaller town __________

Name: Date:
Instructor: Section:

4. There are a total of 49 students in the two second grade classes at Jefferson School. If Carla has 7 more students in her class than Linda, find the number of students in each class.

4.
Carla's class __________
Linda's class__________

5. A rope 82 centimeters long is cut into two pieces with one piece four more than twice as long as the other. Find the length of each piece.

5.
longer piece __________
shorter piece __________

6. The perimeter of a rectangular room is 50 feet. The length is three feet greater than the width. Find the dimensions of the rectangle.

6.
length ______________
width______________

7. The perimeter of a triangular pennant is 116 centimeters. If two sides are of equal length, and the third side is 20 centimeters longer than each of the equal sides, what are the lengths of the three sides?

7.
side 1______________
side 2______________
side 3______________

Name: Date:
Instructor: Section:

Objective 2 Solve problems about quantities and their costs.

Write a system of equations for each problem, then solve the problem.

8. There were 411 tickets sold for a soccer game, some for students and some for nonstudents. Student tickets cost \$4.25 and nonstudent tickets cost \$8.50 each. The total receipts were \$3021.75. How many of each type were sold?

8.
student tix ___________
nonstudent tix ___________

9. A cashier has some \$5 bills and some \$10 bills. The total value of the money is \$750. If the number of tens is equal to twice the number of fives, how many of each type are there?

9.
\$5 bills ___________
\$10 bills ___________

10. The total receipts for a basketball game were \$4690.50. There were 723 tickets sold, some for children and some for adults. If the adult tickets cost \$9.50 and the children's tickets cost \$4, how many of each type were there?

10.
adult tix ___________
children tix ___________

11. Wendy has \$10,000 to invest, part at 7% and part at 4%. If the total annual income from simple interest is to be \$580, how much should she invest at each rate?

11.
7% amount ___________
4% amount ___________

Name: Date:
Instructor: Section:

12. Twice as many general admission tickets to a basketball game were sold as reserved seat tickets. General admission tickets cost $10 and reserved seat tickets cost $15. If the total value of both kinds of tickets was $26,250, how many tickets of each kind were sold?

12.
general admission______

reserved seats _________

13. Carla has $12,000 to invest at 7% and 9%. She wants the income from simple interest on the two investments to total $1000 yearly. How much should she invest at each rate?

13.
7% amount __________

9% amount __________

14. Stan has 14 bills in his wallet worth $95 altogether. If the wallet contains only $5 and $10 bills, how many bills of each denomination does he have?

14.
$5 bills ____________

$10 bills ___________

15. Luke plans to buy 10 ties with exactly $162. If some ties cost $14, and the others cost $25, how many ties of each price should he buy?

15.
$14 ties _____________

$25 ties _____________

Name: Date:
Instructor: Section:

Objective 3 Solve problems about mixtures.

Write a system of equations for each problem, then solve the problem.

16. Jorge wishes to make 150 pounds of coffee blend that can be sold for $8 per pound. The blend will be a mixture of coffee worth $6 per pound and coffee worth $12 per pound. How many pounds of each kind of coffee should be used in the mixture?

16.
$6 coffee__________
$12 coffee__________

17. How many liters of water should be added to 25% antifreeze solution to get 30 liters of a 20% solution? How many liters of 25% solution are needed?

17.
water__________
25% solution__________

18. Ben wishes to blend candy selling for $1.60 a pound with candy selling for $2.50 a pound to get a mixture that will be sold for $1.90 a pound. How many pounds of the $1.60 and the $2.50 candy should be used to get 30 pounds of the mixture?

18.
$1.60 candy __________
$2.50 candy __________

19. How many bags of coffee worth $90 a bag must be mixed with coffee worth $75 a bag to get 50 bags worth $87 a bag?

19.
$90 coffee__________
$75 coffee__________

Name: Date:
Instructor: Section:

20. How many liters of 75% solution should be mixed with a 55% solution to get 70 liters of 63% solution? How many liters of the 55% and 75% solutions should be used?

20.
55% solution__________
75% solution__________

21. A pharmacist wants to add water to a solution that contains 80% medicine. She wants to obtain 12 oz. of a solution that is 20% medicine. How much water and how much of the 80% solution should she use?

21.
water__________
80% solution__________

22. Three quarts of a 24% iodine solution were mixed with a 52% solution to make a 40% iodine solution. How many quarts of the 52% solution were needed?

22.
52% solution__________
42% solution__________

Objective 4 Solve problems about distance, rate (or speed), and time.

Write a system of equations for each problem, then solve the problem.

23. Bill and Hillary start in Washington and fly in opposite directions. At the end of 4 hours, they are 4896 kilometers apart. If Bill flies 60 kilometers per hour faster than Hillary, what are their speeds?

23.
Bill __________
Hillary__________

Name: Date:
Instructor: Section:

24. It takes Carla's boat $\frac{1}{2}$ hour to go 8 miles downstream and 1 hour to make the return trip upstream. Find the speed of the current and the speed of Carla's boat in still water.

24.
boat speed__________
current speed__________

25. Enid leaves Cherry Hill, driving by car toward New York, which is 90 miles away. At the same time, Jerry, riding his bicycle, leaves New York cycling toward Cherry Hill. Enid is traveling 28 miles per hour faster than Jerry. They pass each other $1\frac{1}{2}$ hours later. What are their speeds?

25.
Enid __________
Jerry __________

26. Two planes left Philadelphia traveling in opposite directions. Plane A left 15 minutes before plane B. After plane B had been flying for 1 hour, the planes were 860 miles apart. What were the speeds of the two planes if plane A was flying 40 miles per hour faster than plane B?

26.
plane A __________
plane B __________

27. John left Louisville at noon on the same day that Mike left Louisville at 1 P.M. Both were traveling in the same direction. At 5 P.M., Mike was 62 miles behind John. If John was traveling 2 miles per hour faster than Mike, what were their speeds?

27.
John __________
Mike __________

Name: Date:
Instructor: Section:

28. It takes a kayak $1\frac{1}{2}$ hours to go 24 miles downstream and 4 hours to return. Find the speed of the current and the speed of the kayak in still water.

28.
kayak speed ________
current speed________

29. A plane can travel 300 miles per hour with the wind and 230 miles per hour against the wind. Find the speed of the wind and the speed of the plane in still air.

29.
plane speed ________
wind speed ________

30. At the beginning of a fund-raising walk, Steve and Vic are 30 miles apart. If they leave at the same time and walk in the same direction, Steve would overtake Vic in 15 hours. If they walked toward each other, they would meet in 3 hours. What are their speeds?

30.
Steve ________
Vic ________

Name: Date:
Instructor: Section:

Chapter 4 SYSTEMS OF LINEAR EQUATIONS AND INEQUALITIES

4.5 Solving Systems of Linear Inequalities

Learning Objectives
1 Solve systems of linear inequalities by graphing.

Key Terms

Use the vocabulary terms listed below to complete each statement in exercises 1–2.

system of linear inequalities

solution set of a system of linear inequalities

1. All points that make all inequalities of the system true at the same time is called the ______________________________________.

2. A ________________________________ contains two or more linear inequalities (and no other kinds of inequalities).

Objective 1 Solve systems of linear inequalities by graphing.

Graph the solution of each system of linear inequalities.

1. $7x+3y \geq 21$
   $x - y \leq 6$

1.

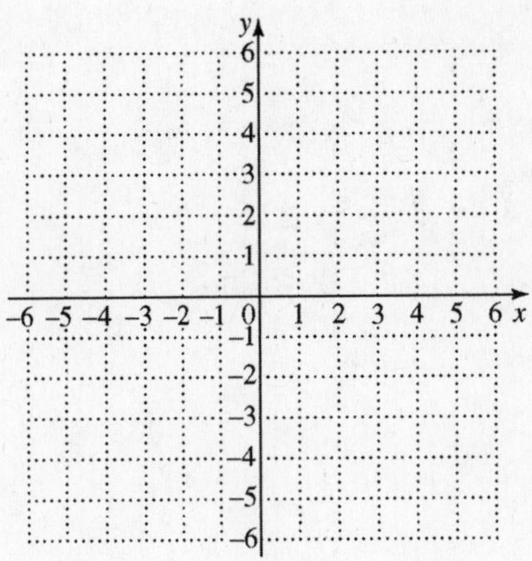

Name: Date:
Instructor: Section:

2. $3x - y \le 3$
$x + y \le 0$

2.

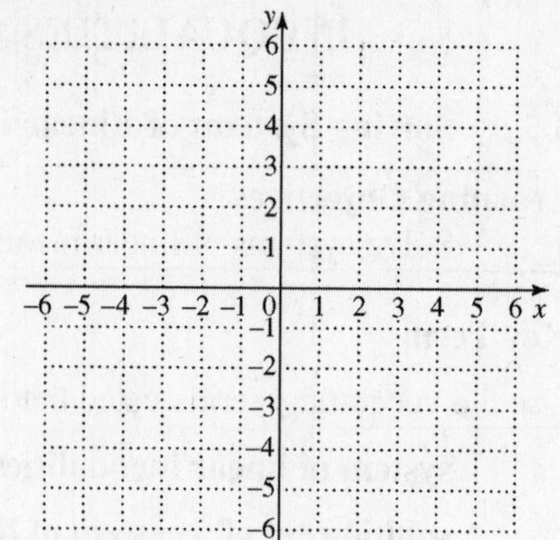

3. $3x - y \le 6$
$3y - 6 \le 2x$

3.

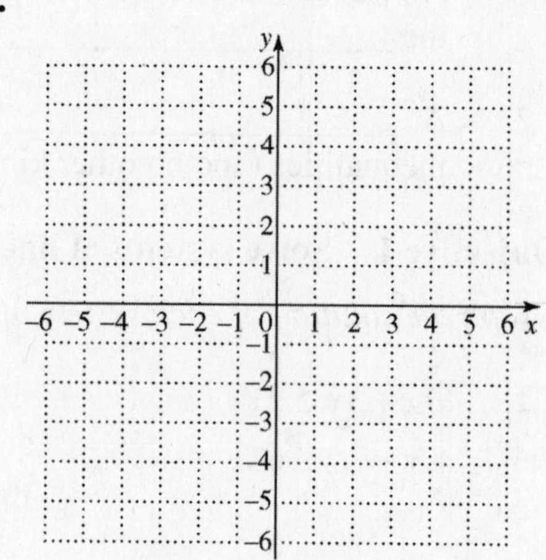

4. 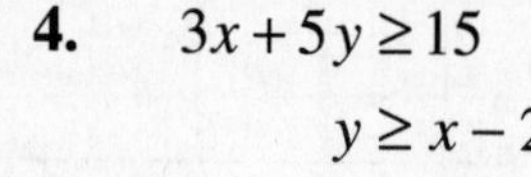
$3x + 5y \ge 15$
$y \ge x - 2$

4.

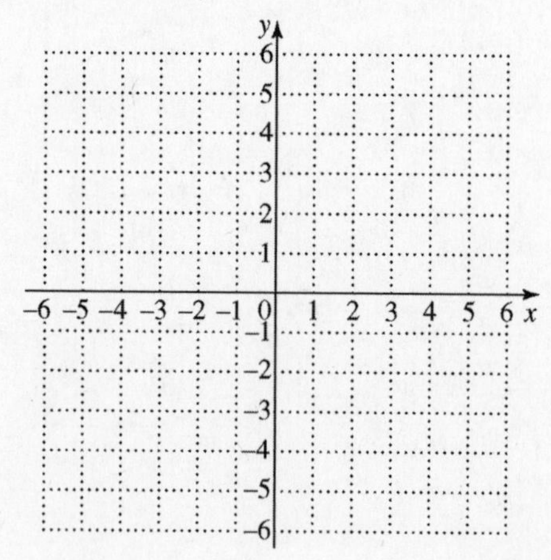

Name: Date:
Instructor: Section:

5. $x + y \leq 3$
$5x - y \geq 5$

5.

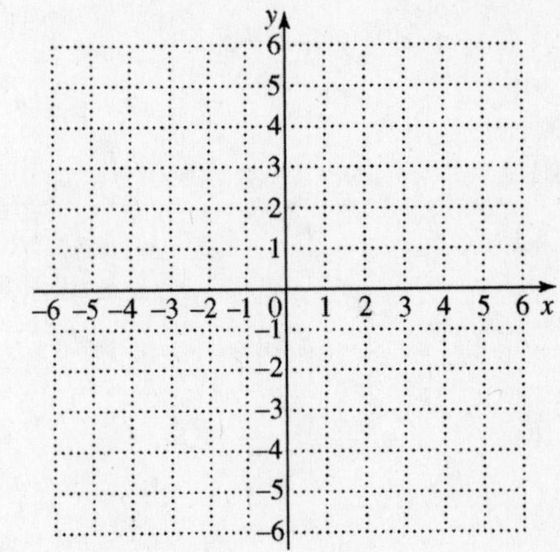

6. $x + 2y \geq -4$
$5x \leq 10 - 2y$

6.

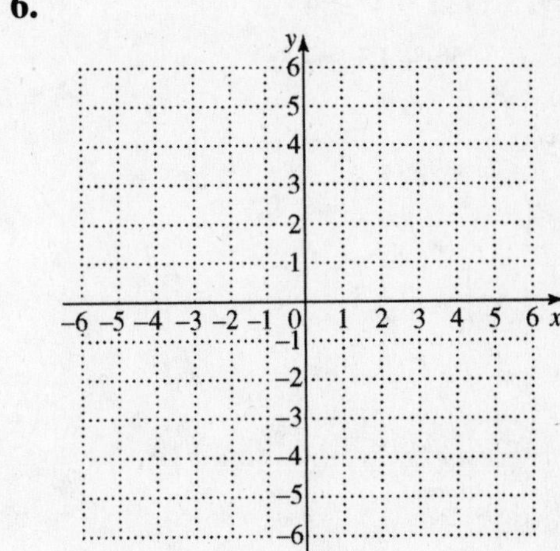

7. $3x - y > 3$
$4x + 3y < 12$

7.

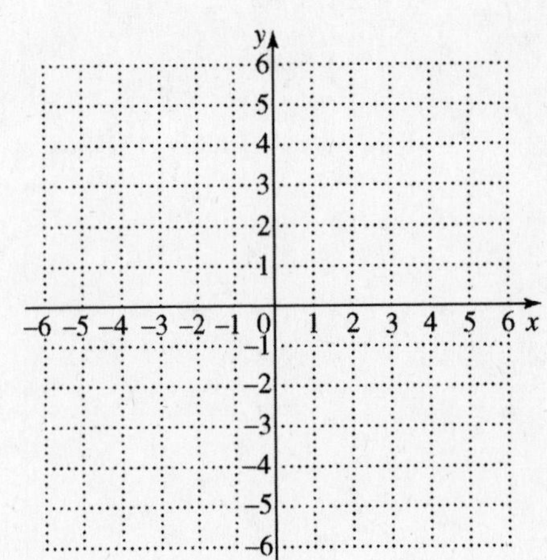

Name: Date:
Instructor: Section:

8. $4x + 5y \leq 20$
$y \leq x + 3$

8.

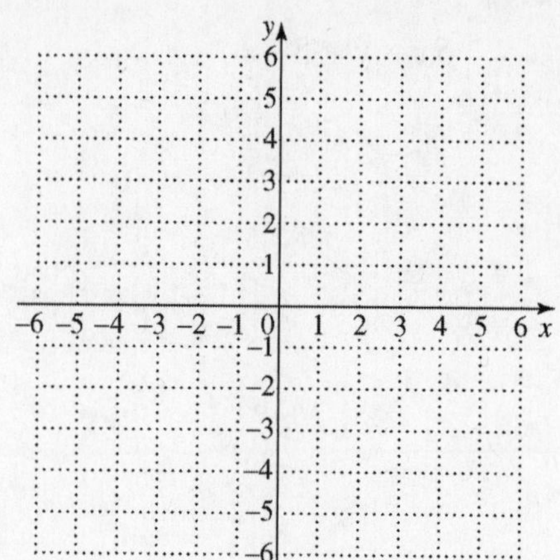

9. $2x - y \geq 4$
$5y + 15 \geq -3x$

9.

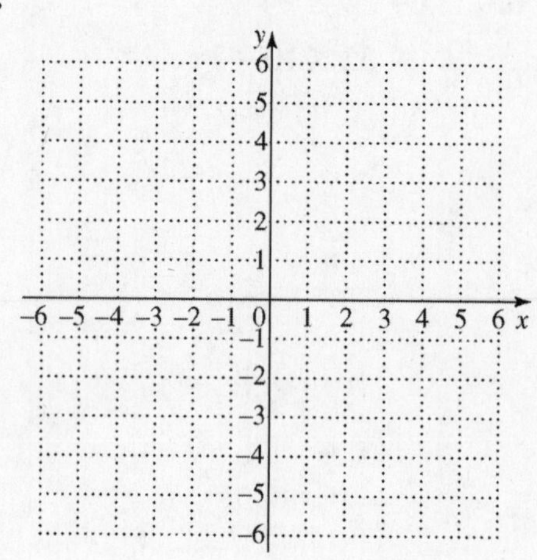

10. $3x - 2y < 8$
$x < 4$

10.

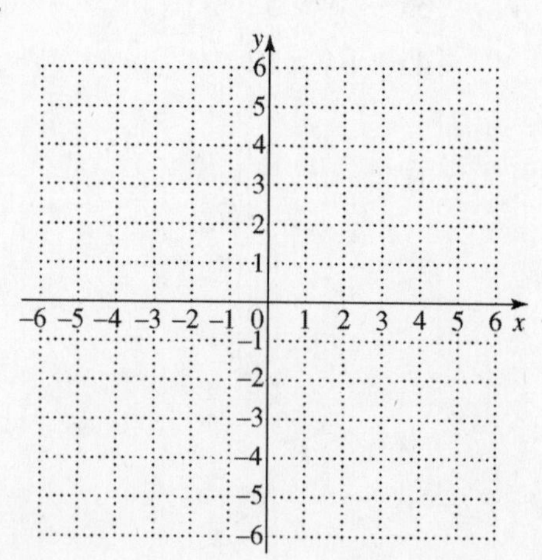

Name: Date:
Instructor: Section:

11. $x - y \le 2$
$y \le 2$

11.

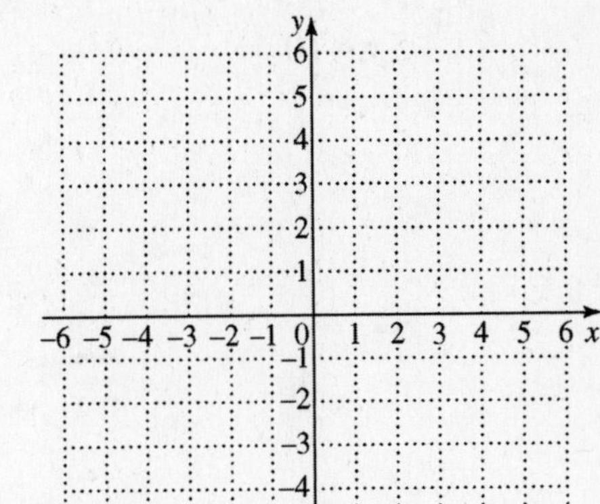

12. $x + y \ge 3$
$x - 2y \le 4$

12.

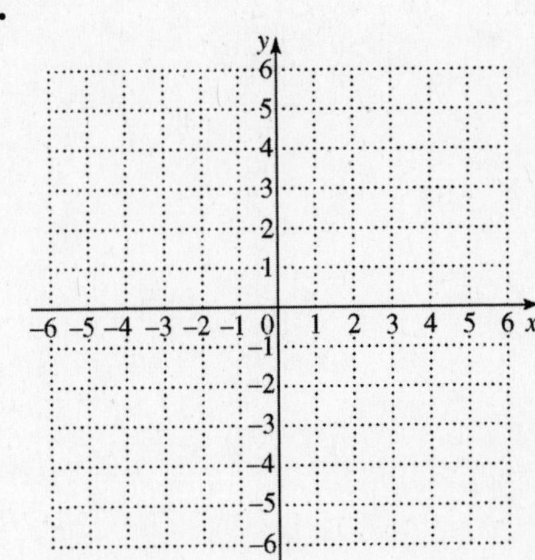

13. $x < 2y + 3$
$0 < x + y$

13.

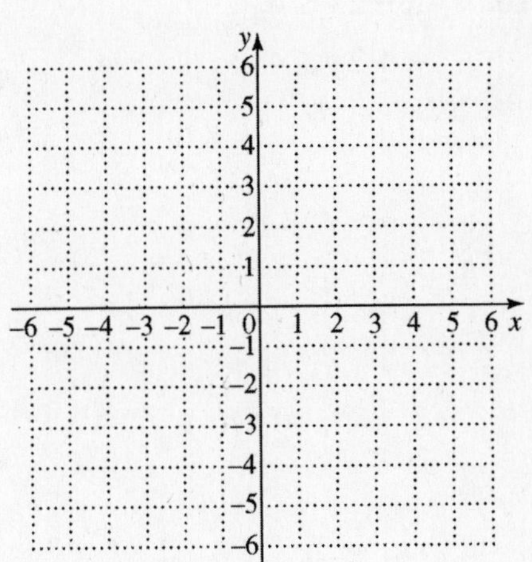

Name: Date:
Instructor: Section:

14. $6x - y > 6$
$2x + 5y < 10$

14.

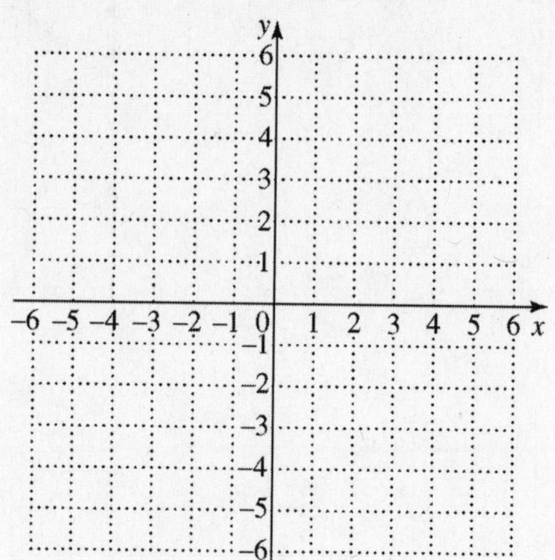

15. $3x - 4y < 12$
$y > -4$

15.

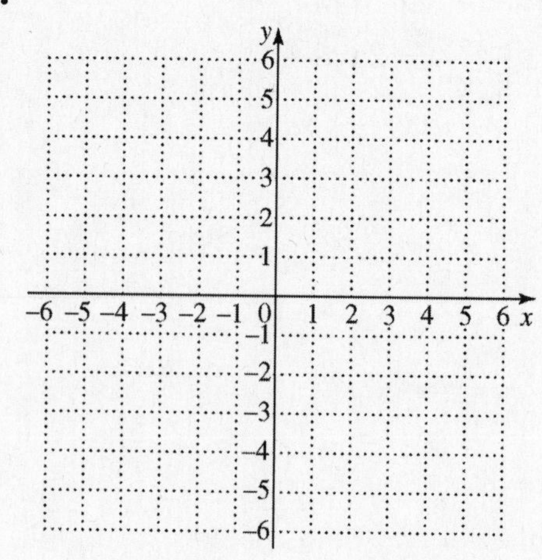

16. $x - 2y \le 4$
$x + 2y \le 4$

16.

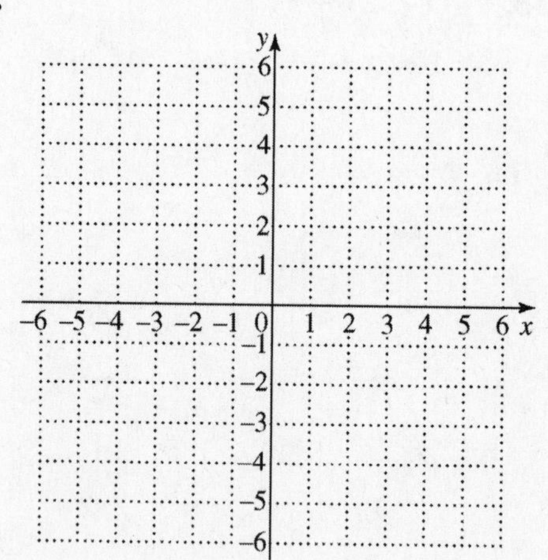

Name: Date:
Instructor: Section:

17. $y > 2$
$4x - 3y < 9$

17.

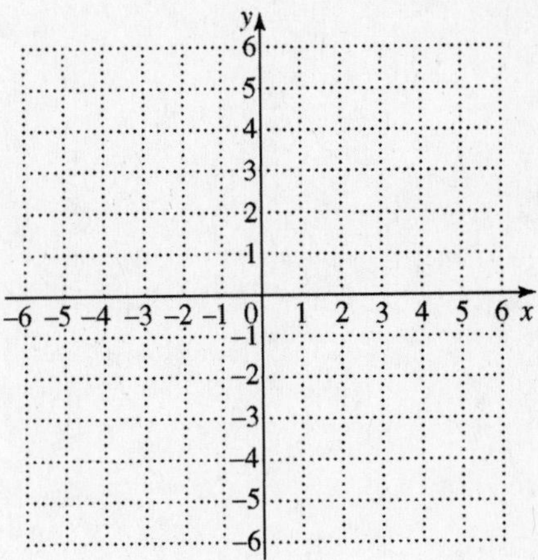

18. $4x - 3y > 12$
$x < 4$

18.

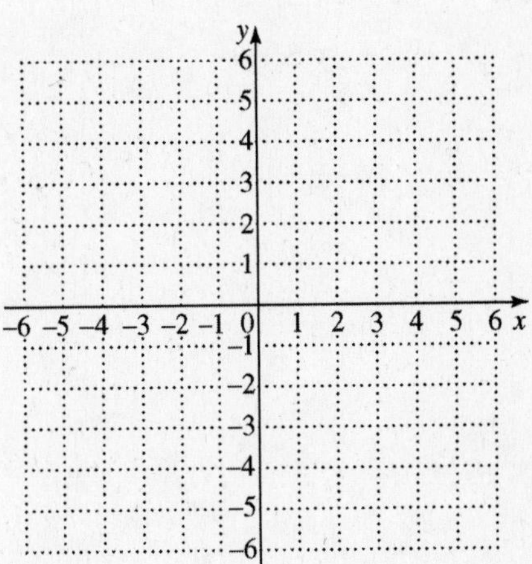

19. $4x - y \le 4$
$7y + 14 \ge -2x$

19.

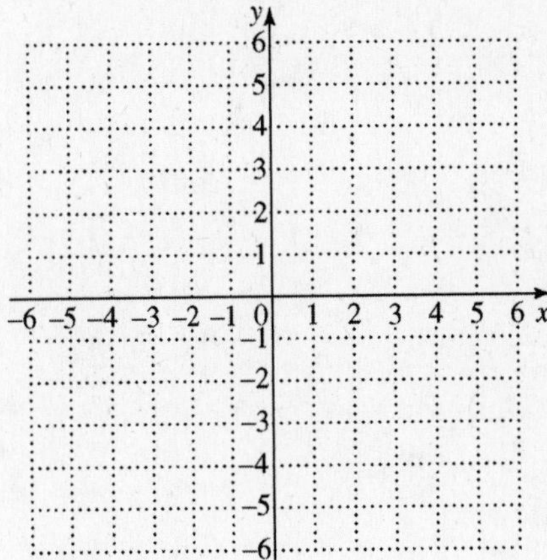

Name: Date:
Instructor: Section:

20. $5x - 2y \le 10$
$y \le -2$

20.

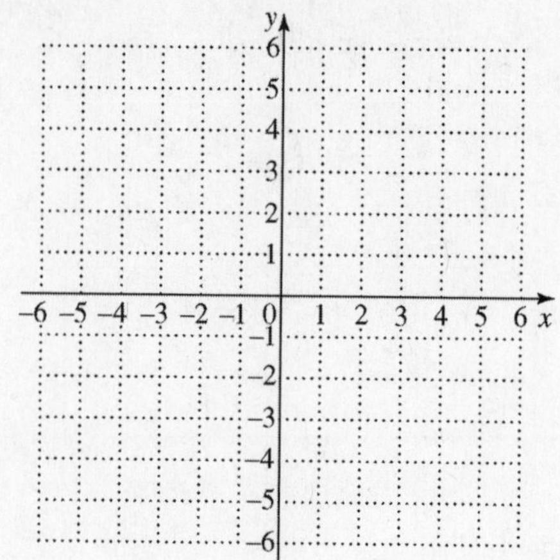

21. $x - 2y \le 3$
$2x + y \le -4$

21.

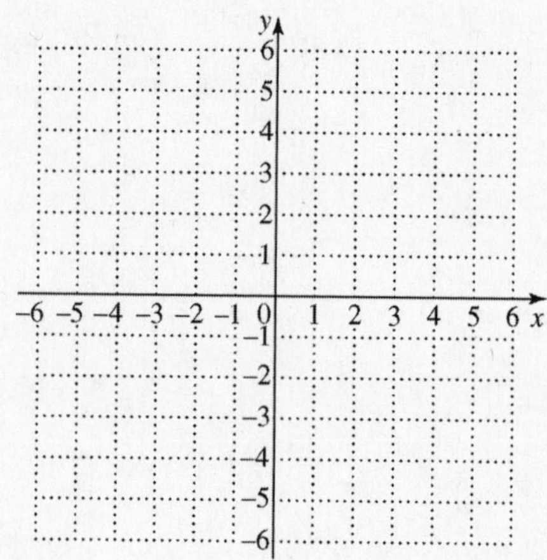

22. $x + y > -3$
$2x - 3y \le -2$

22.

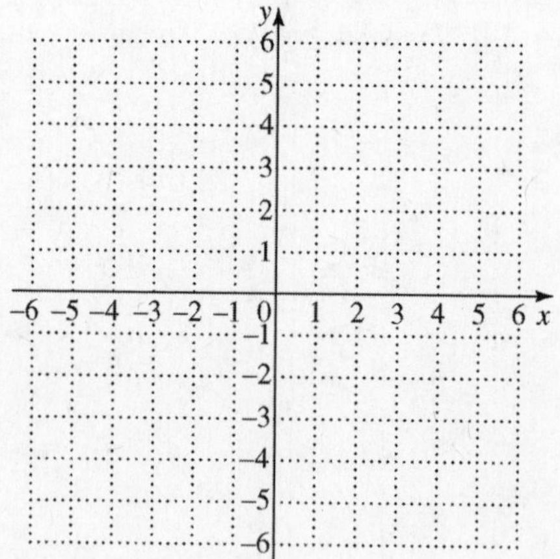

Name: Date:
Instructor: Section:

23. $y < 4$
$x \geq -3$

23.

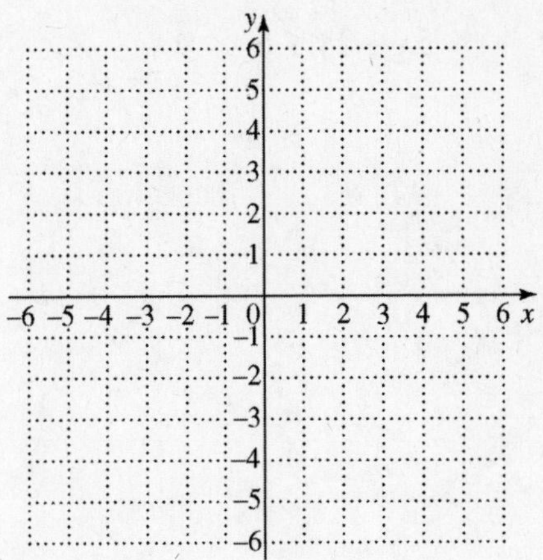

24. $x - 2y \geq -7$
$x - 2y < 2$

24.

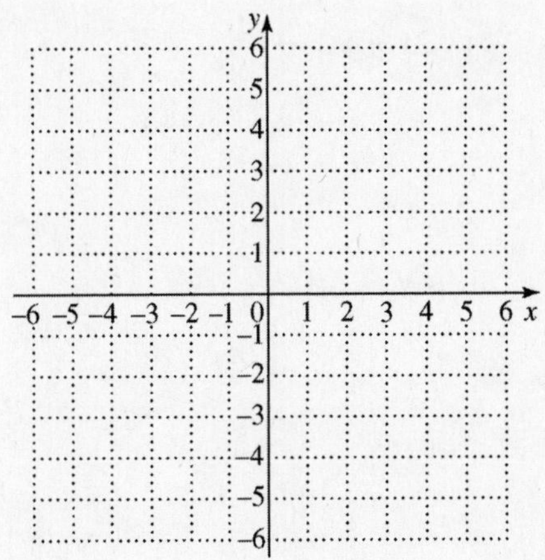

25. $4x - y > 2$
$y > -x - 2$

25.

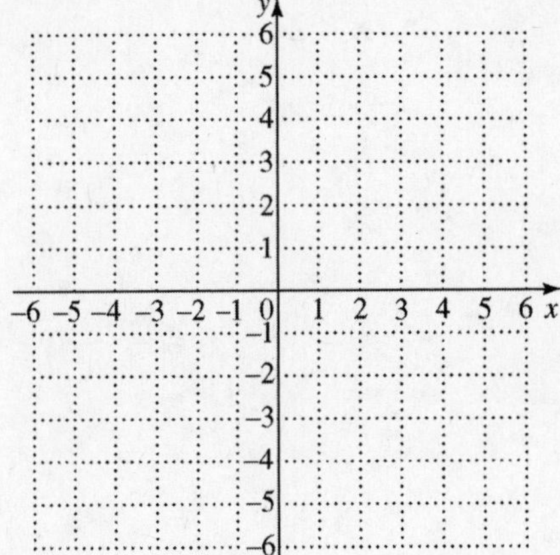

Name: Date:
Instructor: Section:

26. $3x + 2y < 10$

$5x - 2y \le 6$

26.

27. $y \ge -1$

$2x - y > -1$

27.

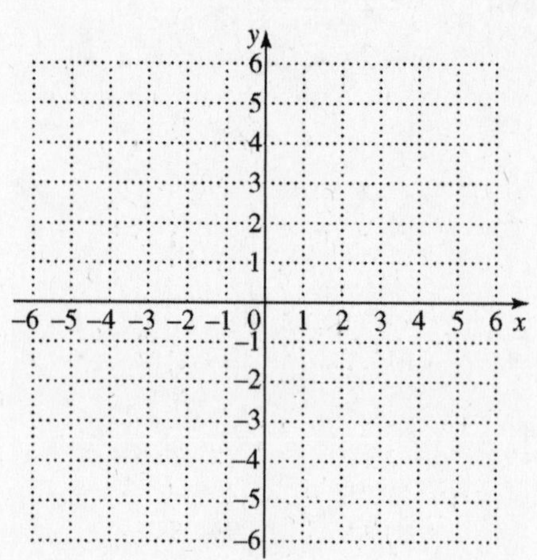

28. $x - 3y \le -7$

$x < 2$

28.

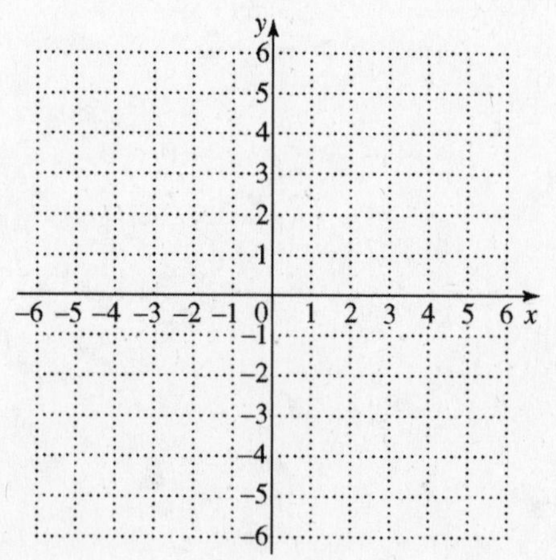

Name: Date:
Instructor: Section:

29. $2x - y > -4$

$2x + y > 0$

29.

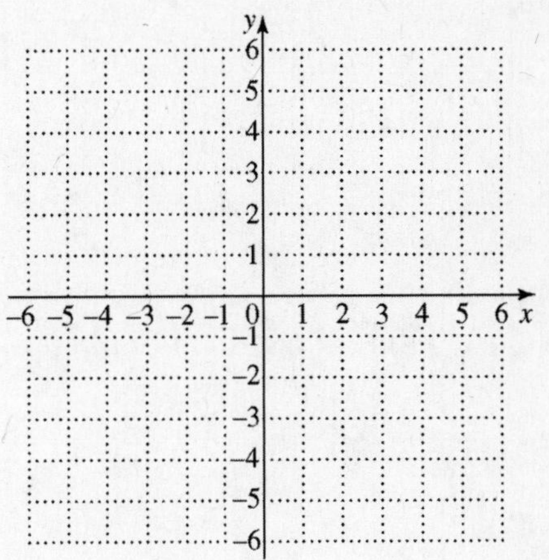

30. $x - 3y \geq -9$

$x + 3y < 3$

30.

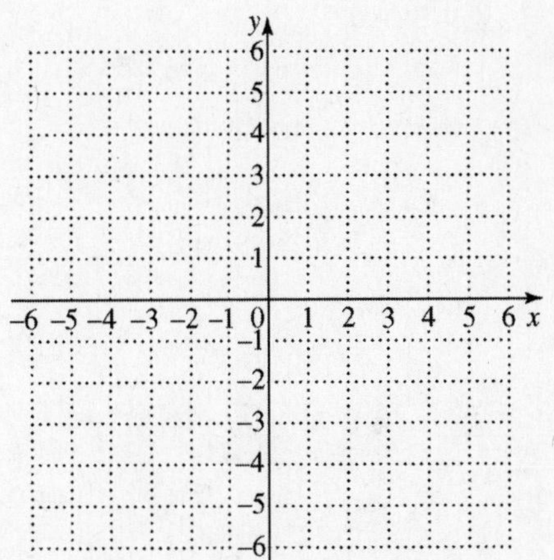

Name: Date:
Instructor: Section:

Chapter 5 EXPONENTS AND POLYNOMIALS

5.1 Adding and Subtracting Polynomials

Learning Objectives

1 Review combining like terms.
2 Know the vocabulary for polynomials.
3 Evaluate polynomials.
4 Add polynomials.
5 Subtract polynomials.
6 Add and subtract polynomials with more than one variable.

Key Terms

Use the vocabulary terms listed below to complete each statement in exercises 1–9.

term	**like terms**	**polynomial**
descending powers		**degree of a term**
degree of a polynomial		**monomial**
binomial		**trinomial**

1. The __________________________ is the sum of the exponents on the variables in that term.

2. A polynomial in x is written in ____________________________ if the exponents on x decrease from left to right.

3. A ________________ is a number, a variable, or a product or quotient of a number and one or more variables raised to powers.

4. A polynomial with exactly three terms is called a ______________________.

5. A _____________________ is a term, or the sum of terms, with whole number exponents.

6. A polynomial with exactly one term is called a ______________________.

7. The ____________________________ is the highest degree of any term of the polynomial.

8. A _____________________ is a polynomial with exactly two terms.

9. Terms with exactly the same variables (including the same exponents) are called ____________________________.

Name: Date:
Instructor: Section:

Objective 1 Review combining like terms.

In each polynomial, combine like terms whenever possible. Write the result with descending powers.

1. $7z^3 - 4z^3 + 5z^3 - 11z^3$ **1.** ________

2. $-1.3z^7 + 0.4z^7 + 2.6z^8$ **2.** ________

3. $-\frac{1}{2}r^3 + \frac{1}{3}r + \frac{1}{4}r^3 - \frac{1}{3}r$ **3.** ________

4. $12x^3 + 7x^2 - 6x^3 + 5x^2$ **4.** ________

5. $6c^3 - 9c^2 - 2c^2 + 14 + 3c^2 - 6c - 8 + 2c^3$ **5.** ________

Objective 2 Know the vocabulary for polynomials.

For each polynomial, first simplify, if possible, and write the resulting polynomial in descending powers of the variable. Then give the degree of this polynomial, and tell whether it is a monomial, *a* binomial, *a* trinomial, *or* none of these.

6. $3n^8 - n^2 - 2n^8$ **6.** ________

degree: ________

type: ________

7. $\frac{7}{8}x^2 - \frac{3}{4}x - \frac{3}{8}x^2 + \frac{1}{4}x$ **7.** ________

degree: ________

type: ________

Name: Date:
Instructor: Section:

8. $10y^4 + 6y^2 - 12y^3 - 5y^4$

8. ______________

degree: __________

type: ____________

9. $-d^2 + 3.2d^3 - 5.7d^8 - 1.1d^5$

9. ______________

degree: __________

type: ____________

10. $-6c^4 - 6c^2 + 9c^4 - 4c^2 + 5c^5$

10. ______________

degree: __________

type: ____________

Objective 3 Evaluate polynomials.

Find the value of each polynomial (a) *when* $x = -2$ *and* (b) *when* $x = 3$.

11. $2x^4 - 47$

11. a. ______________

b. ______________

12. $3x^3 + 4x - 19$

12. a. ______________

b. ______________

13. $-4x^3 + 10x^2 - 1$

13. a. ______________

b. ______________

Name: Date:
Instructor: Section:

14. $x^4 - 3x^2 - 8x + 9$

14. a. ______________

b. ______________

15. $2x^4 + 7x^3 + x - 2$

15. a. ______________

b. ______________

Objective 4 Add polynomials.

Add.

16.
$$\begin{array}{r} 5m^4 + 2m^3 - 4 \\ \underline{-3m^4 + 5m^3 - 3} \end{array}$$

16. ______________

17.
$$\begin{array}{r} 9m^3 + 4m^2 - 2m + 3 \\ \underline{-4m^3 - 6m^2 - 2m + 1} \end{array}$$

17. ______________

18. $(x^2 + 6x - 8) + (3x^2 - 10)$

18. ______________

19. $(3r^3 + 5r^2 - 6) + (2r^2 - 5r + 4)$

19. ______________

Name: Date:
Instructor: Section:

20. $\left(3x^2+2x^4-3\right)+\left(8x^3-5x^4-6x^2\right)$ 20. ________

Objective 5 Subtract polynomials.

Subtract.

21. $\left(-8w^3+11w^2-12\right)-\left(-10w^2+3\right)$ 21. ________

22. $\left(5a^4-6a^2+9a\right)-\left(a^3-19a-1\right)$ 22. ________

23. $\left(8b^4-4b^3+7\right)-\left(2b^2+b+9\right)$ 23. ________

24. $\left(9x^3+7x^2-6x+3\right)-\left(6x^3-6x+1\right)$ 24. ________

25. $\left(7d^4+7d^2-8d+12\right)-\left(10d^2+11d\right)$ 25. ________

Name: Date:
Instructor: Section:

Objective 6 Add and subtract polynomials with more than one variable.

Add or subtract as indicated.

26. $\left(-2a^6+8a^4b-b^2\right)-\left(a^6+7a^4b+2b^2\right)$ **26.** ________________

27. $\left(-4m^2n+3n-6m\right)-\left(2m+7n+4nm^2\right)$ **27.** ________________

28. $(4ab+2bc-9ac)+(3ca-2cb-9ba)$ **28.** ________________

29. $\left(2x^2y+2xy-4xy^2\right)+\left(6xy+9xy^2\right)-\left(9x^2y+5xy\right)$ **29.** ________________

30. $\left(.01ab+.03a^2-.05b^2\right)-\left(-.08a^2+.02b^2+.01ab\right)$ **30.** ________________

Name: Date:
Instructor: Section:

Chapter 5 EXPONENTS AND POLYNOMIALS

5.2 The Product Rule and Power Rules for Exponents

Learning Objectives

1 Use exponents.
2 Use the product rule for exponents.
3 Use the rule $\left(a^m\right)^n = a^{mn}$.
4 Use the rule $(ab)^m = a^m b^m$.
5 Use the rule $\left(\frac{a}{b}\right)^m = \frac{a^m}{b^m}$.
6 Use combinations of the rules for exponents.
7 Use the rules for exponents in a geometry application.

Key Terms

Use the vocabulary terms listed below to complete each statement in exercises 1–3.

exponential expression **base** **power**

1. 2^5 is read "2 to the fifth ______________________".

2. A number written with an exponent is called a(n) ______________________.

3. The ________________ is the number being multiplied repeatedly.

Objective 1 Use exponents.

Write each expression in exponential form and evaluate, if possible.

1. $\left(\frac{1}{3}\right)\left(\frac{1}{3}\right)\left(\frac{1}{3}\right)\left(\frac{1}{3}\right)\left(\frac{1}{3}\right)$ 1. ________________

2. $(.5st)(.5st)(.5st)(.5st)$ 2. ________________

Evaluate each exponential expression. Name the base and the exponent.

3. $(-4)^4$ 3. ________________

base ____________

exponent ________

Name: Date:
Instructor: Section:

4. -3^8

4. ______

base ______

exponent ______

Objective 2 Use the product rule for exponents.

Use the product rule to simplify each expression, if possible. Write each answer in exponential form.

5. $7^4 \cdot 7^3$

5. ______

6. $\left(\frac{1}{2}\right)^9 \cdot \left(\frac{1}{2}\right)$

6. ______

7. $\left(-2c^7\right)\left(-4c^8\right)$

7. ______

8. $\left(3k^7\right)\left(-8k^2\right)\left(-2k^9\right)$

8. ______

Objective 3 Use the rule $\left(a^m\right)^n = a^{mn}$.

Simplify each expression. Write all answers in exponential form.

9. $\left(7^3\right)^4$

9. ______

10. $-\left(v^4\right)^9$

10. ______

11. $\left[\left(\frac{1}{3}\right)^3\right]^5$

11. ______

Name: Date:
Instructor: Section:

12. $\left[(-3)^3\right]^7$ 12. ____________

Objective 4 Use the rule $(ab)^m = a^m b^m$.

Simplify each expression.

13. $\left(\frac{1}{3}x^4\right)^2$ 13. ____________

14. $\left(5r^3t^2\right)^4$ 14. ____________

15. $\left(-.2a^4b\right)^3$ 15. ____________

16. $\left(-2w^3z^7\right)^4$ 16. ____________

Objective 5 Use the rule $\left(\frac{a}{b}\right)^m = \frac{a^m}{b^m}$.

Simplify each expression.

17. $\left(-\frac{2x}{5}\right)^3$ 17. ____________

18. $\left(\frac{xy}{z^2}\right)^4$ 18. ____________

19. $\left(\frac{-2a}{b^2}\right)^7$ 19. ____________

Name: Date:
Instructor: Section:

20. $-\left(\frac{2a^3c}{5b^2}\right)^5$

20. ________________

Objective 6 Use combinations of the rules for exponents.

Simplify. Write all answers in exponential form.

21. $\left(-x^3\right)^2\left(-x^5\right)^4$

21. ________________

22. $\left(2ab^2c\right)^5(ab)^4$

22. ________________

23. $\left(5x^2y^3\right)^7\left(5xy^4\right)^4$

23. ________________

24. $-\left(\frac{2a^3c}{5b^2}\right)^5 \quad (b \neq 0)$

24. ________________

25. $\left(7a^2b^2c\right)^3\left(ab^3c^4\right)^4$

25. ________________

Objective 7 Use the rules for exponents in a geometry application.

Find a polynomial that represents the area of each figure.

26.

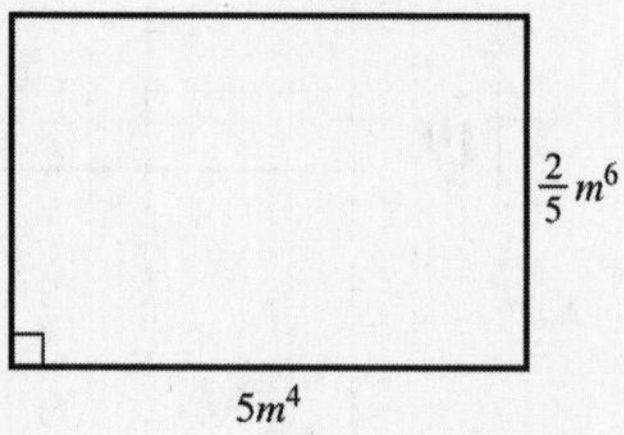

26. ________________

Name: Date:
Instructor: Section:

27.

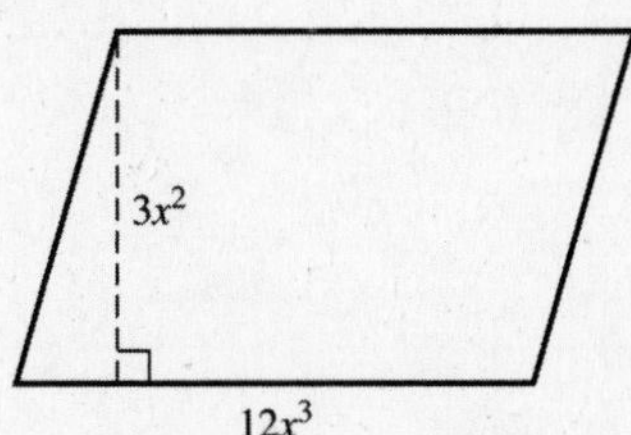

27. ________________

28.

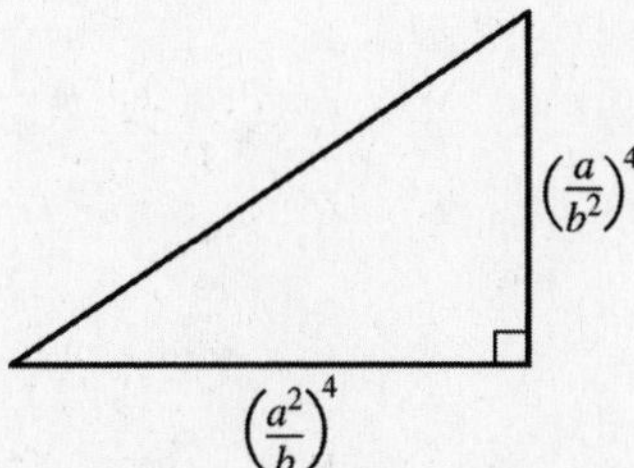

28. ________________

29.

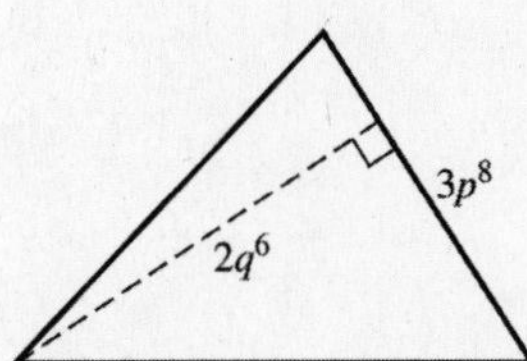

29. ________________

30.

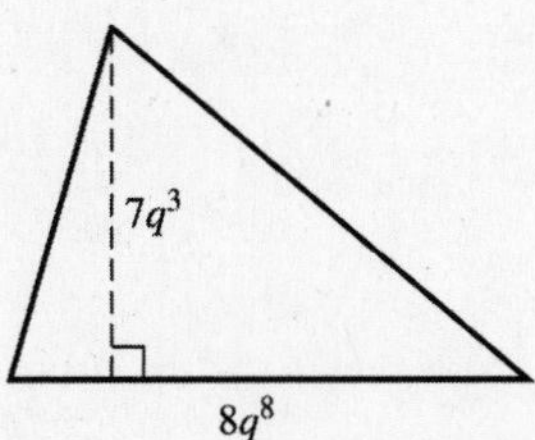

30. ________________

Name: Date:
Instructor: Section:

Chapter 5 EXPONENTS AND POLYNOMIALS

5.3 Multiplying Polynomials

Learning Objectives

1. Multiply a monomial and a polynomial.
2. Multiply two polynomials.
3. Multiply binomials by the FOIL method.

Key Terms

Use the vocabulary terms listed below to complete each statement in exercises 1–3.

FOIL **outer product** **inner product**

1. The ______________________________ of $(2y-5)(y+8)$ is $-5y$.

2. ____________________ is a shortcut method for finding the product of two binomials.

3. The ______________________________ of $(2y-5)(y+8)$ is $16y$.

Objective 1 Multiply a monomial and a polynomial.

Find each product.

1. $(-2y^3)(-8y^4)$ 1. ________________

2. $7z(5z^3+2)$ 2. ________________

3. $2m(3+7m^2+3m^3)$ 3. ________________

4. $4k^2(3+2k^3+6k^4)$ 4. ________________

5. $-6z(z^5+3z^3+4z+2)$ 5. ________________

Name: Date:
Instructor: Section:

6. $-3y^2\left(2y^3+3y^2-4y+11\right)$ 6. ________

7. $7b^2\left(-5b^2+1-4b\right)$ 7. ________

8. $-4r^4\left(2r^2-3r+2\right)$ 8. ________

9. $8mn\left(4m^2+2mn+7n^2\right)$ 9. ________

10. $-3r^2s^3\left(8r^2s^2-4rs+2rs^2\right)$ 10. ________

Objective 2 Multiply two polynomials.

Find each product.

11. $(2x+5)(3x+4)$ 11. ________

12. $(3m-5)(2m+4)$ 12. ________

13. $(x+3)\left(x^2-3x+9\right)$ 13. ________

14. $(y+4)\left(y^2-4y+16\right)$ 14. ________

Name: Date:
Instructor: Section:

15. $(r+3)(2r^2-3r+5)$ 15. ______

16. $(3y-4)(3y^3-2y^2-y+4)$ 16. ______

17. $(2m^2+1)(3m^3+2m^2-4m)$ 17. ______

18. $(3x^2+x)(2x^2+3x-4)$ 18. ______

19. $(y^2-2y-3)(y^2-2y-3)$ 19. ______

20. $(3x^3+3x^2-2x+1)(2x^2-x+2)$ 20. ______

Objective 3 Multiply binomials by the FOIL method.

Find each product.

21. $(3x+2y)(2x-3y)$ 21. ______

Name: Date:
Instructor: Section:

22. $(5a-b)(4a+3b)$ 22. ______________

23. $(3+4a)(1+2a)$ 23. ______________

24. $(4x-3y)(x+2y)$ 24. ______________

25. $(2m+3n)(-3m+4n)$ 25. ______________

26. $\left(2v^2+w^2\right)\left(v^2-3w^2\right)$ 26. ______________

27. $(x-.5)(x+.3)$ 27. ______________

28. $(2y+.1)(2y-.5)$ 28. ______________

29. $\left(x-\frac{1}{3}\right)\left(x-\frac{4}{3}\right)$ 29. ______________

30. $\left(z+\frac{4}{5}\right)\left(z-\frac{2}{5}\right)$ 30. ______________

Name: Date:
Instructor: Section:

Chapter 5 EXPONENTS AND POLYNOMIALS

5.4 Special Products

Learning Objectives
1 Square binomials.
2 Find the product of the sum and difference of two terms.
3 Find greater powers of binomials.

Key Terms

Use the vocabulary terms listed below to complete each statement in exercises 1–2.

conjugate **binomial**

1. A polynomial with two terms is called a ______________________.
2. The ______________________ of $a + b$ is $a - b$.

Objective 1 Square binomials.

Find each square by using the pattern for the square of a binomial.

1. $(5y-3)^2$ 1. ______________
2. $(2m+5)^2$ 2. ______________
3. $(7+x)^2$ 3. ______________
4. $(5-3y)^2$ 4. ______________
5. $(2p+3q)^2$ 5. ______________
6. $(2m-3p)^2$ 6. ______________

Name: Date:
Instructor: Section:

7. $(4y-.7)^2$ 7. ________

8. $\left(4x-\frac{1}{4}y\right)^2$ 8. ________

9. $\left(3x+\frac{1}{3}y\right)^2$ 9. ________

10. $\left(3a+\frac{1}{2}b\right)^2$ 10. ________

Objective 2 Find the product of the sum and difference of two terms.

Find each product by using the pattern for the sum and difference of two terms.

11. $(12+x)(12-x)$ 11. ________

12. $(8k+5p)(8k-5p)$ 12. ________

13. $(7x-3y)(7x+3y)$ 13. ________

14. $(2+3x)(2-3x)$ 14. ________

15. $(x+.2)(x-.2)$ 15. ________

16. $(9-.4y)(9+.4y)$ 16. ________

Name: Date:
Instructor: Section:

17. $\left(7m-\frac{3}{4}\right)\left(7m+\frac{3}{4}\right)$ 17. ______________

18. $\left(\frac{3}{4}s+\frac{7}{5}t\right)\left(\frac{3}{4}s-\frac{7}{5}t\right)$ 18. ______________

19. $\left(4x+\frac{7}{4}\right)\left(4x-\frac{7}{4}\right)$ 19. ______________

20. $\left(\frac{4}{7}t+2u\right)\left(\frac{4}{7}t-2u\right)$ 20. ______________

Objective 3 Find greater powers of binomials.

Find each product.

21. $(a-3)^3$ 21. ______________

22. $(2x+4)^3$ 22. ______________

23. $(4x+y)^3$ 23. ______________

24. $(.3x-.2y)^3$ 24. ______________

Name: Date:
Instructor: Section:

25. $\left(\frac{1}{2}t+2u\right)^3$ **25.** ______________

26. $(j+3)^4$ **26.** ______________

27. $(x+2y)^4$ **27.** ______________

28. $(3b-2)^4$ **28.** ______________

29. $(2x-y)^4$ **29.** ______________

30. $(4s+3t)^4$ **30.** ______________

Name: Date:
Instructor: Section:

Chapter 5 EXPONENTS AND POLYNOMIALS

5.5 Integer Exponents and the Quotient Rule

Learning Objectives

1. Use 0 as an exponent.
2. Use negative numbers as exponents.
3. Use the quotient rule for exponents.
4. Use combinations of rules.

Key Terms

Use the vocabulary terms listed below to complete each statement in exercises 1–3.

exponent **base** **product rule for exponents**

power rule for exponents

1. The statement "If m and n are any integers, then $(a^m)^n = a^{mn}$" is an example of the ______________________________.

2. In the expression a^m, a is the ______________ and m is the ______________.

3. The statement "If m and n are any integers, then $a^m \cdot a^n = a^{m+n}$ is an example of the ______________________________.

Objective 1 Use 0 as an exponent.

Evaluate each expression.

1. $-(-8)^0$ 1. ______________

2. -12^0 2. ______________

3. $2^0 + 6^0$ 3. ______________

4. $\left(\frac{2}{3}\right)^0 + \left(\frac{1}{3}\right)^0 - 2^0$ 4. ______________

5. $-15^0 - (-15)^0$ 5. ______________

6. $-r^0 \ (r \neq 0)$ 6. ______________

Name: Date:
Instructor: Section:

7. $\frac{0^8}{8^0}$ 7. ____________

Objective 2 Use negative numbers as exponents.

Evaluate or simplify each expression, and write it using only positive exponents. Assume that all variables represent nonzero real numbers.

8. 4^{-2} 8. ____________

9. $\frac{2}{r^{-7}}$ 9. ____________

10. $-2k^{-4}$ 10. ____________

11. $-(-4)^{-4}$ 11. ____________

12. $(m^2n)^{-9}$ 12. ____________

13. $3x^{-2} - \frac{6}{x^2}$ 13. ____________

14. $\frac{2x^{-4}}{3y^{-7}}$ 14. ____________

15. $\frac{2^{-4}}{8^{-2}}$ 15. ____________

Objective 3 Use the quotient rule for exponents.

Use the quotient rule to simplify each expression, and write it using only positive exponents. Assume that all variables represent nonzero real numbers.

16. $\frac{12^{-7}}{12^{-6}}$ 16. ____________

Name: Date:
Instructor: Section:

17. $\dfrac{2^4 \cdot x^2}{2^5 \cdot x^8}$

17. ____________

18. $\dfrac{4k^7m^{10}}{8k^3m^5}$

18. ____________

19. $\dfrac{12x^9y^5}{12^4x^3y^7}$

19. ____________

20. $\dfrac{a^4b^3}{a^{-2}b^{-3}}$

20. ____________

21. $\dfrac{3^{-1}m^{-4}p^6}{3^4m^{-1}p^{-2}}$

21. ____________

22. $\dfrac{8b^{-3}c^4}{8^{-4}b^{-7}c^{-3}}$

22. ____________

Objective 4 Use combinations of rules.

Simplify each expression, and write it using only positive exponents. Assume that all variables represent nonzero real numbers.

23. $\dfrac{(2y)^{-4}}{(3y)^{-2}}$

23. ____________

24. $\left(2p^{-3}q^2\right)^2\left(4p^4q^{-1}\right)^{-1}$

24. ____________

Name: Date:
Instructor: Section:

25. $(9xy)^7(9xy)^{-8}$

25. ________________

26. $\dfrac{c^{10}\left(c^2\right)^3}{\left(c^3\right)^3\left(c^2\right)^{-9}}$

26. ________________

27. $\dfrac{\left(a^{-1}b^{-2}\right)^{-4}\left(ab^2\right)^6}{\left(a^3b\right)^{-2}}$

27. ________________

28. $\left(\dfrac{k^3t^4}{k^2t^{-1}}\right)^{-4}$

28. ________________

29. $\dfrac{\left(3^{-2}x^{-5}y\right)^{-4}\left(2x^2y^{-4}\right)^2}{\left(2x^{-2}y^2\right)^{-2}}$

29. ________________

30. $\dfrac{\left(4a^{-1}b^4\right)^{-3}\left(4a^2b\right)^{-1}}{\left(2a^{-3}b^2\right)^{-2}}$

30. ________________

Name: Date:
Instructor: Section:

Chapter 5 EXPONENTS AND POLYNOMIALS

5.6 Dividing a Polynomial by a Monomial

Learning Objectives
1 Divide a polynomial by a monomial.

Key Terms

Use the vocabulary terms listed below to complete each statement in exercises 1–3.

quotient **dividend** **divisor**

1. In the division $\dfrac{5x^5-10x^3}{5x^2}=x^3-2x$, the expression $5x^5-10x^3$ is the ________________.

2. In the division $\dfrac{5x^5-10x^3}{5x^2}=x^3-2x$, the expression x^3-2x is the ________________.

3. In the division $\dfrac{5x^5-10x^3}{5x^2}=x^3-2x$, the expression $5x^2$ is the ________________.

Objective 1 Divide a polynomial by a monomial.

Perform each division.

1. $\dfrac{16a^5-24a^3}{8a^2}$ **1.** ________________

2. $\left(20x^4-10x^2\right)\div(2x)$ **2.** ________________

3. $\left(20a^3-9a\right)\div(4a)$ **3.** ________________

4. $\dfrac{12x^6+28x^5+20x^3}{4x^2}$ **4.** ________________

Name: Date:
Instructor: Section:

5. $\dfrac{6p^4+18p^7}{6p^4}$ 5. ________

6. $\dfrac{16x^4-12x^3+8x^2}{4x^3}$ 6. ________

7. $\dfrac{6y^5+9y^8-21y^{10}}{3y^5}$ 7. ________

8. $\dfrac{24w^8+12w^6-18w^4}{-6w^5}$ 8. ________

9. $\dfrac{40p^4-35p^3-15p}{5p^2}$ 9. ________

10. $\dfrac{2.4a^2b^2+.6ab^2+3a^2b}{.3ab}$ 10. ________

11. $\dfrac{9r^2s+18rs^2-27s^3}{-27rs^2}$ 11. ________

12. $\dfrac{30x^3y^3-45x^2y^2+15xy}{-75xy^2}$ 12. ________

13. $\dfrac{32a^5-6a^4b+24a^3b^2}{-8a^2}$ 13. ________

Name: Date:
Instructor: Section:

14. $\dfrac{3.2st^2 + .8s^2t + .4s^2t^2}{-.4st}$

14. ______________

15. $\dfrac{-28p^5 - 21p^3 - 35p^2 + p}{7p^3}$

15. ______________

16. $\dfrac{5x^2y^4 - 30x^4y^3 + 30x^5y^2}{-5x^2y^2}$

16. ______________

17. $\left(6z^5 + 27z^3 - 12z + 10\right) \div (3z)$

17. ______________

18. $\left(9x^4 + 24x^3 - 48x + 12\right) \div (3x)$

18. ______________

19. $\left(m^2 + 7m - 42\right) \div (2m)$

19. ______________

20. $\dfrac{70q^4 - 40q^2 + 10q}{10q^2}$

20. ______________

21. $\dfrac{2y^9 + 8y^6 - 41y^3 - 12}{y^3}$

21. ______________

22. $\dfrac{12z^5 + 28z^4 - 8z^3 + 3z}{4z^3}$

22. ______________

Name: Date:
Instructor: Section:

23. $\dfrac{48x+64x^4+2x^8}{4x}$

23. ______________

24. $\dfrac{-25u^3v+20u^2v^2-45uv^3}{5uv}$

24. ______________

25. $\dfrac{21y^2-14y+42}{-7y^2}$

25. ______________

26. $\dfrac{39m^4-12m^3+15}{-3m^2}$

26. ______________

27. $\dfrac{-20d^4-8d^3+14d^2+8}{-2d^2}$

27. ______________

28. $\left(15x^5-10x^4-10x^2+4\right)\div(-5x)$

28. ______________

29. $\dfrac{-12y^4-15y^3+2y}{-3y^2}$

29. ______________

30. $\dfrac{18r^4-12r^3+36r^2-12}{6r}$

30. ______________

Name: Date:
Instructor: Section:

Chapter 5 EXPONENTS AND POLYNOMIALS

5.7 Dividing a Polynomial by a Polynomial

Learning Objectives
1 Divide a polynomial by a polynomial.
2 Apply division to a geometry problem.

Key Terms

Use the vocabulary terms listed below to complete each statement in exercises 1–3.

quotient **dividend** **divisor**

1. In the division $\frac{6x^2-9x-12}{2x-5}=3x+3+\frac{3}{2x-5}$, the expression $2x-5$ is the ________________.

2. In the division $\frac{6x^2-9x-12}{2x-5}=3x+3+\frac{3}{2x-5}$, the expression $3x+3+\frac{3}{2x-5}$ is the ____________________.

3. In the division $\frac{6x^2-9x-12}{2x-5}=3x+3+\frac{3}{2x-5}$, the expression $6x^2-9x-12$ is the ____________________.

Objective 1 Divide a polynomial by a polynomial.

Perform each division.

1. $\frac{18a^2-9a-5}{3a+1}$ 1. ________________

2. $\left(r^2-r-20\right)\div(r-5)$ 2. ________________

Name: Date:
Instructor: Section:

3. $\dfrac{p^2+5p-24}{p-3}$ 3. ______________

4. $\left(9w^2+12w+4\right)\div(3w+2)$ 4. ______________

5. $\dfrac{81a^2-1}{9a+1}$ 5. ______________

6. $\dfrac{4x^2-25}{2x-5}$ 6. ______________

7. $\left(2a^2-11a+16\right)\div(2a+3)$ 7. ______________

Name: Date:
Instructor: Section:

8. $\dfrac{5w^2 - 22w + 4}{w - 4}$

8. ______________

9. $\dfrac{9m^2 - 18m + 16}{3m - 4}$

9. ______________

10. $\dfrac{-6x^2 + 23x - 20}{2x - 5}$

10. ______________

11. $\dfrac{5b^2 + 32b + 3}{b + 7}$

11. ______________

Name: Date:
Instructor: Section:

12. $\dfrac{12p^3 - 28p^2 + 21p - 5}{6p - 5}$

12. ______________

13. $\dfrac{12y^3 - 11y^2 + 9y + 18}{4y + 3}$

13. ______________

14. $\dfrac{2z^3 - 7z^2 + 3z + 2}{2z + 3}$

14. ______________

15. $\dfrac{6m^3 + 7m^2 - 13m + 16}{3m + 2}$

15. ______________

Name: Date:
Instructor: Section:

16. $\left(27p^4 - 36p^3 - 6p^2 + 23p - 20\right) \div (3p - 4)$

16. ________________

17. $\left(3x^3 - 11x^2 + 25x - 25\right) \div \left(x^2 - 3x - 5\right)$

17. ________________

18. $\dfrac{6x^4 - 12x^3 + 13x^2 - 5x - 1}{2x^2 + 3}$

18. ________________

19. $\dfrac{12y^5 - 8y^4 - y^3 + 2y^2 - 5}{4y^2 - 3}$

19. ________________

Name: Date:
Instructor: Section:

20. $\dfrac{2a^4+5a^2+3}{2a^2+3}$ 20. ________________

21. $\dfrac{y^3+1}{y+1}$ 21. ________________

22. $\dfrac{b^4-1}{b^2-1}$ 22. ________________

23. $\dfrac{3x^4+2x^3-2x^2-2x-1}{x^2-1}$ 23. ________________

Name: Date:
Instructor: Section:

24. $\dfrac{6x^5+7x^4-7x^3+7x+2}{3x+2}$

24. ____________

25. $\dfrac{32x^5-243}{2x-3}$

25. ____________

26. $\dfrac{6y^6+8y^5-8y^4-3y^3-4y^2+4y+3}{2y^3-1}$

26. ____________

Objective 2 Apply division to a geometry problem.

Work each problem.

27. The area of a rectangle is given by $6r^3-5r^2+16r-5$ square units, and the width is $3r-1$ units. What is the length of the rectangle?

27. ____________

Name: Date:
Instructor: Section:

28. The area of a rectangle is given by $12p^3 - 7p^2 + 5p - 1$ square units, and the width is $4p - 1$ units. What is the length of the rectangle?

28. ______________

29. The area of a parallelogram is given by $4y^3 - 44y - 600$ square units, and the height is $y - 6$ units. What is the base of the parallelogram?

29. ______________

30. The area of a parallelogram is given by $3t^3 + 16t^2 - 32t - 64$ square units, and the base is $t^2 + 4t - 16$ units. What is the height of the parallelogram?

30. ______________

Name: Date:
Instructor: Section:

Chapter 5 EXPONENTS AND POLYNOMIALS

5.8 An Application of Exponents: Scientific Notation

Learning Objectives
1 Express numbers in scientific notation.
2 Convert numbers in scientific notation to numbers without exponents.
3 Use scientific notation in calculations.

Key Terms

Use the vocabulary terms listed below to complete each statement in exercises 1–3.

scientific notation **quotient rule** **power rule**

1. A number written as $a \times 10^n$, where $1 \le |a| < 10$ and n is an integer, is written in ______________________.

2. The statement "If m and n are any integers and $b \neq 0$, then $\left(\frac{a}{b}\right)^m = \frac{a^m}{b^m}$" is an example of the ______________________.

3. The statement "If m and n are any integers and $b \neq 0$, then $\frac{a^m}{a^n} = a^{m-n}$" is an example of the ______________________.

Objective 1 Express numbers in scientific notation.

Write each number in scientific notation.

1. 325 1. ______________
2. 4579 2. ______________
3. 23,651 3. ______________
4. −38,600,000 4. ______________
5. 9,540,000 5. ______________
6. −429,600,000,000 6. ______________
7. 0.0503 7. ______________

Name: Date:
Instructor: Section:

8. 0.007068 **8.** ________

9. –0.0002208 **9.** ________

10. 0.00000476 **10.** ________

Objective 2 Convert numbers in scientific notation to numbers without exponents.

Write each number in scientific notation.

11. 7.2×10^7 **11.** ________

12. -2.45×10^6 **12.** ________

13. 2.3×10^4 **13.** ________

14. 4.5×10^7 **14.** ________

15. 6.4×10^{-3} **15.** ________

16. 7.24×10^{-4} **16.** ________

17. 4.007×10^{-2} **17.** ________

18. 4.752×10^{-1} **18.** ________

19. -4.02×10^0 **19.** ________

20. -9.11×10^{-4} **20.** ________

Objective 3 Use scientific notation in calculations.

Perform the indicated operations, and write the answers in scientific notation.

21. $\left(2.3 \times 10^4\right) \times \left(1.1 \times 10^{-2}\right)$ **21.** ________

Name: Date:
Instructor: Section:

22. $\dfrac{9.39\times10^1}{3\times10^3}$ 22. ______________

23. $(6\times10^4)\times(3\times10^5)\div(9\times10^7)$ 23. ______________

24. $(3\times10^4)\times(4\times10^2)\div(2\times10^3)$ 24. ______________

25. $\dfrac{(7.5\times10^6)\times(4.2\times10^{-5})}{(6\times10^4)\times(2.5\times10^{-3})}$ 25. ______________

26. $\dfrac{(2.1\times10^{-3})\times(4.8\times10^4)}{(1.6\times10^6)\times(7\times10^{-6})}$ 26. ______________

Work each problem. Give answers without exponents.

27. There are about 6×10^{23} atoms in a mole of atoms. How many atoms are there in 8.1×10^{-5} mole? 27. ______________

Name: Date:
Instructor: Section:

28. The earth has a mass of 6×10^{24} kilograms and a volume of 1.1×10^{21} cubic meters. What is the Earth's density in kilograms per cubic meter? Round to the nearest hundred.

28. ______________

29. A light-year is the distance that light travels in one year. The speed of light is about 3×10^{5} km per second. How many kilometers are in a light-year?

29. ______________

30. The Sahara desert covers approximately 3.5×10^{6} square miles. Its sand is, on average, 12 feet deep.

a. Find the volume, in cubic feet, of sand in the Sahara. $\left(\text{Hint: } 1 \text{ mi}^2 = 5280^2 \text{ ft}^2\right)$ Round your answer to two decimal places.

b. The volume of a single grain of sand is approximately 1.3×10^{-9} cubic feet. About how many grains of sand are in the Sahara?

30. a. ______________

b. ______________

Name: Date:
Instructor: Section:

Chapter 6 FACTORING AND APPLICATIONS

6.1 Factors; The Greatest Common Factor

Learning Objectives

1. Find the greatest common factor of a list of numbers.
2. Find the greatest common factor of a list of variable terms.
3. Factor out the greatest common factor.
4. Factor by grouping.

Key Terms

Use the vocabulary terms listed below to complete each statement in exercises 1–4.

factor **factored form** **greatest common factor (GCF)** **factoring**

1. The process of writing a polynomial as a product is called ________________.

2. An expression is in ________________________ when it is written as a product.

3. The __________________________ is the largest quantity that is a factor of each of a group of quantities.

4. An expression A is a ____________________ of an expression B if B can be divided by A with 0 remainder.

Objective 1 Find the greatest common factor of a list of numbers.

Find the greatest common factor for each group of numbers.

1. 60, 75, 120 — 1. ________________

2. 108, 48, 84 — 2. ________________

3. 9, 18, 24, 48 — 3. ________________

4. 70, 126, 42, 56 — 4. ________________

5. 84, 280, 112 — 5. ________________

Name: Date:
Instructor: Section:

6. 56, 21, 49 **6.** ______

7. 42, 48, 72 **7.** ______

Objective 2 Find the greatest common factor of a list of variable terms.

Find the greatest common factor for each list of terms.

8. $-15ab^2,\ -45a^3b^4,\ 70ab^3$ **8.** ______

9. $12ab^3,\ 18a^2b^4,\ 26ab^2,\ 32a^2b^2$ **9.** ______

10. $6k^2m^4n^5,\ 8k^3m^7n^4,\ k^4m^8,n^7$ **10.** ______

11. $29w^3x^7y^4,\ w^4x^5y^7,\ 58w^2x^9,y^5$ **11.** ______

12. $45a^7y^4,\ 75a^3y^2,\ -90a^2y,\ 30a^4y^3$ **12.** ______

13. $9xy^4, 72x^4y^7,\ 27xy^2,\ 108x^2y^5$ **13.** ______

14. $-72u^2v^3,\ -54uv^2,\ -63uv^4$ **14.** ______

Name: Date:
Instructor: Section:

Objective 3 Factor out the greatest common factor.

Factor out the greatest common factor or a negative common factor if the coefficient of the term of greatest degree is negative.

15. $24ab - 8a^2 + 40ac$ **15.** ______________

16. $45a^2b^3 - 90ab + 15ab^2$ **16.** ______________

17. $20x^2 + 40x^2y - 70xy^2$ **17.** ______________

18. $2a(x-2y) + 9b(x-2y)$ **18.** ______________

19. $26x^8 - 13x^{12} + 52x^{10}$ **19.** ______________

20. $27r^2 - 54r^4 - 81r^5$ **20.** ______________

21. $56x^2y^4 - 24xy^3 + 32xy^2$ **21.** ______________

22. $x^2(r-4s) + z^2(r-4s)$ **22.** ______________

Name: Date:
Instructor: Section:

Objective 4 Factor by grouping.

Factor each polynomial by grouping.

23. $1 + p - q - pq$ **23.** ____________

24. $15 - 5x - 3y + xy$ **24.** ____________

25. $8x^2 + 12xy - 2xy - 3y^2$ **25.** ____________

26. $6x^3 + 9x^2y^2 - 2xy^3 - 3y^5$ **26.** ____________

27. $12x^3 - 4xy - 3x^2y^2 + y^3$ **27.** ____________

28. $2x^2 - 14xy + xy - 7y^2$ **28.** ____________

29. $3r^3 - 2r^2s + 3s^2r - 2s^3$ **29.** ____________

30. $2a^3 + 3a^2b + 8ab^2 + 12b^3$ **30.** ____________

Name: Date:
Instructor: Section:

Chapter 6 FACTORING AND APPLICATIONS

6.2 Factoring Trinomials

Learning Objectives
1 Factor trinomials with a coefficient of 1 for the squared term.
2 Factor trinomials after factoring out the greatest common factor.

Key Terms

Use the vocabulary terms listed below to complete each statement in exercises 1–3.

prime polynomial **factoring** **greatest common factor**

1. ______________________ is the process of writing a polynomial as a product.

2. The ______________________ of a polynomial is the greatest term that is a factor of all the terms in the polynomial.

3. A ______________________ is a polynomial that cannot be factored using only integers.

Objective 1 Factor trinomials with a coefficient of 1 for the squared term.

Factor completely. If a polynomial cannot be factored, write prime.

1. $r^2 + r + 3$ 1. ____________

2. $a^2 - 10a + 21$ 2. ____________

3. $s^2 - 4s - 32$ 3. ____________

4. $n^2 - 16n + 64$ 4. ____________

5. $x^2 + 11x + 18$ 5. ____________

Name: Date:
Instructor: Section:

6. $x^2 - 11x + 28$ 6. ______________

7. $x^2 - x - 2$ 7. ______________

8. $x^2 + 14x - 49$ 8. ______________

9. $x^2 - 2x - 35$ 9. ______________

10. $x^2 - 8x - 33$ 10. ______________

11. $x^2 - 15xy + 56y^2$ 11. ______________

12. $a^2 + 10ab + 21b^2$ 12. ______________

13. $q^2 - 4q - 12$ 13. ______________

14. $b^2 + 10bc + 25c^2$ 14. ______________

15. $a^2 - 10ab + 16b^2$ 15. ______________

Name: Date:
Instructor: Section:

Objective 2 Factor trinomials after factoring out the greatest common factor.

Factor completely. If a polynomial cannot be factored, write **prime**.

16. $3d^2 - 18d + 27$ **16.** ______________

17. $2m^3 - 2m^2 - 4m$ **17.** ______________

18. $2n^4 - 16n^3 + 30n^2$ **18.** ______________

19. $4a^2 - 24b + 5$ **19.** ______________

20. $2a^3b - 10a^2b^2 + 12ab^3$ **20.** ______________

21. $3p^6 + 18p^5 + 24p^4$ **21.** ______________

22. $3h^3k - 21h^2k - 54hk$ **22.** ______________

23. $10k^6 + 70k^5 + 100k^4$ **23.** ______________

Name: Date:
Instructor: Section:

24. $3xy^2 - 24xy + 36x$ **24.** ________________

25. $x^5 - 3x^4 + 2x^3$ **25.** ________________

26. $a^2b - 12ab^2 + 35b^3$ **26.** ________________

27. $2x^2y^2 - 2xy^3 - 12y^4$ **27.** ________________

28. $2s^2t - 16st - 40t$ **28.** ________________

29. $qr^3 - 4q^2r^2 - 21q^3r$ **29.** ________________

30. $2x^3 - 14x^2y + 20xy^2$ **30.** ________________

Name: Date:
Instructor: Section:

Chapter 6 FACTORING AND APPLICATIONS

6.3 Factoring Trinomials by Grouping

Learning Objectives
1 Factor trinomials by grouping when the coefficient of the squared term is not 1.

Key Terms

Use the vocabulary terms listed below to complete each statement in exercises 1–2.

coefficient **trinomial**

1. In the term $6x^2y$, 6 is the ____________________.

2. A polynomial with three terms is a ____________________.

Objective 1 Factor trinomials by grouping when the coefficient of the squared term is not 1.

Complete the factoring.

1. $2x^2 + 5x - 3 = (2x - 1)(\quad)$ 1. ____________

2. $6x^2 + 19x + 10 = (3x + 2)(\quad)$ 2. ____________

3. $16x^2 + 4x - 6 = (4x + 3)(\quad)$ 3. ____________

4. $24y^2 - 17y + 3 = (3y - 1)(\quad)$ 4. ____________

Factor each trinomial by grouping.

5. $8b^2 + 18b + 9$ 5. ____________

6. $3x^2 + 13x + 14$ 6. ____________

Name: Date:
Instructor: Section:

7. $15a^2 + 16a + 4$

7. ________________

8. $6n^2 + 11n + 4$

8. ________________

9. $3b^2 + 8b + 4$

9. ________________

10. $3m^2 - 5m - 12$

10. ________________

11. $3p^3 + 8p^2 + 4p$

11. ________________

12. $8m^2 + 26mn + 6n^2$

12. ________________

13. $7a^2b + 18ab + 8b$

13. ________________

14. $2s + 5st - 3st^2$

14. ________________

Name: Date:
Instructor: Section:

15. $9c^2 + 24cd + 12d^2$ **15.** ________________

16. $25a^2 + 30ab + 9b^2$ **16.** ________________

17. $10c^2 - 29ct + 21t^2$ **17.** ________________

18. $24s^2 - 14st - 5t^2$ **18.** ________________

19. $12x^2 + 32xy - 35y^2$ **19.** ________________

20. $24c^2 + 90cd - 81d^2$ **20.** ________________

21. $6m^3 + 2m^2n - 8mn^2$ **21.** ________________

22. $40x^2 + 18xy - 9y^2$ **22.** ________________

Name: Date:
Instructor: Section:

23. $18f^2 + 27fg - 5g^2$ **23.** ____________

24. $16p^2 + 8pq + q^2$ **24.** ____________

25. $40a^3 - 82a^2 + 40a$ **25.** ____________

26. $4x^2 + 32x + 55$ **26.** ____________

27. $8x^2 - 4xy - 4y^2$ **27.** ____________

28. $7m^2 + 3mn - 22n^2$ **28.** ____________

29. $10c^2 + 39cd + 36d^2$ **29.** ____________

30. $9x^3 - 30x^2y + 24xy^2$ **30.** ____________

Name: Date:
Instructor: Section:

Chapter 6 FACTORING AND APPLICATIONS

6.4 Factoring Trinomials Using FOIL

Learning Objectives
1 Factor trinomials using FOIL.

Key Terms

Use the vocabulary terms listed below to complete each statement in exercises 1–3.

FOIL **outer product** **inner product**

1. The ______________________________ of $(2y-5)(y+8)$ is $-5y$.

2. ____________________ is a shortcut method for finding the product of two binomials.

3. The ______________________________ of $(2y-5)(y+8)$ is $16y$.

Objective 1 Factor trinomials using FOIL.

Factor each trinomial completely.

1. $10x^2+19x+6$ 1. __________________

2. $4y^2+3y-10$ 2. __________________

3. $2a^2+13a+6$ 3. __________________

4. $8q^2+10q+3$ 4. __________________

5. $8m^2-10m-3$ 5. __________________

Name: Date:
Instructor: Section:

6. $14b^2 + 3b - 2$ 6. ____________

7. $15q^2 - 2q - 24$ 7. ____________

8. $3a^2 + 8ab + 4b^2$ 8. ____________

9. $9w^2 + 12wz + 4z^2$ 9. ____________

10. $10c^2 - cd - 2d^2$ 10. ____________

11. $6x^2 + xy - 12y^2$ 11. ____________

12. $18x^2 - 27xy + 4y^2$ 12. ____________

13. $12y^2 + 11y - 15$ 13. ____________

14. $3x^2 - 11x - 4$ 14. ____________

Name: Date:
Instructor: Section:

15. $2p^2 + 11p + 5$ **15.** ____________

16. $6y^2 + y - 1$ **16.** ____________

17. $9y^2 - 16y - 4$ **17.** ____________

18. $3p^2 + 17p + 10$ **18.** ____________

19. $9r^2 + 12r - 5$ **19.** ____________

20. $7x^2 + 27x - 4$ **20.** ____________

21. $4c^2 + 14cd - 8d^2$ **21.** ____________

22. $2x^4 + 5x^3 - 12x^2$ **22.** ____________

23. $27r^2 + 6rt - 8t^2$ **23.** ____________

Name: Date:
Instructor: Section:

24. $6x^4y^2 + x^2y - 15$

24. ____________

25. $28c^2 + 23cd - 15d^2$

25. ____________

26. $8x^3 - 10x^2y + 3xy^2$

26. ____________

27. $6n^2 + 13ns - 63s^2$

27. ____________

28. $-30a^4b + 3a^3b + 6a^2b$

28. ____________

29. $12a^3 + 26a^2b + 12ab^2$

29. ____________

30. $2y^5z^2 - 5y^4z^3 - 3y^3z^4$

30. ____________

Name: Date:
Instructor: Section:

Chapter 6 FACTORING AND APPLICATIONS

6.5 Special Factoring Techniques

Learning Objectives

1. Factor a difference of squares.
2. Factor a perfect square trinomial.
3. Factor a difference of cubes.
4. Factor a sum of cubes

Key Terms

Use the vocabulary terms listed below to complete each statement in exercises 1–2.

perfect square trinomial **difference of squares**

1. A ____________________ is a binomial that can be factored as the product of the sum and difference of two terms.

2. A ____________________ is a trinomial that can be factored as the square of a binomial.

Objective 1 Factor a difference of squares.

Factor each binomial completely. If a binomial cannot be factored, write prime.

1. $25a^2 - 36$ 1. __________

2. $x^2 + 16$ 2. __________

3. $9j^2 - \frac{16}{49}$ 3. __________

4. $121m^2 - 9n^2$ 4. __________

5. $16y^4 - 81$ 5. __________

Name: Date:
Instructor: Section:

6. $q^2 - (2r+3)^2$ 6. ______________

7. $m^4n^2 - m^2$ 7. ______________

8. $(r-s)^2 - (r+s)^2$ 8. ______________

Objective 2 Factor a perfect square trinomial.

Factor each polynomial completely.

9. $4w^2 + 12w + 9$ 9. ______________

10. $z^2 - \frac{4}{3}z + \frac{4}{9}$ 10. ______________

11. $16q^2 - 40q + 25$ 11. ______________

12. $64p^4 + 48p^2q^2 + 9q^4$ 12. ______________

13. $100p^2 - \frac{25}{2}pr + \frac{25}{64}r^2$ 13. ______________

14. $9m^2 + .6m + .01$ 14. ______________

Name: Date:
Instructor: Section:

15. $(p-q)^2-20(p-q)+100$ **15.** ______________

16. $(m-n)^2-12(m-n)+36$ **16.** ______________

Objective 3 Factor a difference of cubes.

Factor.

17. x^3-y^3 **17.** ______________

18. $8r^3-27s^3$ **18.** ______________

19. $216m^3-125p^6$ **19.** ______________

20. $8a^3-125b^3$ **20.** ______________

21. $216x^3-8y^3$ **21.** ______________

22. $(m+n)^3-(m-n)^3$ **22.** ______________

Name:
Instructor:

Date:
Section:

23. $x^3 - (x-1)^3$

23. ______________

Objective 4 Factor a sum of cubes.

Factor.

24. $x^3 + y^3$

24. ______________

25. $27r^3 + 8s^3$

25. ______________

26. $8a^3 + 64b^3$

26. ______________

27. $125p^3 + q^3$

27. ______________

28. $64x^3 + 343y^3$

28. ______________

29. $(x-y)^3 + (x+y)^3$

29. ______________

30. $t^3 + (t+2)^3$

30. ______________

Name: Date:
Instructor: Section:

Chapter 6 FACTORING AND APPLICATIONS

6.6 A General Approach to Factoring

Learning Objectives

1. Factor out any common factor.
2. Factor binomials.
3. Factor trinomials.
4. Factor polynomials with more than three terms.

Key Terms

Use the vocabulary terms listed below to complete each statement in exercises 1–2.

FOIL **factoring by grouping**

1. When there are more than three terms in a polynomial, use a process called ____________________ to factor the polynomial.

2. ____________________ is a shortcut method for finding the product of two binomials.

Objective 1 Factor out any common factor.

Factor completely.

1. $-12x^2 - 6x$ **1.** ________________

2. $12a^2b^2 + 3a^2b - 9ab^2$ **2.** ________________

3. $5r^2t - 10rt + 5rt^2$ **3.** ________________

Name: Date:
Instructor: Section:

4. $12x^4 - 8x^2 + 20x$ **4.** ________

5. $3a^2 + 6a(x - y)$ **5.** ________

6. $(x+1)(2x+3) - (x+1)$ **6.** ________

7. $2m(m-n) - (m+n)(m-n)$ **7.** ________

Objective 2 Factor binomials.

Factor completely.

8. $16s^4 - r^4$ **8.** ________

9. $2x^3y^4 - 72xy^2$ **9.** ________

Name: Date:
Instructor: Section:

10. $128x^3 - 2y^3$ **10.** ______________

11. $(r+s)^3 + 8$ **11.** ______________

12. $(a-b)^2 - (a+b)^2$ **12.** ______________

13. $y^6 + 1$ **13.** ______________

14. $32 - 2(x-y)^2$ **14.** ______________

15. $(3a-1)^2 - y^6$ **15.** ______________

Name: Date:
Instructor: Section:

Objective 3 Factor trinomials.

Factor completely.

16. $4x^2+12xy+9y^2$ **16.** ______________

17. $2a^2-17a+30$ **17.** ______________

18. $4y^4+8y^2-45$ **18.** ______________

19. $25x^2-5xy-2y^2$ **19.** ______________

20. $3(a+1)^2+19(a+1)-14$ **20.** ______________

21. $12m^2+11m-5$ **21.** ______________

Name: Date:
Instructor: Section:

22. $3k + 42k^2 + 147k^3$ **22.** ____________

23. $4b^2(b+2) - 3b(b+2) - (b+2)$ **23.** ____________

Objective 4 Factor polynomials with more than three terms.

Factor completely.

24. $14w^2 + 6wx - 35wx - 15x^2$ **24.** ____________

25. $bx - by - ay + ax$ **25.** ____________

26. $x^2 + 7xy + 2x + 14y$ **26.** ____________

27. $x^3 - 21 - 3x^2 + 7x$ **27.** ____________

Name: Date:
Instructor: Section:

28. $3x^2 - 2x + 6x - 4$ **28.** ______________

29. $a^2 - 6ab + 9b^2 - 25$ **29.** ______________

30. $r^3 - s^3 - sr^2 + rs^2$ **30.** ______________

Name: Date:
Instructor: Section:

Chapter 6 FACTORING AND APPLICATIONS

6.7 Solving Quadratic Equations by Factoring

Learning Objectives
1 Solve quadratic equations by factoring.
2 Solve other equations by factoring.

Key Terms

Use the vocabulary terms listed below to complete each statement in exercises 1–2.

quadratic equation **standard form**

1. An equation written in the form $ax^2 + bx + c = 0$ is written in the ____________________ of a quadratic equation.

2. An equation that can written in the form $ax^2 + bx + c = 0$, with $a \neq 0$, is a ____________________.

Objective 1 Solve quadratic equations by factoring.

Solve each equation and check your solutions.

1. $x^2 + 7x + 10 = 0$ 1. ____________________

2. $3x^2 + 7x + 2 = 0$ 2. ____________________

3. $b^2 - 49 = 0$ 3. ____________________

Name: Date:
Instructor: Section:

4. $2x^2 - 3x - 20 = 0$ **4.** ______________

5. $x^2 - 2x - 63 = 0$ **5.** ______________

6. $8r^2 = 24r$ **6.** ______________

7. $3x^2 - 7x - 6 = 0$ **7.** ______________

8. $3 - 5x = 8x^2$ **8.** ______________

9. $9x^2 + 12x + 4 = 0$ **9.** ______________

10. $25x^2 = 20x$ **10.** ______________

Name: Date:
Instructor: Section:

11. $9y^2 = 16$ **11.** ____________

12. $12x^2 + 7x - 12 = 0$ **12.** ____________

13. $14x^2 - 17x - 6 = 0$ **13.** ____________

14. $c(5c + 17) = 12$ **14.** ____________

15. $3x(x + 3) = (x + 2)^2 - 1$ **15.** ____________

Objective 2 Solve other equations by factoring.

Solve each equation and check your solutions.

16. $3x(x + 7)(x - 2) = 0$ **16.** ____________

Name: Date:
Instructor: Section:

17. $x\left(2x^2 - 7x - 15\right) = 0$

17. ______________

18. $z\left(4z^2 - 9\right) = 0$

18. ______________

19. $z^3 - 49z = 0$

19. ______________

20. $25a = a^3$

20. ______________

21. $x^3 + 2x^2 - 8x = 0$

21. ______________

22. $2m^3 + m^2 - 6m = 0$

22. ______________

Name: Date:
Instructor: Section:

23. $\left(4x^2-9\right)(x-2)=0$ **23.** ________________

24. $z^4+8z^3-9z^2=0$ **24.** ________________

25. $3z^3+z^2-4z=0$ **25.** ________________

26. $(x+4)\left(x^2+7x+10\right)=0$ **26.** ________________

27. $\left(y^2-5y+6\right)\left(y^2-36\right)=0$ **27.** ________________

28. $15x^2=x^3+56x$ **28.** ________________

Name: Date:
Instructor: Section:

29. $(y-7)\left(2y^2+7y-15\right)=0$

29. ________________

30. $\left(x-\frac{3}{2}\right)\left(2x^2+11x+15\right)=0$

30. ________________

Name: Date:
Instructor: Section:

Chapter 6 FACTORING AND APPLICATIONS

6.8 Applications of Quadratic Equations

Learning Objectives

1. Solve problems about geometric figures.
2. Solve problems about consecutive integers.
3. Solve problems using the Pythagorean formula.
4. Solve problems using given quadratic models.

Key Terms

Use the vocabulary terms listed below to complete each statement in exercises 1–2.

hypotenuse **legs**

1. In a right triangle, the sides that form the right angle are the ______________.

2. The longest side of a right triangle is the ___________________________.

Objective 1 Solve problems about geometric figures.

Solve each problem. Check your answers to be sure they are reasonable.

1. A book is three times as long as it is wide. Find the length and width of the book in inches if its area is numerically 128 more than its perimeter.

1. width__________

length __________

2. The length of a rectangle is three times its width. If the width were increased by 4 and the length remained the same, the resulting rectangle would have an area of 231 square inches. Find the dimensions of the original rectangle.

2. width__________

length __________

Name: Date:
Instructor: Section:

3. Two rectangles with different dimensions have the same area. The length of the first rectangle is three times its width. The length of the second rectangle is 4 meters more than the width of the first rectangle, and its width is 2 meters more than the width of the first rectangle. Find the lengths and widths of the two rectangles.

3. Rectangle 1:

width __________

length __________

Rectangle 2:

width __________

length __________

4. Each side of one square is 1 meter less than twice the length of each side of a second square. If the difference between the areas of the two squares is 16 meters, find the lengths of the sides of the two rectangles.

4. square 1__________

square 2__________

5. The area of a triangle is 42 square centimeters. The base is 2 centimeters less than twice the height. Find the base and height of the triangle.

5. base__________

height __________

Name: Date:
Instructor: Section:

6. The volume of a box is 192 cubic feet. If the length of the box is 8 feet and the width is 2 feet more than the height, find the height and width of the box.

6. height ____________

width ____________

7. Mr. Fixxall is building a box which will have a volume of 60 cubic meters. The height of the box will be 4 meters, and the length will be 2 meters more than the width. Find the width and length of the box.

7. width ____________

length ____________

Objective 2 Solve problems about consecutive integers.

Solve each problem.

8. Find two consecutive integers such that the sum of the squares of the two integers is 3 more than the opposite (additive inverse) of the smaller integer.

8. ____________

Name: Date:
Instructor: Section:

9. If the square of the sum of two consecutive integers is reduced by twice their product, the result is 25. Find the integers.

9. ____________

10. Find all possible pairs of consecutive odd integers whose sum is equal to their product decreased by 47.

10. ____________

11. Find two consecutive positive even integers whose product is 168.

11. ____________

12. Find two consecutive positive even integers whose product is six more than three times its sum.

12. ____________

Name: Date:
Instructor: Section:

13. The product of two consecutive even positive integers is 10 more than seven times the larger. Find the integers.

13. ______________

14. Find three consecutive positive odd integers such that four times the sum of all three equals 13 more than the product of the smaller two.

14. ______________

Objective 3 Solve problems using the Pythagorean formula.

Solve each problem.

15. The hypotenuse of a right triangle is 4 inches longer than the shorter leg. The longer leg is 4 inches shorter than twice the shorter leg. Find the lengths of the three sides.

15. ______________

Name: Date:
Instructor: Section:

16. A field is in the shape of a right triangle. The shorter leg measures 45 meters. The hypotenuse measures 45 meters less than twice the longer the leg. Find the dimensions of the lot.

16. ____________

17. A train and a car leave a station at the same time, the train traveling due north and the car traveling west. When they are 100 miles apart, the train has traveled 20 miles farther than the car. Find the distance each has traveled.

17. car ____________

train ____________

18. Penny and Carla started biking from the same corner. Penny biked east and Carla biked south. When they were 26 miles apart, Carla had biked 14 miles further than Penny. Find the distance each biked.

18. Penny ____________

Carla ____________

Name: Date:
Instructor: Section:

19. Two trains leave New York City at the same time. One train travels due north and the other travels due east. When they are 75 miles apart, the train going north has gone 30 miles less than twice the distance traveled by the train going east. Find the distance traveled by the train going east.

19. ______________

20. Mark is standing directly beneath a kite attached to a string which Nina is holding, with her hand touching the ground. The height of the kite at that instant is 12 feet less than twice the distance between Mark and Nina. The length of the kite string is 12 feet more than the distance between Mark and Nina. Find the length of the kite string.

20. ______________

21. Two ships left a dock at the same time. When they were 25 miles apart, the ship that sailed due south had gone 10 miles less than twice the distance traveled by the ship that sailed due west. Find the distance traveled by the ship that sailed due south.

21. ______________

Name: Date:
Instructor: Section:

22. A ladder is leaning against a building. The distance from the bottom of the ladder to the building is 8 feet less than the length of the ladder. How high up the side of the building is the top of the ladder if that distance is 4 feet less than the length of the ladder?

22. ____________

Objective 4 Solve problems using given quadratic models.

Solve each problem.

23. A ball is dropped from the roof of a 19.6 meter high building. Its height h (in meters) t seconds later is given by the equation $h = -4.9t^2 + 19.6$.

(a) After how many seconds is the height 18.375 meters?

(b) After how many seconds is the height 14.7 meters?

(c) After how many seconds does the ball hit the ground?

23. a. ____________

b. ____________

c. ____________

Name: Date:
Instructor: Section:

24. If an object is propelled upward from a height of 16 feet with an initial velocity of 48 feet per second, its height h (in feet) t seconds later is given by the equation $h = -16t^2 + 48t + 16$.

(a) After how many seconds is the height 52 feet?

(b) After how many seconds is the height 48 feet?

24. a. ______________

b. ______________

25. The total cost of a product can be modeled by the equation $C = 400 - 100x + x^2$ where x represents the number of items produced. How many items can be produced at a cost of $1500?

25. ______________

26. Jeff threw a stone straight upward at 46 feet per second from a dock 6 feet above a lake. The height of the stone above the lake t seconds after it is thrown is given by $h = -16t^2 + 46t + 6$. How long will it take for the stone to reach a height of 39 feet?

26. ______________

Name: Date:
Instructor: Section:

27. A company determines that its daily revenue R (in dollars) for selling x items is modeled by the equation $R = x(150 - x)$. How many items must be sold for its revenue to be $4400?

27. ______________

28. If a ball is batted at an angle of 35°, the distance that the ball travels is given approximately by $D = 0.029v^2 + 0.021v - 1$, where v is the bat speed in miles per hour and D is the distance traveled in feet. Find the distance a batted ball will travel if the ball is batted with a velocity of 90 miles per hour. Round your answer to the nearest whole number.

28. ______________

Name: Date:
Instructor: Section:

29. Altitude affects the distance a batted ball travels. The data shown below can be modeled by the equation $D = -0.000000234a^2 + .0069a + 400$, where D is the distance the ball travels in feet and a is the altitude above sea level in feet. Use the model to find how far a ball will travel in Wrigley Field where the altitude is approximately 580 feet. Round your answer to the nearest whole number.
(Source: Robert K. Adair, *The Physics of Baseball*, (Harper Perennial, 1994), 18–19)

29. ______________

Stadium	**Altitude (ft)**	**Distance (ft)**
Yankee Stadium (New York)	0	400
Kauffman Stadium (Kansas City)	740	405
Turner Field (Atlanta)	1050	407
Coors Field (Denver)	5280	430

30. The unemployment rate in a certain community can be modeled by the equation $y = 0.0248x^2 - 0.4810x + 7.8543$, where y is the unemployment rate (percent) and x is the month ($x = 1$ represents January, $x = 2$ represents February, etc.) Use the model to find the unemployment rate in August. Round your answer to the nearest tenth.

30. ______________

Name: Date:
Instructor: Section:

Chapter 7 RATIONAL EXPRESSIONS AND FUNCTIONS

7.1 Rational Expressions and Functions; Multiplying and Dividing

Learning Objectives

1. Define rational expressions.
2. Define rational functions and describe their domains.
3. Write rational expressions in lowest terms.
4. Multiply rational expressions.
5. Find reciprocals for rational expressions.
6. Divide rational expressions.

Key Terms

Use the vocabulary terms listed below to complete each statement in exercises 1–2.

rational expression **rational function**

1. A ______________________ is a function that is defined by rational expression in the form $f(x)=\dfrac{P(x)}{Q(x)}$, where $Q(x)\neq 0$.

2. The quotient of two polynomials with denominator not 0 is called a ______________________.

Objective 1 Define rational expressions.

Objective 2 Define rational functions and describe their domains.

Find all numbers that are not in the domain of each function. Then give the domain using set notation.

1. $f(a)=\dfrac{2a-3}{4a-7}$ 1. ______________ ______________

2. $f(s)=\dfrac{8s+7}{3s-2}$ 2. ______________ ______________

3. $f(x)=\dfrac{x-6}{x^2+1}$ 3. ______________ ______________

Name: Date:
Instructor: Section:

4. $f(r)=\dfrac{r+7}{r^2-25}$ 4. ________

____

5. $f(q)=\dfrac{q+7}{q^2-3q+2}$ 5. ________

____

6. $f(x)=\dfrac{2x-5}{x^2-10x+25}$ 6. ________

____

Objective 3 Write rational expressions in lowest terms.

Write each rational expression in lowest terms.

7. $\dfrac{16-x^2}{2x-8}$ 7. ________

8. $\dfrac{12k^3+12k^2}{3k^2+3k}$ 8. ________

9. $\dfrac{9x^2-9x-108}{2x-8}$ 9. ________

10. $\dfrac{3y^2-13y-10}{2y^2-9y-5}$ 10. ________

Name: Date:
Instructor: Section:

11. $\dfrac{a^2-3a}{3a-a^2}$

11. ______________

12. $\dfrac{x^2-y^2}{x^3-y^3}$

12. ______________

Objective 4 Multiply rational expressions.

Multiply. Write each answer in lowest terms.

13. $\dfrac{3x+12}{6x-30}\cdot\dfrac{x^2-x-20}{x^2-16}$

13. ______________

14. $\dfrac{x^2+x-12}{x^2+7x+10}\cdot\dfrac{x^2+3x-10}{x^2+2x-8}$

14. ______________

15. $\dfrac{2x^2+5x-12}{x^2-2x-24}\cdot\dfrac{x^2-9x+18}{9-4x^2}$

15. ______________

Name: Date:
Instructor: Section:

16. $\dfrac{x^2+10x+21}{x^2+14x+49}\cdot\dfrac{x^2+12x+35}{x^2-6x-27}$

16. ________

17. $\dfrac{3m^2-m-10}{2m^2-7m-4}\cdot\dfrac{4m^2-1}{6m^2+7m-5}$

17. ________

18. $\dfrac{x^2-x-6}{x^2-2x-8}\cdot\dfrac{x^2+7x+12}{9-x^2}$

18. ________

Objective 5 Find reciprocals for rational expressions.

Find the reciprocal.

19. $\dfrac{8-s}{s-8}$

19. ________

20. $\dfrac{r^2+2r}{5+r}$

20. ________

Name: Date:
Instructor: Section:

21. $\dfrac{x^2+4}{3x-6}$ 21. ______________

22. $\dfrac{7z+7}{z^2-9}$ 22. ______________

23. 0 23. ______________

24. $\dfrac{x^2-3x+4}{x^2+x+2}$ 24. ______________

Objective 6 Divide rational expressions.

Divide. Write each answer in lowest terms.

25. $\dfrac{4m-12}{2m+10} \div \dfrac{9-m^2}{m^2-25}$ 25. ______________

26. $\dfrac{27-3k^2}{3k^2+8k-3} \div \dfrac{k^2-6k+9}{6k^2-19k+3}$ 26. ______________

Name: Date:
Instructor: Section:

27. $\dfrac{y^2+7y+10}{3y+6} \div \dfrac{y^2+2y-15}{4y-4}$

27. ____________

28. $\dfrac{2k^2+5k-12}{2k^2+k-3} \div \dfrac{k^2+8x+16}{2k^2+11k+12}$

28. ____________

29. $\dfrac{z^4+2z^3+z^2}{z^5-4z^3} \div \dfrac{9z+9}{6z+12}$

29. ____________

30. $\dfrac{2a^2-5a-12}{a^2-10a+24} \div \dfrac{4a^2-9}{a^2-9a+18}$

30. ____________

Name: Date:
Instructor: Section:

Chapter 7 RATIONAL EXPRESSIONS AND FUNCTIONS

7.2 Adding and Subtracting Rational Expressions

Learning Objectives

1. Add and subtract rational expressions with the same denominator.
2. Find a least common denominator.
3. Add and subtract rational expressions with different denominators.

Key Terms

Use the vocabulary terms listed below to complete each statement in exercises 1–2.

least common denominator (LCD) **equivalent expressions**

1. $\frac{24x-8}{9x^2-1}$ and $\frac{8}{3x+1}$ are ________________________.

2. The simplest expression that is divisible by all denominators is called the ________________________.

Objective 1 Add and subtract rational expressions with the same denominator.

Add or subtract as indicated. Write each answer in lowest terms.

1. $\frac{z^2}{z-y}-\frac{y^2}{z-y}$ **1.** ____________

2. $\frac{-4x+3}{x-7}+\frac{2x+11}{x-7}$ **2.** ____________

3. $\frac{n}{m+3}-\frac{-3n+7}{m+3}$ **3.** ____________

Name: Date:
Instructor: Section:

4. $\dfrac{3x-2}{x^2+x-2}-\dfrac{x}{x^2+x-2}$ 4. ______

5. $\dfrac{r}{r^2-s^2}+\dfrac{s}{r^2-s^2}$ 5. ______

6. $\dfrac{2x+3}{x^2+3x-10}+\dfrac{2-x}{x^2+3x-10}$ 6. ______

7. $\dfrac{x}{x^2-7x+10}-\dfrac{2}{x^2-7x+10}$ 7. ______

8. $\dfrac{k}{k^2-6k+8}-\dfrac{2}{k^2-6k+8}$ 8. ______

9. $\dfrac{1}{q^2-6q-7}+\dfrac{q}{q^2-6q-7}$ 9. ______

Name: Date:
Instructor: Section:

10. $\frac{b}{a^2 - b^2} - \frac{a}{a^2 - b^2}$

10. ______________

Objective 2 Find a least common denominator.

Assume that the expressions given are denominators of fractions. Find the least common denominator (LCD) for each group.

11. $5a + 10, a^2 + 2a$

11. ______________

12. $q^2 - 36, (q + 6)^2$

12. ______________

13. $x^2 + 5x + 6, 3x + 6$

13. ______________

14. $m^3 - 2m^2, m^2 + 5n - 14$

14. ______________

Name: Date:
Instructor: Section:

15. $a^2 - 2a,\ 2a^2 + a - 10$

15. ______________

16. $r^2 + 5r + 4,\ r^2 + r$

16. ______________

17. $3n + n^2,\ 3 - n$

17. ______________

18. $z^4 + 2z^3 - 8z^2,\ z^3 + 8z^2 + 16z$

18. ______________

19. $p - 4,\ p^2 - 16,\ (p+4)^2$

19. ______________

20. $2z^2 + 7z - 4,\ 2z^2 - 7z + 3$

20. ______________

Name: Date:
Instructor: Section:

Objective 3 Add and subtract rational expressions with different denominators.

Add or subtract as indicated. Write each answer in lowest terms..

21. $\frac{-4}{x^2-4}-\frac{3}{4-2x}$ **21.** ________________

22. $\frac{3z}{z^2-4}+\frac{4z-3}{z^2-4z+4}$ **22.** ________________

23. $\frac{1}{9m-3}-\frac{m-2}{3m^2+11m-4}$ **23.** ________________

24. $\frac{1-3x}{4x^2-1}+\frac{3x-5}{2x^2+5x+2}$ **24.** ________________

25. $\frac{3}{9b^2-16}+\frac{2}{3b^2+2b-8}$ **25.** ________________

Name: Date:
Instructor: Section:

26. $\dfrac{1}{9m-3}-\dfrac{m-2}{3m^2+11m-4}$ 26. ______________

27. $\dfrac{3}{n^2-16}-\dfrac{6n}{n^2+8n+16}$ 27. ______________

28. $\dfrac{4}{2c^2+c-3}+\dfrac{c}{2c^2-5c-12}$ 28. ______________

29. $\dfrac{4z}{z^2+6z+8}+\dfrac{2z-1}{z^2+5z+6}$ 29. ______________

30. $\dfrac{4y}{y^2+4y+3}-\dfrac{3y+1}{y^2-y-2}$ 30. ______________

Name: Date:
Instructor: Section:

Chapter 7 RATIONAL EXPRESSIONS AND FUNCTIONS

7.3 Complex Fractions

Learning Objectives

1. Simplify complex fractions by simplifying the numerator and denominator (Method 1).
2. Simplify complex fractions by multiplying by a common denominator (Method 2).
3. Compare the two methods of simplifying complex fractions.
4. Simplify rational expressions with negative exponents.

Key Terms

Use the vocabulary terms listed below to complete each statement in exercises 1–2.

complex fraction **LCD**

1. A ________________________ is a rational expression with one or more fractions in the numerator, denominator, or both.

2. To simplify a complex fraction, multiply the numerator and denominator by the ____________________________ of all the fractions within the complex fraction.

Objective 1 Simplify complex fractions by simplifying the numerator and denominator (Method 1).

Use Method 1 to simplify each complex fraction.

1. $\dfrac{\frac{rs}{3r^2}}{\frac{s^2}{3}}$ 1. ______________

2. $\dfrac{\frac{3a+4}{a}}{\frac{1}{a}+\frac{2}{5}}$ 2. ______________

Name: Date:
Instructor: Section:

3. $\dfrac{\dfrac{5}{rs^2}-\dfrac{2}{rs}}{\dfrac{3}{rs}-\dfrac{4}{r^2 s}}$

3. ______________

4. $\dfrac{\dfrac{2}{a+2}-4}{\dfrac{1}{a+2}-3}$

4. ______________

5. $\dfrac{\dfrac{a+2}{a-2}}{\dfrac{1}{a^2-4}}$

5. ______________

6. $\dfrac{\dfrac{3}{w-4}-\dfrac{3}{w+4}}{\dfrac{1}{w+4}+\dfrac{1}{w^2-16}}$

6. ______________

Name: Date:
Instructor: Section:

7. $\dfrac{q+\dfrac{1}{q+1}}{q-\dfrac{1}{q}}$

7. ______________

Objective 2 Simplify complex fractions by multiplying by a common denominator (Method 2).

Use Method 2 to simplify each complex fraction.

8. $\dfrac{\dfrac{16r^2}{11s^3}}{\dfrac{8r^4}{22s}}$

8. ______________

9. $\dfrac{\dfrac{9}{x^2}-1}{\dfrac{3}{x}-1}$

9. ______________

10. $\dfrac{2x-y^2}{x+\dfrac{y^2}{x}}$

10. ______________

Name: Date:
Instructor: Section:

11. $$\frac{r+\frac{3}{r}}{\frac{5}{r}+rt}$$

11. ______________

12. $$\frac{\frac{x-2}{x+2}}{\frac{x}{x-2}}$$

12. ______________

13. $$\frac{\frac{15}{10k+10}}{\frac{5}{3k+3}}$$

13. ______________

14. $$\frac{\frac{r}{r+1}+1}{\frac{2r+1}{r-1}}$$

14. ______________

Name: Date:
Instructor: Section:

Objective 3 Compare the two methods of simplifying complex fractions.

Use either method to simplify each complex fraction.

15. $$\frac{\frac{1}{r}}{\frac{1+r}{1-r}}$$

15. ______________

16. $$\frac{\frac{1}{t+1}-1}{\frac{1}{t-1}+1}$$

16. ______________

17. $$\frac{\frac{25k^2-m^2}{4k}}{\frac{5k+m}{7k}}$$

17. ______________

18. $$\frac{\frac{4}{p}-2p}{\frac{3-p^2}{6}}$$

18. ______________

Name: Date:
Instructor: Section:

19. $$\frac{\frac{4}{x}-\frac{1}{2}}{\frac{5}{x}+\frac{1}{3}}$$

19. ________________

20. $$\frac{\frac{1}{m-1}+4}{\frac{2}{m-1}-4}$$

20. ________________

21. $$\frac{\frac{4}{s+3}-\frac{2}{s-3}}{\frac{5}{s^2-9}}$$

21. ________________

22. $$\frac{\frac{6}{k+1}-\frac{5}{k-3}}{\frac{3}{k-3}+\frac{2}{k+2}}$$

22. ________________

Name: Date:
Instructor: Section:

Objective 4 Simplify rational expressions with negative exponents.

Simplify each expression, using only positive exponents in the answer.

23. $\dfrac{x^{-1}}{x^{-1}+5}$ **23.** ______________

24. $\dfrac{1+x^{-1}}{(x+y)^{-1}}$ **24.** ______________

25. $\dfrac{2x^{-1}+y^{2}}{z^{-3}}$ **25.** ______________

26. $\left(x^{-2}-y^{-2}\right)^{-1}$ **26.** ______________

Name: Date:
Instructor: Section:

27. $\dfrac{4x^{-2}}{2+6y^{-3}}$

27. ______________

28. $\dfrac{s^{-1}+r}{r^{-1}+s}$

28. ______________

29. $\dfrac{x^{-1}}{y-x^{-1}}$

29. ______________

30. $\dfrac{(m+n)^{-2}}{m^{-2}-n^{-2}}$

30. ______________

Name: Date:
Instructor: Section:

Chapter 7 RATIONAL EXPRESSIONS AND FUNCTIONS

7.4 Equations with Rational Expressions and Graphs

Learning Objectives	
1	Determine the domain of the variable in a rational equation.
2	Solve rational equations.
3	Recognize the graph of a rational function.

Key Terms

Use the vocabulary terms listed below to complete each statement in exercises 1–4.

domain of the variable in a rational equation **discontinuous**

vertical asymptote **horizontal asymptote**

1. A rational function in simplest form $f(x)=\dfrac{P(x)}{x-a}$ has the line $x=a$ as a
________________________________.

2. The ____ is the intersection of the domains of the rational expressions in the equation.

3. A graph of a function is ______________________ if there are one or more breaks in the graph.

4. A horizontal line that a graph approaches as $|x|$ gets larger without bound is called a __________________________________.

Objective 1 Determine the domain of the variable in a rational equation.

(a) Without actually solving the equations below, list all possible numbers that would have to be rejected if they appeared as potential solutions. (b) Then give the domain using set notation.

1. $\dfrac{1}{x}+\dfrac{2}{x+1}=0$

1. a.______________

b.______________

Name: Date:
Instructor: Section:

2. $\frac{4}{2x-5}-\frac{6}{3x+1}=\frac{1}{2}$

2. a. ________________

b. ________________

3. $\frac{10}{x-7}+\frac{7}{8+x}=0$

3. a. ________________

b. ________________

4. $\frac{x}{6}-\frac{1}{2}=\frac{3}{x+1}$

4. a. ________________

b. ________________

5. $\frac{4}{x^2-2x}+\frac{x}{2}=0$

5. a. ________________

b. ________________

6. $\frac{3}{3+x}+\frac{5}{x^2-x}=\frac{9}{x}$

6. a. ________________

b. ________________

Name: Date:
Instructor: Section:

Objective 2 Solve rational equations.

Solve each equation.

7. $\frac{3}{k}-\frac{2}{k+2}=\frac{7}{3}$ **7.** ______________

8. $\frac{z+4}{z-7}-6=\frac{5}{z-7}$ **8.** ______________

9. $\frac{8}{2m+4}+\frac{2}{3m+6}=\frac{7}{9}$ **9.** ______________

10. $\frac{2}{z-1}+\frac{3}{z+1}-\frac{17}{24}=0$ **10.** ______________

Name: Date:
Instructor: Section:

11. $\frac{5a+1}{2a+2}=\frac{5a-5}{5a+5}+\frac{3a+1}{a+1}$ **11.** ______________

12. $\frac{x}{2x+2}=\frac{-2x}{4x+4}+\frac{2x-3}{x+1}$ **12.** ______________

13. $\frac{4}{n}-\frac{2}{n+1}=3$ **13.** ______________

14. $x-11+\frac{18}{x}=0$ **14.** ______________

Name: Date:
Instructor: Section:

15. $\dfrac{1}{b+2}-\dfrac{5}{b^2+9b+14}=\dfrac{-3}{b+7}$

15. ______________

16. $\dfrac{-16}{n^2-8n+12}=\dfrac{3}{n-2}+\dfrac{n}{n-6}$

16. ______________

17. $\dfrac{q+12}{q^2-16}-\dfrac{3}{q-4}=\dfrac{1}{q+4}$

17. ______________

18. $\dfrac{9}{x^2-x-12}=\dfrac{3}{x-4}-\dfrac{x}{x+3}$

18. ______________

Name: Date:
Instructor: Section:

19. $\dfrac{4}{y+2}-\dfrac{3}{y+3}=\dfrac{8}{y^2+5y+6}$

19. ______________

20. $\dfrac{-13}{r^2+6r+8}+\dfrac{4}{r+2}=\dfrac{3}{r+4}$

20. ______________

21. $\dfrac{1}{q^2+5q+6}+\dfrac{1}{q^2-2q-8}=\dfrac{-1}{12q+24}$

21. ______________

22. $-\dfrac{17}{z^2+5z-6}-\dfrac{3}{1-z}=\dfrac{z+2}{z+6}$

22. ______________

Name: Date:
Instructor: Section:

23. $\dfrac{1}{x^2+5x+6}=\dfrac{2}{x^2-x-6}-\dfrac{3}{9-x^2}$

23. ____________

24. $\dfrac{3p}{p^2+5p+6}=\dfrac{5p}{p^2+2p-3}-\dfrac{2}{p^2+p-2}$

24. ____________

Objective 3 Recognize the graph of a rational function.

Graph each rational function. Give the equations of the vertical and horizontal asymptotes.

25. $f(x)=\dfrac{5}{x}$

25.

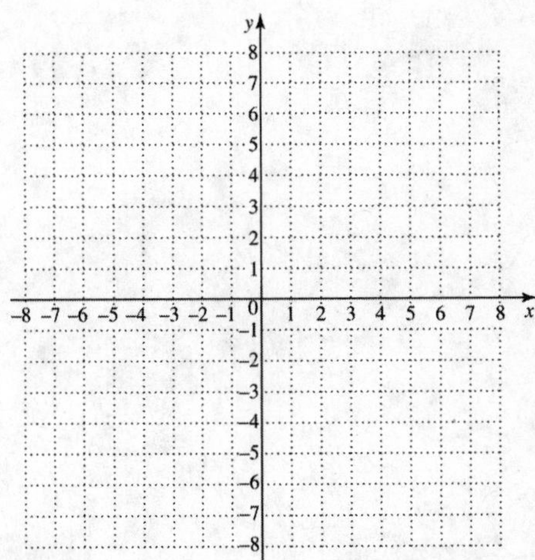

Vertical asymptote:__________

Horizontal asymptote:__________

Name: Date:
Instructor: Section:

26. $f(x) = \dfrac{2}{x-1}$

26.

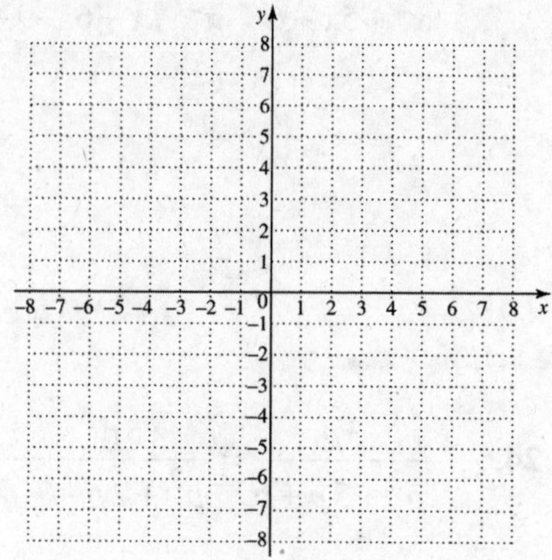

Vertical asymptote:___________

Horizontal asymptote:_________

27. $f(x) = \dfrac{1}{x-3}$

27.

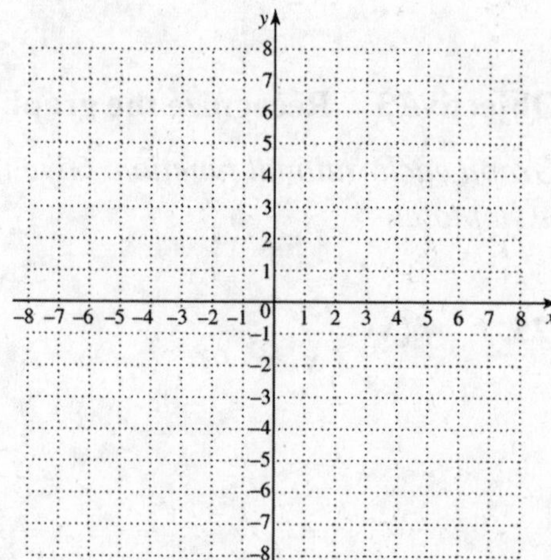

Vertical asymptote:___________

Horizontal asymptote:_________

Name: Date:
Instructor: Section:

28. $f(x) = \dfrac{1}{x-4}$

28.

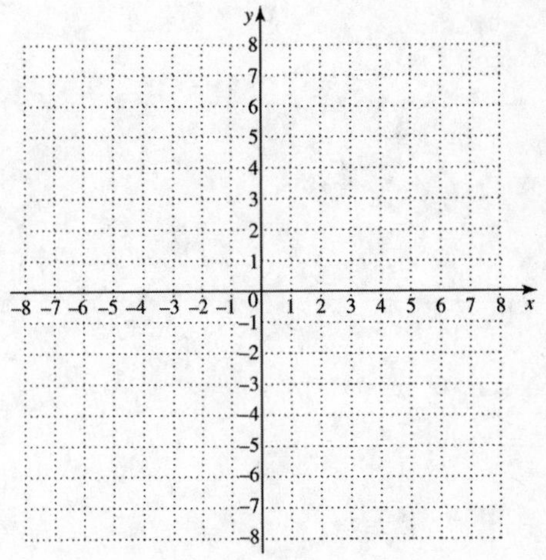

Vertical asymptote:___________

Horizontal asymptote:_________

29. $f(x) = \dfrac{3}{x+2}$

29.

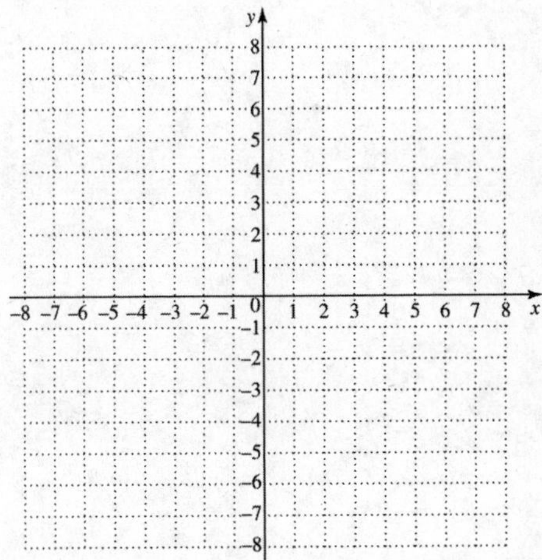

Vertical asymptote:___________

Horizontal asymptote:_________

Name: Date:
Instructor: Section:

30. $f(x) = -\frac{1}{x}$

30.

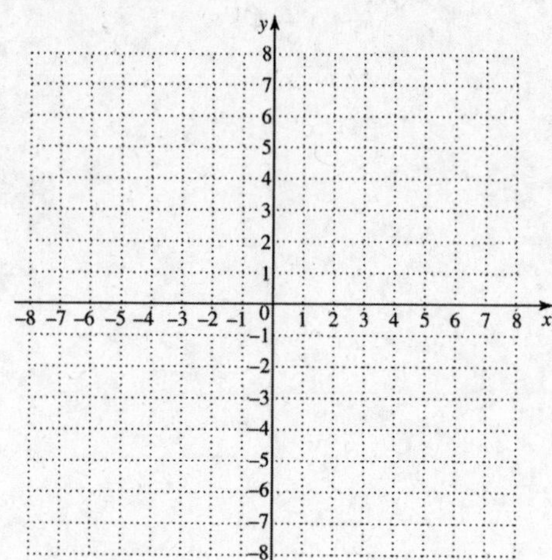

Vertical asymptote:___________

Horizontal asymptote:_________

Name: Date:
Instructor: Section:

Chapter 7 RATIONAL EXPRESSIONS AND FUNCTIONS

7.5 Applications of Rational Expressions

Learning Objectives
1. Find the value of an unknown variable in a formula.
2. Solve a formula for a specified variable.
3. Solve applications using proportions.
4. Solve applications about distance, rate, and time.
5. Solve applications about work rates.

Key Terms

Use the vocabulary terms listed below to complete each statement in exercises 1–2.

ratio **proportion**

1. A comparison of two quantities using a quotient is a ________________.

2. A statement that two ratios are equal is a ________________.

Objective 1 Find the value of an unknown variable in a formula.

Find the value of the variable indicated.

1. If $F = \dfrac{GmM}{d^2}$, $F = 150$, $G = 32$, $M = 50$, and $d = 10$, find m. 1. ________________

2. If $\dfrac{1}{R} = \dfrac{1}{R_1} + \dfrac{1}{R_2}$, $R = 10$, and $R_1 = 20$, find R_2. 2. ________________

Name: Date:
Instructor: Section:

3. If $\frac{1}{f}=\frac{1}{d_0}+\frac{1}{d_i}$, $f=10$, and $d_0=25$, find d_i. **3.** ______________

4. If $h=\frac{2A}{B+b}$, $A=40$, $h=8$, and $b=3$, find B. **4.** ______________

5. If $c=\frac{100b}{L}$, $c=80$, and $b=16$, find L. **5.** ______________

6. If $\frac{mn}{p}=\frac{qr}{s}$, $m=4$, $n=2.5$, $q=7.5$, $r=2.5$, and $s=1.5$, find p. **6.** ______________

Objective 2 Solve a formula for a specified variable.

Solve each formula for the specified variable.

7. $\frac{V_1P_1}{T_1}=\frac{V_2P_2}{T_2}$ for T_2 **7.** ______________

Name: Date:
Instructor: Section:

8. $A = \dfrac{R_1 R_2}{R_1 + R_2}$ for R_1 8. ______________

9. $C = \dfrac{5}{9}(F - 32)$ for F 9. ______________

10. $h = \dfrac{2A}{B + b}$ for B 10. ______________

11. $F = f\left[\dfrac{v + v_0}{v - v_s}\right]$ for v_s 11. ______________

12. $E = \dfrac{e(R + r)}{r}$ for r 12. ______________

Name:	Date:
Instructor:	Section:

Objective 3 Solve applications using proportions.

Use proportions to solve each problem.

13. Holly drove 315 miles in 7 hours. How long would it take her to drive 225 miles if she traveled at the same rate?

13. ________________

14. Ryan's car travels 24 miles using 1 gallon of gas. If Ryan has 4 gallons of gas in his car, how much more gas will he need to travel 288 miles?

14. ________________

15. Connie paid $0.90 in sales tax on a purchase of $12.00. Later that day she purchased an item for which she paid $55.90 including sales tax, at the same tax rate. How much was the sales tax on that item?

15. ________________

16. Mike paid $5.12 sales tax on an item priced $64. Later he made a purchase that cost him a total of $51.84, including sales tax at the same rate. How much sales tax did he pay on the second item?

16. ________________

Name: Date:
Instructor: Section:

17. A certain city with a population of 400,000 had 1600 burglaries during the last year. If the number of burglaries increased by 200 this year and its burglary rate remained the same, by what number did its population increase?

17. ________________

18. Megan has invested $5000 in an account that earns $200 in income per year. If she wishes to earn $240 per year at the same rate, by how much must she increase her investment?

18. ________________

Objective 4 Solve applications about distance, rate, and time.

Solve each problem.

19. Lauren's boat can go 9 miles per hour in still water. How far downstream can Lauren go if the river has a current of 3 miles per hour and she must be back in 4 hours?

19. ________________

Name: Date:
Instructor: Section:

20. Pauline and Pete agree to meet in Columbia. Pauline travels 120 miles, while Pete travels 80 miles. If Pauline's speed is 20 miles per hour greater than Pete's and they both spend the same amount of time traveling, at what speed does each travel?

20. ______________

21. Leo can get to the lake using either the old road at 40 miles per hour, or the new road at 60 miles per hour. If both roads are the same length and he gets there 1 hour sooner on the new road, how far is it to the lake?

21. ______________

22. Olivia can ride her bike 4 miles per hour faster than Ted can ride his bike. If Olivia can go 30 miles in the same time that Ted can go 15 miles, what are their speeds?

22. ______________

Name: Date:
Instructor: Section:

23. Carl traveled to his destination at an average speed of 70 miles per hour. Coming home, his average speed was 50 miles per hour and the trip took 2 hours longer. How far did he travel each way?

23. ______________

24. A plane traveling 450 miles per hour can go 1000 miles with the wind in $\frac{1}{2}$ hour less than when traveling against the wind. Find the speed of the wind.

24. ______________

Objective 5 Solve applications about work rates.

Solve each problem.

25. Kelly can clean the house in 6 hours, but it takes Linda 4 hours. How long would it take them to clean the house if they worked together?

25. ______________

Name: Date:
Instructor: Section:

26. One pipe can fill a swimming pool in 8 hours and another pipe can fill the pool in 12 hours. How long will it take to fill the pool if both pipes are open?

26. ______________

27. Chuck can weed the garden in $\frac{1}{2}$ hour, but David takes 2 hours. How long does it take them to weed the garden if they work together?

27. ______________

28. A swimming pool can be filled by an inlet pipe in 18 hours and emptied by an outlet pipe in 24 hours. How long will it take to fill the empty pool if the outlet pipe is accidentally left open at the same time as the inlet pipe is opened?

28. ______________

Name: Date:
Instructor: Section:

29. Fred can seal an asphalt driveway in $\frac{1}{3}$ the time it takes John. Working together, it takes them $1\frac{1}{2}$ hours. How long would it have taken Fred working alone?

29. ______________

30. A cold water faucet can fill a sink in 12 minutes, and a hot water faucet can fill it in 15 minutes. The drain can empty the sink in 25 minutes. If both faucets are on and the drain is open, how long would it take to fill the sink?

30. ______________

Name: Date:
Instructor: Section:

Chapter 7 RATIONAL EXPRESSIONS AND FUNCTIONS

7.6 Variation

Learning Objectives

1. Write an equation expressing direct variation.
2. Find the constant of variation, and solve direct variation problems.
3. Solve inverse variation problems.
4. Solve joint variation problems.
5. Solve combined variation problems.

Key Terms

Use the vocabulary terms listed below to complete each statement in exercises 1–3.

varies directly **varies inversely** **constant of variation**

1. In the equations for direct and inverse variation, k is the ________________.

2. If there exists a real number k such that $y = \frac{k}{x}$, then y ________________ as x.

3. If there exists a real number k such that $y = kx$, then y ________________ as x.

Objective 1 Write an equation expressing direct variation.

Objective 2 Find the constant of variation, and solve direct variation problems.

Find the constant of variation, and write a direct variation equation.

1. $y = 15$ when $x = 5$ **1.** ________________

2. $y = 12$ when $x = 8$ **2.** ________________

3. $y = 13.75$ when $x = 55$ **3.** ________________

Name: Date:
Instructor: Section:

4. $y = 8$ when $x = 6$ **4.** ____________

5. $y = 14$ when $x = 9$ **5.** ____________

6. $y = 11$ when $x = 22$ **6.** ____________

Solve each problem.

7. If m varies directly as p, and $m = 40$ when $p = 5$, find m when p is 9. **7.** ____________

8. If m varies directly as the square of p, and $m = 1$ when $p = 2$, find m when p is 7. **8.** ____________

9. The circumference of a circle varies directly as the radius. A circle with a radius of 7 centimeters has a circumference of 43.96 centimeters. Find the circumference of the circle if the radius changes to 11 centimeters. **9.** ____________

Name: Date:
Instructor: Section:

10. The pressure exerted by a certain liquid at a given point varies directly as the depth of the point beneath the surface of the liquid. The pressure at 10 feet is 50 pounds per square inch (psi). What is the pressure at 20 feet?

10. ________

11. The force required to compress a spring varies directly as the change in length of the spring. If a force of 20 newtons is required to compress a spring 2 centimeters in length, how much force is required to compress a spring of length 10 centimeters?

11. ________

12. The surface area of a sphere varies directly as the square of its radius. If the surface area of a sphere with a radius of 12 inches is 576π square inches, find the surface area of a sphere with a radius of 3 inches.

12. ________

Objective 3 Solve inverse variation problems.

Solve each problem.

13. If y varies inversely as x, and $y = 10$ when $x = 3$, find y when $x = 12$.

13. ________

14. If z varies inversely as x^2, and $z = 9$ when $x = \frac{2}{3}$, find z when $x = \frac{5}{4}$.

14. ________

Name: Date:
Instructor: Section:

15. The current in a simple electrical circuit varies inversely as the resistance. If the current is 50 amperes (an ampere is a unit for measuring current) when the resistance is 10 ohms (an ohm is a unit for measuring resistance), find the current if the resistance is 5 ohms.

15. ______________

16. The illumination produced by a light source varies inversely as the square of the distance from the source. If the illumination produced 4 feet from a light source is 75 footcandles, find the illumination produced 9 feet from the same source.

16. ______________

17. The weight of an object varies inversely as the square of its distance from the center of the earth. If an object 8000 miles from the center of the earth weighs 90 pounds, find its weight when it is 12,000 miles from the center of the earth.

17. ______________

18. The speed of a pulley varies inversely as its diameter. One kind of pulley, with a diameter of 3 inches, turns at 150 revolutions per minute. Find the speed of a similar pulley with diameter of 5 inches.

18. ______________

Name: Date:
Instructor: Section:

Objective 4 Solve joint variation problems.

Solve each problem.

19. If q varies jointly as p and r^2, and $q = 27$ when $p = 9$ and $r = 2$, find q when $p = 8$ and $r = 4$.

19. ______________

20. Suppose y varies jointly as x^2 and z^2, and $y = 72$ when $x = 2$ and $z = 3$. Find y when $x = 4$ and $z = 2$.

20. ______________

21. Suppose d varies jointly as f^2 and g^2, and $d = 384$ when $f = 3$ and $g = 8$. Find d when $f = 6$ and $g = 2$.

21. ______________

22. For a fixed interest rate, interest varies jointly as the principal and the time in years. If $5000 invested for 4 years earns $900, how much interest will $6000 invested for 3 years earn at the same interest rate?

22. ______________

Name: Date:
Instructor: Section:

23. The work w (in joules) done when lifting an object is jointly proportional to the product of the mass m (in kg) of the object and the height h (in meter) the object is lifted. If the work done when a 120 kg object is lifted 1.8 meters above the ground is 2116.8 joules, how much work is done when lifting a 100kg object 1.5 meters above the ground?

23. ______________

24. The absolute temperature of an ideal gas varies jointly as its pressure and its volume. If the absolute temperature is 250° when the pressure is 25 pounds per square centimeter and the volume is 50 cubic centimeters, find the absolute temperature when the pressure is 50 pounds per square centimeter and the volume is 75 cubic centimeters.

24. ______________

Objective 5 Solve combined variation problems.

Solve each problem.

25. If y varies directly as x and inversely as z and $y = 3$ when $x = 72$ and $z = 8$, find y when $x = 1$ and $z = 4$.

25. ______________

26. If m varies directly as w and inversely as r^2, and $m = 1845$ when $w = 4.5$ and $r = .1$, find m when $w = 2.5$ and $r = .2$.

26. ______________

Name: Date:
Instructor: Section:

27. p varies jointly as P, V, and t and inversely as v and T. Suppose $p = 65.625$ when $P = 50$, $V = 9$, $t = 350$, $v = 8$, and $T = 300$. Find p when $P = 60$, $V = 8$, $t = 300$, $v = 6$, and $T = 200$.

27. ______________

28. The volume of a gas varies inversely as the pressure and directly as the temperature. If a certain gas occupies a volume of 1.3 liters at 300 K and a pressure of 18 kilograms per square centimeter, find the volume at 340 K and a pressure of 24 kilograms per square centimeter.

28. ______________

29. The time required to lay a sidewalk varies directly as its length and inversely as the number of people who are working on the job. If three people can lay a sidewalk 100 feet long in 15 hours, how long would it take two people to lay a sidewalk 40 feet long?

29. ______________

30. When an object is moving in a circular path, the centripetal force varies directly as the square of the velocity and inversely as the radius of the circle. A stone that is whirled at the end of a string 50 centimeters long at 900 centimeters per second has a centripetal force of 3,240,000 dynes. Find the centripetal force if the stone is whirled at the end of a string 75 centimeters long at 1500 centimeters per second.

30. ______________

Name: Date:
Instructor: Section:

Chapter 8 EQUATIONS, INEQUALITIES AND SYSTEMS REVISITED

8.1 Review of Solving Equations and Linear Inequalities

Learning Objectives
1. Solve linear equations.
2. Solve linear inequalities.
3. Solve three-part linear inequalities

Key Terms

Use the vocabulary terms listed below to complete each statement in exercises 1–8.

linear (first-degree) equation in one variable **solution**

solution set **equivalent equations**

conditional equation **identity** **contradiction**

linear inequality in one variable

1. Equations that have exactly the same solution sets are called ______________________________.

2. An equation that can be written in the form $Ax + B = C$, where A, B, and C are real numbers and $A \neq 0$, is called a ____.

3. The set of all numbers that satisfy an equation is called its ______________________________.

4. An equation with no solution is called a(n) ______________________________.

5. A(n) ________________________________ is an equation that is true for some values of the variable and false for other values.

6. An equation that is true for all values of the variable is called a(n) ________________________.

7. A ____________________ of an equation is a number that makes the equation true when substituted for the variable.

8. A(n) ________________________ can be written in the form $Ax + B > C$, $Ax + B \geq C$, $Ax + B < C$, or $Ax + B \leq C$, where A, B, and C are real numbers with $A \neq 0$.

Name: Date:
Instructor: Section:

Objective 1 Solve linear equations.

Solve and check each equation.

1. $11x - 14x - 7 + 8 = 4x + 5 - 2$ 1. ________

2. $18 - 3y + 11 = 7y + 6y - 27 - 9y$ 2. ________

3. $5(2p + 1) - (p + 3) = 7$ 3. ________

4. $-[p - (4p + 2)] = 3 + (4p + 7)$ 4. ________

5. $4t - 3(4 - 2t) = 2(t - 3) + 6t + 2$ 5. ________

6. $6x - (8 - x) = 9[-2 - (5 + 2x) - 12]$ 6. ________

Name: Date:
Instructor: Section:

7. $\frac{2x+5}{5}-\frac{3x+1}{2}=\frac{7-x}{2}$ **7.** ________________

8. $.35(140)+.15w=.05(w+1100)$ **8.** ________________

9. $\frac{x-5}{2}-\frac{x+6}{3}=-4$ **9.** ________________

Decide whether each equation is a **conditional equation***, an* **identity***, or a* **contradiction***. Give the solution set.*

10. $7(2-5b)-32=10b-3(6+15b)$ **10.** ________________

11. $7(3-4q)-10(q-2)=19(5-2q)$ **11.** ________________

Name: Date:
Instructor: Section:

12. $13p - 9(3 - 2p) = 3(10p - 9) + 1$ **12.** ________

Objective 2 Solve linear inequalities.

Solve each inequality, giving its solution set in both interval and graph forms. Check your answers.

13. $5a + 3 \le 6a$ **13.** ________

14. $-2 + 8b \ge 7b - 1$ **14.** ________

15. $6 + 3x < 4x + 4$ **15.** ________

16. $3 + 5p \le 4p + 3$ **16.** ________

17. $9 + 8b > 9b + 11$ **17.** ________

18. $6 < -3t + 4t$ **18.** ________

Name: Date:
Instructor: Section:

19. $-2s < 4$

19. ______________

20. $4k \geq -16$

20. ______________

21. $\frac{3}{5}n \geq 0$

21. ______________

22. $-5t \leq -35$

22. ______________

23. $-9m > -36$

23. ______________

24. $12 - 2(p - 3) \geq -8p$

24. ______________

Name: Date:
Instructor: Section:

Objective 3 Solve three-part linear inequalities.

Solve each inequality, giving its solution set in both interval and graph forms. Check your answers.

25. $9 < 2x + 1 \le 15$

25. ______________

26. $-17 \le 3x - 2 < -11$

26. ______________

27. $6 < 2x - 4 < 8$

27. ______________

28. $-5 < -2 - x \le 4$

28. ______________

29. $1 < 3z + 4 < 19$

29. ______________

30. $-3 \le \frac{3y - 3}{-4} \le 1$

30. ______________

Name: Date:
Instructor: Section:

Chapter 8 EQUATIONS, INEQUALITIES AND SYSTEMS REVISITED

8.2 Set Operations and Compound Inequalities

Learning Objectives
1 Find the intersection of two sets.
2 Solve compound inequalities with the word *and*.
3 Find the union of two sets.
4 Solve compound inequalities with the word *or*.

Key Terms

Use the vocabulary terms listed below to complete each statement in exercises 1–3.

intersection **compound inequality** **union**

1. The ______________________ of two sets, A and B, is the set of elements that belong to either A or B or both.

2. A ____________________________________ is formed by joining two inequalities with a connective word such as *and* or *or*.

3. The ______________________ of two sets, A and B, is the set of elements that belong to both A or B.

Objective 1 Find the intersection of two sets.

Let $A = \{0, 1, 2, 3, 4, 5\}$, $B = \{2, 4, 6, 8, 10\}$, $C = \{1, 3, 5, 7, 9\}$, $D = \{0, 2, 4\}$, *and* $E = \{0\}$. *Specify each set.*

1. $A \cap B$ 1. ________________

2. $A \cap D$ 2. ________________

3. $B \cap C$ 3. ________________

4. $A \cap C$ 4. ________________

5. $A \cap E$ **5.** ________

6. $B \cap D$ **6.** ________

7. $C \cap E$ **7.** ________

Objective 2 Solve compound inequalities with the word *and*.

For each compound inequality, give the solution set in both interval and graph forms.

8. $2q > -2$ and $q + 3 < 5$ **8.** ________

9. $x - 3 \leq 6$ and $x + 2 \geq 7$ **9.** ________

10. $1 - 2s \leq -3$ and $2s + 7 \geq 11$ **10.** ________

11. $2z + 1 < 3$ and $3z - 3 > 3$ **11.** ________

12. $3x + 2 < 11$ and $2 - 3x \leq 14$ **12.** ________

Name: Date:
Instructor: Section:

13. $5t > 0$ and $5t + 4 \leq 9$

13. ______________

14. $-2x + 1 < 3$ and $3x \leq 12$

14. ______________

15. $7y + 5 \leq 3$ and $-3y \geq -9$

15. ______________

Objective 3 Find the union of two sets.

Let $A = \{0, 1, 2, 3, 4, 5\}$, $B = \{2, 4, 6, 8, 10\}$, $C = \{1, 3, 5, 7, 9\}$, $D = \{0, 2, 4\}$, *and* $E = \{0\}$. *Specify each set.*

16. $A \cup B$

16. ______________

17. $C \cup D$

17. ______________

18. $A \cup D$

18. ______________

19. $B \cup C$

19. ______________

Name: Date:
Instructor: Section:

20. $A \cup C$ **20.** ____________

21. $A \cup E$ **21.** ____________

22. $B \cup D$ **22.** ____________

Objective 4 Solve compound inequalities with the word *or*.

For each compound inequality, give the solution set in both interval and graph forms.

23. $s - 5 > 0$ or $s + 7 < 6$ **23.** ____________

24. $r - 2 \leq 0$ or $r - 2 \geq 4$ **24.** ____________

25. $q + 3 > 7$ or $q + 1 \leq -3$ **25.** ____________

26. $4x < x - 5$ or $6x > 2x + 3$ **26.** ____________

Name: Date:
Instructor: Section:

27. $4x - 9 > 3$ or $8 - x > 10$

27. ______________________

28. $4t < 2t + 10$ or $t - 3 > 3$

28. ______________________

29. $3 > 4m + 2$ or $4m - 3 \geq -2$

29. ______________________

30. $2r + 4 \geq 8$ or $4r - 3 < 1$

30. ______________________

Name: Date:
Instructor: Section:

Chapter 8 EQUATIONS, INEQUALITIES AND SYSTEMS REVISITED

8.3 Absolute Value Equations and Inequalities

Learning Objectives

1. Use the distance definition of absolute value.
2. Solve equations of the form $|ax + b| = k$, for $k > 0$.
3. Solve inequalities of the form $|ax + b| < k$ and of the form $|ax + b| > k$, for $k > 0$.
4. Solve absolute value equations that involve rewriting.
5. Solve equations of the form $|ax + b| = |cx + d|$.
6. Solve special cases of absolute value equations and inequalities.

Key Terms

Use the vocabulary terms listed below to complete each statement in exercises 1–2.

absolute value equation **absolute value inequality**

1. An ______________________________ is an equation that involves the absolute value of a variable expression.

2. An ______________________________ is an inequality that involves the absolute value of a variable expression.

Objective 1 Use the distance definition of absolute value.

Graph the solution set of each equation or inequality.

1. $|m| = 7$

1.

2. $|p| < 3$

2.

3. $|r| > 2$

3.

4. $|x| \geq 6$

4.

5. $|t| \leq 0$

5.

Name: Date:
Instructor: Section:

Objective 2 Solve equations of the form $|ax + b| = k$, for $k > 0$.

Solve each equation.

6. $|2x+3|=10$ **6.** ____________

7. $|3k-1|=6$ **7.** ____________

8. $|5r-15|=0$ **8.** ____________

9. $\left|\frac{1}{2}x-3\right|=4$ **9.** ____________

10. $\left|5-\frac{4}{3}x\right|=9$ **10.** ____________

Name: Date:
Instructor: Section:

Objective 3 Solve inequalities of the form $|ax + b| < k$ and of the form $|ax + b| > k$, for $k > 0$.

Solve each inequality and graph the solution set.

11. $|x-2|>8$

11. ______________

12. $|3r-9|\le 10$

12. ______________

13. $|4y-1|\le 2$

13. ______________

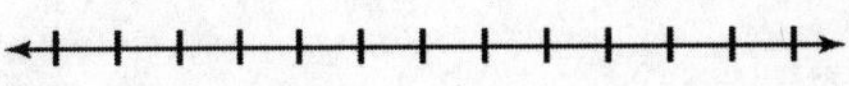

14. $|2r-9|\ge 23$

14. ______________

15. $|5r+2|<18$

15. ______________

Name: Date:
Instructor: Section:

Objective 4 Solve absolute value equations that involve rewriting.

Solve each equation.

16. $|7t+5|+6=14$ **16.** ______________

17. $|5-2w|+7=5$ **17.** ______________

18. $|2w-1|+7=12$ **18.** ______________

19. $\left|2-\frac{1}{2}x\right|-5=18$ **19.** ______________

20. $|4t+3|+8=10$ **20.** ______________

Name: Date:
Instructor: Section:

Objective 5 Solve equations of the form $|ax + b| = |cx + d|$.

Solve each problem.

21. $|2x-8|=|6x+7|$ **21.** ________

22. $|y+5|=|3y+1|$ **22.** ________

23. $|2p-4|=|7-p|$ **23.** ________

24. $|3x-2|=|5x+8|$ **24.** ________

25. $\left|y-\frac{1}{4}\right|=\left|\frac{1}{2}y+1\right|$ **25.** ________

Name: Date:
Instructor: Section:

Objective 6 Solve special cases of absolute value equations and inequalities.

Solve each problem.

26. $\left|7+\frac{1}{2}x\right|=0$ **26.** ________________

27. $|m-2|\geq -1$ **27.** ________________

28. $|k+5|\leq -2$ **28.** ________________

29. $|3-2x|+5\leq 1$ **29.** ________________

30. $|3p+4|>-7$ **30.** ________________

Name: Date:
Instructor: Section:

Chapter 8 EQUATIONS, INEQUALITIES AND SYSTEMS REVISITED

8.4 Review of Systems of Linear Equations in Two Variables

Learning Objectives
1 Solve linear systems with two equations and two variables.
2 Solve special systems.

Key Terms

Use the vocabulary terms listed below to complete each statement in exercises 1–7.

system of equations **linear system**
solution set of a system **consistent system** **inconsistent system**
independent equations **dependent equations**

1. Equations of a system that have different graphs are called ______________________________.

2. A system of equations with at least one solution is a ______________________________.

3. Two or more equations that are to be solved at the same time form a ____.

4. The ______________________________________ of linear equations includes all the ordered pairs that make all the equations of the system true at the same time.

5. Equations of a system that have the same graph (because they are different forms of the same equation) are called ______________________________.

6. A system with no solution is called a(n) ___________________________________.

7. A(n) _____________________________________ consists of two or more linear equations with the same variables.

Name: Date:
Instructor: Section:

Objective 1 Solve linear systems with two equations and two variables.

Solve each system by graphing.

1. $x - 2y = 6$
$2x + y = 2$

1.

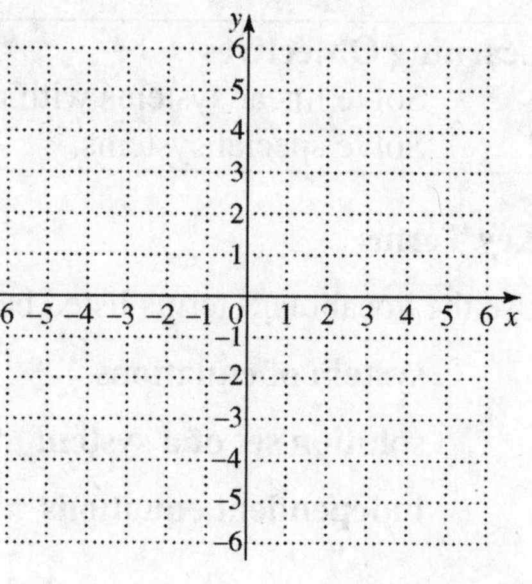

2. $2x + 3y = 5$
$3x - y = 13$

2.

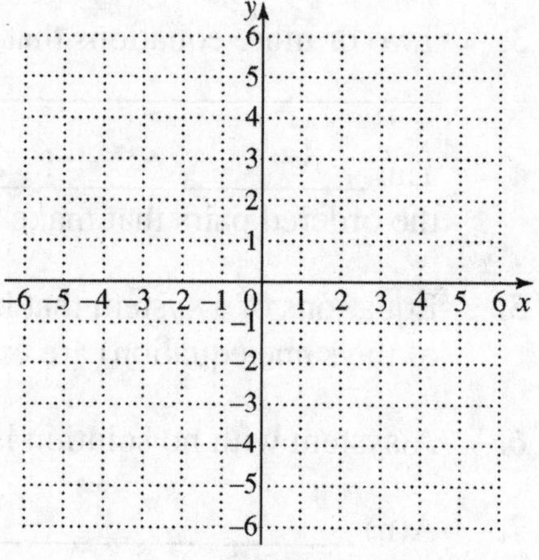

Name: Date:
Instructor: Section:

3. $6x - 5y = 4$

$2x - 5y = 8$

3.

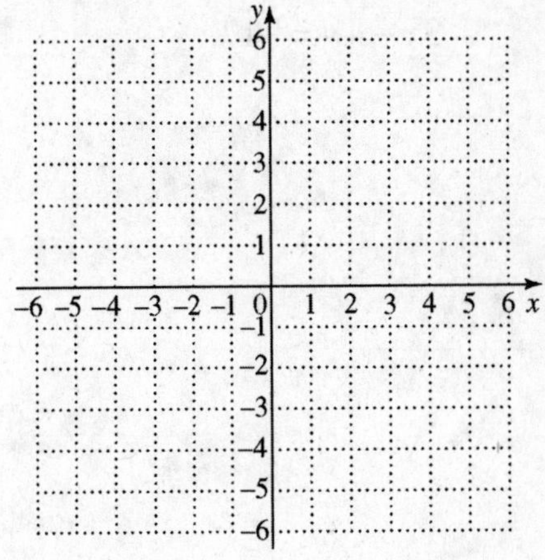

4. $3x - y = -7$

$2x + y = -3$

4.

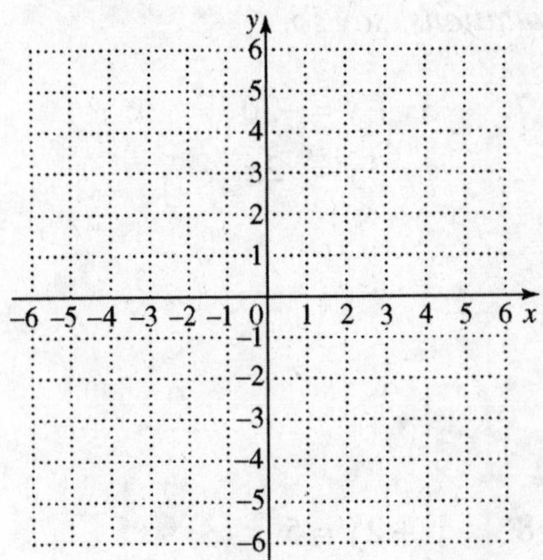

5. $2x = y$

$5x + 3y = 0$

5.

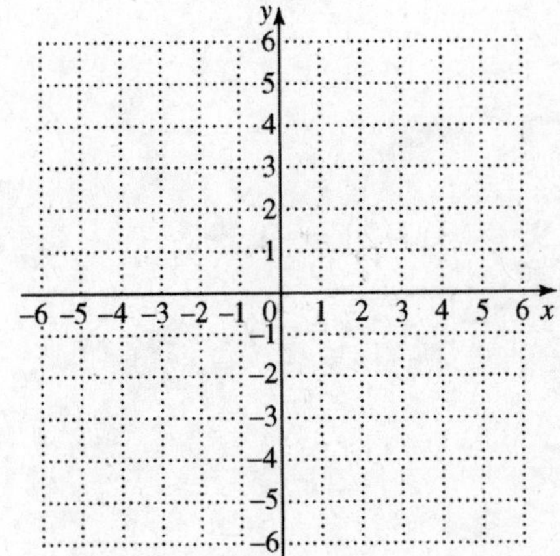

Name: Date:
Instructor: Section:

6. $y-2=0$

$3x-4y=-17$

6.

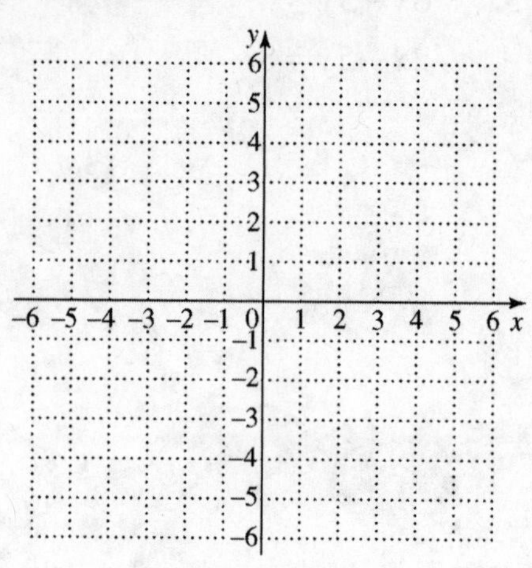

Solve each system by substitution. If the system is inconsistent or has dependent equations, say so.

7. $3x+y=-20$

$y=2x$

7. ____________

8. $x+2y=5$

$x=2y+1$

8. ____________

9. $3x-2y=-1$

$x=\frac{3}{4}y$

9. ____________

Name: Date:
Instructor: Section:

10. $2x + y = 6$
$y = 5 - 3x$

10. ______________

11. $y = 11 - 2x$
$x = 18 - 3y$

11. ______________

12. $x + 3y = 9$
$x - 2y = -1$

12. ______________

Solve each system by elimination. If the system is inconsistent or has dependent equations, say so.

13. $x + 2y = 7$
$x - y = -2$

13. ______________

Name: Date:
Instructor: Section:

14. $3x - y = 11$
$x + y = 5$

14. ____________

15. $3x + 4y = -13$
$5x - 2y = -13$

15. ____________

16. $2x + 8y = 3$
$4x - 12y = -1$

16. ____________

17. $\frac{1}{2}x + \frac{1}{4}y = 5$
$\frac{1}{2}x - \frac{3}{4}y = -3$

17. ____________

Name: Date:
Instructor: Section:

18. $\frac{x}{2}+\frac{y}{3}=-2$
$\frac{3x}{2}+\frac{5y}{3}=-8$

18. ______________

Solve each system using any method. If the system is inconsistent or has dependent equations, say so.

19. $2x+3y=11$
$3x-2y=9$

19. ______________

20. $2x-6y=6$
$3x+2y=14$

20. ______________

21. $5x-3y=-12$
$2x+3y=-9$

21. ______________

22. $4x+3y=-16$
$x-4y=-4$

22. ______________

23. $x-y=-1$
$4x+y=-24$

23. ______________

24. $-5x-3y=8$
$2x-3y=24$

24. ______________

Name: Date:
Instructor: Section:

Objective 2 Solve special systems.

Solve each system of equation using any method.

25. $8x + 4y = -1$
$4x + 2y = 3$

25. ______________

26. $x + 2y = 4$
$8y = -4x + 16$

26. ______________

27. $4x + 3y = 12$
$6y + 8x = -24$

27. ______________

28. $2x + 3y = 0$
$6x = -9y$

28. ______________

Name: Date:
Instructor: Section:

29. $-3x + 2y = 6$
$-6x + 4y = 12$

29. ________________

30. $3x + 3y = 8$
$x = 4 - y$

30. ________________

Name: Date:
Instructor: Section:

Chapter 8 EQUATIONS, INEQUALITIES AND SYSTEMS REVISITED

8.5 Systems of Linear Equations in Three Variables; Applications

Learning Objectives

1. Understand the geometry of systems of three equations in three variables.
2. Solve linear systems (with three equations and three variables) by elimination.
3. Solve linear systems (with three equations and three variables) in which some of the equations have missing terms.
4. Solve special systems.
5. Solve application problems with three variables using a system of three equations.

Key Terms

Use the vocabulary terms listed below to complete each statement in exercises 1–3.

ordered triple **inconsistent system** **dependent system**

1. The solution of a linear system of equations in three variables is written as a(n) ________________________.

2. A system of equations in which all solutions of the first equation are also solutions of the second equation is a(n) ________________________________.

3. A system of equations that has no common solution is called a(n) ________________________.

Objective 1 Understand the geometry of systems of three equations in three variables.

Answer each question.

1. If a system of linear equations in three variables has a single solution, how do the planes that are the graphs of the equations intersect? **1.** ________________

2. If a system of linear equations in three variables has no solution, how do the planes that are the graphs of the equations intersect? **2.** ________________

Name: Date:
Instructor: Section:

Objective 2 Solve linear systems (with three equations and three variables) by elimination.

Solve each system of equations

3.
$$\begin{aligned} x+y+z &= 0 \\ x-y+z &= -2 \\ x-y-z &= -4 \end{aligned}$$

3. ____________

4.
$$\begin{aligned} 3x-y+2z &= -6 \\ 2x+y+2z &= -1 \\ 3x+y-z &= -10 \end{aligned}$$

4. ____________

5.
$$\begin{aligned} 4x+2y+3z &= 11 \\ 2x+y-4z &= -22 \\ 3x+3y+z &= -1 \end{aligned}$$

5. ____________

6.
$$\begin{aligned} x-2y+5z &= -7 \\ 2x+3y-4z &= 14 \\ 3x-5y+z &= 7 \end{aligned}$$

6. ____________

Name: Date:
Instructor: Section:

7. $2x - 5y + 2z = 30$

$x + 4y + 5z = -7$

$\frac{1}{2}x - \frac{1}{4}y + z = 4$

7. ______________

8. $\frac{1}{3}x + \frac{1}{6}y - \frac{2}{3}z = -1$

$\frac{3}{4}x + \frac{1}{3}y + \frac{1}{4}z = -3$

$\frac{1}{2}x + \frac{3}{2}y + \frac{3}{4}z = 21$

8. ______________

9. $\frac{2}{3}x - \frac{1}{4}y + \frac{5}{8}z = 0$

$\frac{1}{5}x + \frac{2}{3}y - \frac{1}{4}z = -7$

$\frac{3}{5}x - \frac{4}{3}y + \frac{7}{8}z = 5$

9. ______________

Name: Date:
Instructor: Section:

Objective 3 Solve linear systems (with three equations and three variables) in which some of the equations have missing terms.

Solve each system of equations.

10.

$$\begin{aligned} x - \phantom{y + {}} z &= -3 \\ y + z &= 4 \\ x - y \phantom{{} + z} &= 3 \end{aligned}$$

10. ______________

11.

$$\begin{aligned} 2x + 3y \phantom{{} - 5z} &= 3 \\ 6y - 5z &= 3 \\ 4x + 9y \phantom{{} - 5z} &= 8 \end{aligned}$$

11. ______________

12.

$$\begin{aligned} x + 5y \phantom{{} - 3z} &= -23 \\ 4y - 3z &= -29 \\ 2x + \phantom{4y - {}} 5z &= 19 \end{aligned}$$

12. ______________

Name: Date:
Instructor: Section:

13.

$$\begin{aligned} 7x + \quad\quad z &= 5 \\ 3y - 2z &= -16 \\ 5x + y \quad\quad &= -2 \end{aligned}$$

13. ______________

14.

$$\begin{aligned} 2x + 5y \quad\quad &= 18 \\ 3y + 2z &= 4 \\ \frac{1}{4}x - y \quad\quad &= -1 \end{aligned}$$

14. ______________

15.

$$\begin{aligned} 5x - \quad\quad 2z &= 8 \\ 4y + 3z &= -9 \\ \frac{1}{2}x + \frac{2}{3}y \quad\quad &= -1 \end{aligned}$$

15. ______________

Name: Date:
Instructor: Section:

16.
$$\begin{aligned} x + 2y \quad &= -2 \\ \frac{1}{2}y + z &= -1 \\ \frac{2}{3}x - \frac{3}{4}y \quad &= 7 \end{aligned}$$

16. ______________

Objective 4 Solve special systems.

Solve each system of equations.

17.
$$\begin{aligned} 8x - 7y + 2z &= 1 \\ 3x + 4y - z &= 6 \\ -8x + 7y - 2z &= 5 \end{aligned}$$

17. ______________

18.
$$\begin{aligned} 3x - 2y + 4z &= 5 \\ -3x + 2y - 4z &= -5 \\ \frac{3}{2}x - y + 2z &= \frac{5}{2} \end{aligned}$$

18. ______________

Name: Date:
Instructor: Section:

19.

$$\begin{aligned} -x + 5y - 2z &= 3 \\ 2x - 10y + 4z &= -6 \\ -3x + 15y - 6z &= 9 \end{aligned}$$

19. ______________

20.

$$\begin{aligned} 8x - 4y + 2z &= 0 \\ 3x + y - 4z &= 0 \\ 5x + y + 2z &= 0 \end{aligned}$$

20. ______________

21.

$$\begin{aligned} 3x - 2y + 5z &= 6 \\ x - 4y - z &= 1 \\ \frac{3}{2}x - y + \frac{5}{2}z &= -3 \end{aligned}$$

21. ______________

Name: Date:
Instructor: Section:

22.
$$2x+7y-8z=3$$
$$5x-y-z=1$$
$$x+\frac{7}{2}y-4z=3$$

22. ______________

23.
$$x-5y+2z=0$$
$$-x+5y-2z=0$$
$$\frac{1}{2}x-\frac{5}{2}y+z=0$$

23. ______________

24.
$$3x-2y+5z=0$$
$$6x-4y+10z=0$$
$$\frac{3}{2}x-y+\frac{5}{2}z=0$$

24. ______________

Name: Date:
Instructor: Section:

Objective 5 Solve problems with three variables using a system of three equations.

Solve each problem involving three unknowns.

25. Three numbers have a sum of 31. The middle number is 1 more than the smallest number. The sum of the smaller two numbers is 7 more than the largest number. Find the three numbers.

25. ______________

26. The sum of the measures of the angles of any triangle is 180°. In a certain triangle, the first angle measures 20° less than the second angle, and the second angle measures 10° more than the third. Find the measures of the three angles.

26. ______________

27. Lance has some \$5, \$10, and \$20-bills. He has a total of 51 bills, worth \$795. The number of \$5-bills is 25 less than the number of \$20-bills. Find the number of each type of bill he has.

27. \$5 ______________

\$10 ______________

\$15 ______________

Name: Date:
Instructor: Section:

28. Sara has $80,000 to invest. She invests part at 5%, one fourth this amount at 6%, and the balance at 7%. Her total annual income from interest is $4700. Find the amount invested at each rate.

28. 5%__________

6%__________

7%__________

29. A box contains $6.25 in nickels, dimes, and quarters. There are 85 coins in all, with three times as many nickels as dimes. How many of each coin are there?

29. nickels__________

dimes__________

quarters__________

30. A merchant wishes to mix gourmet coffee selling for $8 per pound, $10 per pound, and $15 per pound to get 50 pounds of a mixture that can be sold for $11.70 per pound. The amount of the $8 coffee must be 3 pounds more than the amount of the $10 coffee. Find the number of pounds of each that must be used.

30. $8 coffee__________

$10 coffee__________

$15 coffee__________

Name: Date:
Instructor: Section:

Chapter 8 EQUATIONS, INEQUALITIES AND SYSTEMS REVISITED

8.6 Solving Systems of Linear Equations by Matrix Methods

Learning Objectives

1. Define a matrix.
2. Write the augmented matrix for a system.
3. Use row operations to solve a system with two equations.
4. Use row operations to solve a system with three equations.
5. Use row operations to solve special systems.

Key Terms

Use the vocabulary terms listed below to complete each statement in exercises 1–5.

matrix **elements of the matrix** **square matrix**

augmented matrix **row echelon form**

1. A(n) ______________________________ is a matrix that has a vertical bar that separates the columns of the matrix into two groups.

2. A(n) ______________________________ is a rectangular array of numbers, consisting of horizontal rows and vertical columns.

3. A(n) ______________________________ is a matrix that has the same number of rows as columns.

4. If a matrix is written with 1s down the diagonal from upper left to lower right and 0s below the 1s, it is said to be in ______________________________.

5. The numbers in a matrix are the ______________________________.

Objective 1 Define a matrix.

Give the dimensions of each matrix.

1. $\begin{bmatrix} -4 & 3 \\ 0 & 7 \\ 6 & 2 \end{bmatrix}$

1. ________________

2. $\begin{bmatrix} 1 & 4 & 7 \\ 6 & 5 & -5 \end{bmatrix}$

2. ________________

Name: Date:
Instructor: Section:

3. $\begin{bmatrix} 8 & 7 & 6 & 4 \\ 4 & -1 & 0 & 6 \\ -5 & 3 & -4 & 7 \end{bmatrix}$ **3.** ________

4. $\begin{bmatrix} 1 & 4 \\ 3 & 2 \\ -2 & 0 \\ 5 & -3 \end{bmatrix}$ **4.** ________

Use the matrix $\begin{bmatrix} 3 & 2 & 6 \\ 6 & 0 & 1 \\ -2 & 10 & -11 \\ 1 & 5 & 2 \end{bmatrix}$ *to answer questions 5 and 6.*

5. What are the elements of the second row? **5.** ________

6. What are the elements of the second column? **6.** ________

Objective 2 Write the augmented matrix for a system.

Write the augmented matrix for each system.

7. $3x - 4y = 7$ **7.** ________
$2x + y = 12$

8. $2x - 3y = 12$ **8.** ________
$7x + 3y = 15$

9. $\frac{1}{2}x + \frac{1}{2}y = -16$ **9.** ________
$-3x + y = 2$

Name: Date:
Instructor: Section:

10. $y+3=-2x$
$x=4y-5$

10. ______________

11. $-2x+3y-5z=7$
$6x+2y-4z=12$
$5x-2y+z=-1$

11. ______________

12. $x+y+z=10$
$2x+y-3z=11$
$x+2z=-2$

12. ______________

Objective 3 Use row operations to solve a system with two equations.

Use row operations to solve each system.

13. $x+3y=-7$
$4x+3y=-1$

13. ______________

14. $x-2y=-1$
$2x+y=8$

14. ______________

Name: Date:
Instructor: Section:

15. $3x - 3y = 15$
$2x + y = 4$

15. ______________

16. $y = x - 2$
$2x = -3y + 9$

16. ______________

17. $x = 3y + 4$
$2y = 5x + 19$

17. ______________

18. $3x + 2y = 12$
$5x - 2y = 12$

18. ______________

Name: Date:
Instructor: Section:

Objective 4 Use row operations to solve a system with three equations.

Use row operations to solve each system.

19. $x - y - z = 6$
$-x + 3y + 2z = -11$
$3x + 2y + z = 1$

19. ______________

20. $x + y + z = 5$
$x - 2y + 3z = 16$
$2x - y + z = 9$

20. ______________

21. $x + 2y + 3z = 1$
$x + y + 2z = 0$
$2x - y - z = 1$

21. ______________

Name: Date:
Instructor: Section:

22. $x + y + z = 6$
$x + 2y - 3z = -11$
$-2x + y - z = -11$

22. ______________

23. $2x + y + z = 8$
$x - y + z = 3$
$3x + y - z = 1$

23. ______________

24. $2x - y + 2z = -1$
$-x - 3y + z = 1$
$x + y + z = 1$

24. ______________

Name: Date:
Instructor: Section:

Objective 5 Use row operations to solve special systems.

Use row operations to solve each system.

25. $x + y = 3$
$3x + 3y = -2$

25. ______________

26. $x - 2y = 3$
$3x - 6y = 9$

26. ______________

27. $2x + y = 10$
$-4x - 2y = -20$

27. ______________

Name: Date:
Instructor: Section:

28.

$$\begin{aligned} x+y+z&=6 \\ x-y-z&=2 \\ 2x+y+z&=8 \end{aligned}$$

28. ______________

29.

$$\begin{aligned} x+3y-z&=1 \\ 2x+y-z&=2 \\ 4x-3y-z&=4 \end{aligned}$$

29. ______________

30.

$$\begin{aligned} x+2y-z&=6 \\ 2x+4y-2z&=12 \\ -3x-6y+3z&=-18 \end{aligned}$$

30. ______________

Name: Date:
Instructor: Section:

Chapter 9 ROOTS, RADICALS, AND ROOT FUNCTIONS

9.1 Radical Expressions and Graphs

Learning Objectives

1. Find square roots.
2. Decide whether a given root is rational, irrational, or not a real number.
3. Find cube, fourth, and other roots.
4. Graph functions defined by radical expressions.
5. Find *n*th roots of *n*th powers.
6. Use a calculator to find roots.

Key Terms

Use the vocabulary terms listed below to complete each statement in exercises 1–5.

square root	**principal square root**	**radicand**
radical	**radical expression**	**perfect square**
irrational number	**cube root**	**index (order)**

1. The number or expression inside a radical sign is called the ______________.

2. A number with a rational square root is called a ______________________.

3. In a radical of the form $\sqrt[n]{a}$, the number n is the ______________________.

4. The number b is a ______________________ of a if $b^2 = a$.

5. The expression $\sqrt[n]{a}$ is called a ______________________________.

6. The positive square root of a number is its ______________________________.

7. A real number that is not rational is called an ____________________________.

8. A ______________________ is a radical sign and the number or expression in it.

9. The number b is a ______________________ of a if $b^3 = a$.

Objective 1 Find square roots.

Find all square roots of each number.

1. 625   1. ______________

Name: Date:
Instructor: Section:

2. $\frac{49}{144}$ 2. ______

3. $\frac{121}{196}$ 3. ______

Find each square root

4. $\sqrt{400}$ 4. ______

5. $-\sqrt{\frac{2500}{6400}}$ 5. ______

Objective 2 Decide whether a given root is rational, irrational, or not a real number.

Tell whether each square root is rational, irrational, *or* not a real number.

6. $\sqrt{72}$ 6. ______

7. $\sqrt{-36}$ 7. ______

8. $-\sqrt{\frac{625}{484}}$ 8. ______

9. $\sqrt{8.1}$ 9. ______

10. $\sqrt{-49}$ 10. ______

Objective 3 Find cube, fourth, and other roots.

Find each root.

11. $\sqrt[3]{-64}$ 11. ______

Name: Date:
Instructor: Section:

12. $\sqrt[4]{-625}$

12. ______________

13. $-\sqrt[5]{243}$

13. ______________

14. $-\sqrt[3]{-343}$

14. ______________

15. $\sqrt[4]{256}$

15. ______________

Objective 4 Graph functions defined by radical expressions.

Graph each function and give its domain and its range.

16. $f(x) = \sqrt{x+1}$

16.

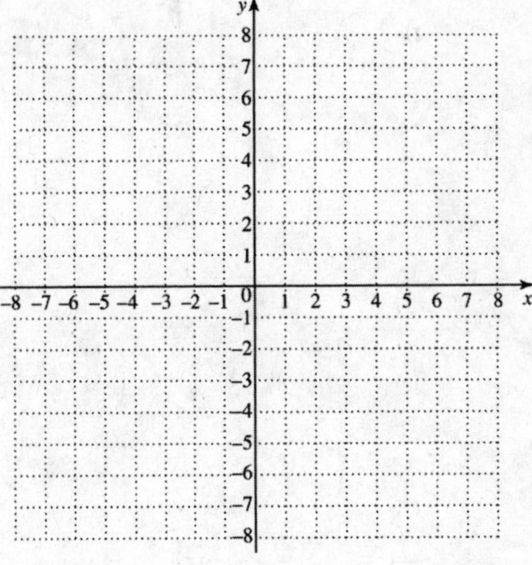

domain: ______________

range: ______________

Name: Date:
Instructor: Section:

17. $f(x) = \sqrt{x} - 1$

17.

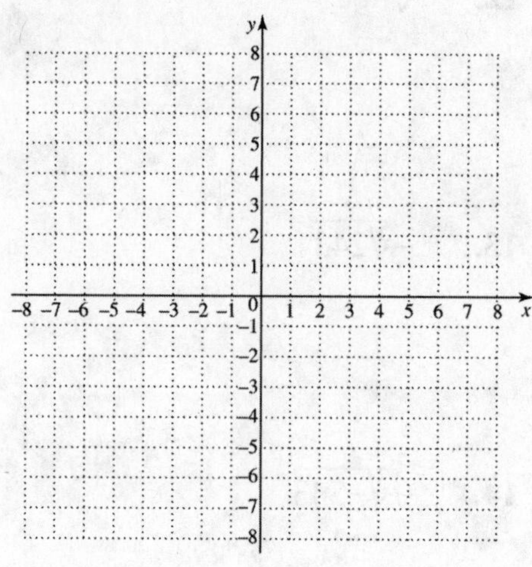

domain: ______________

range: ______________

18. $f(x) = \sqrt{x} + 2$

18.

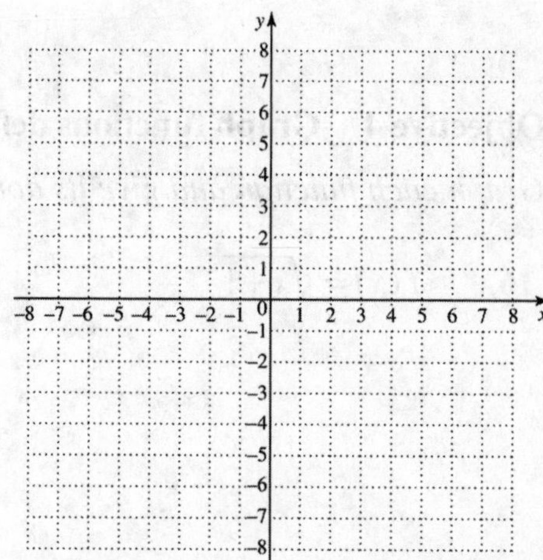

domain: ______________

range: ______________

Name: Date:
Instructor: Section:

19. $f(x)=\sqrt[3]{x}-2$

19.

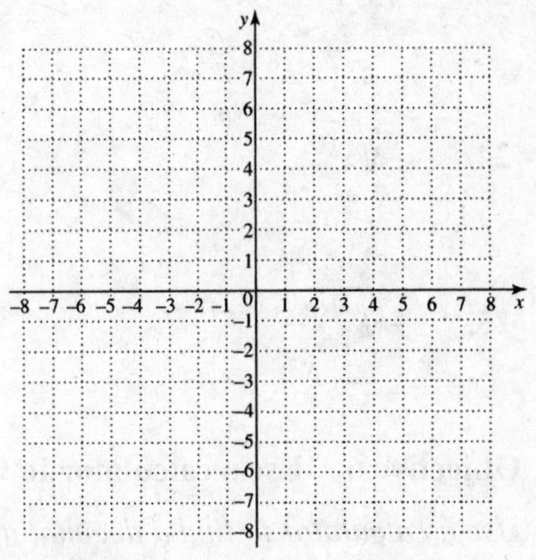

domain: ______________

range: ______________

20. $f(x)=\sqrt[3]{x}+2$

20.

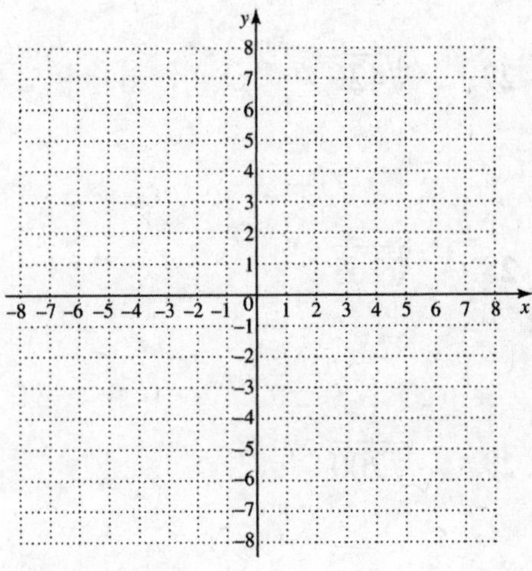

domain: ______________

range: ______________

Objective 5 Find *n*th roots of *n*th powers.

Simplify each root.

21. $\sqrt[5]{x^{25}}$

21. ______________

22. $-\sqrt[5]{x^5}$

22. ______________

Name: Date:
Instructor: Section:

23. $\sqrt[6]{x^{12}}$ **23.** ________________

24. $-\sqrt[5]{x^{10}}$ **24.** ________________

25. $-\sqrt[4]{x^{16}}$ **25.** ________________

Objective 6 Use a calculator to find roots.

Use a calculator to find a decimal approximation for each radical. Round answers to three decimal places if necessary.

26. $-\sqrt{87}$ **26.** ________________

27. $\sqrt[4]{42}$ **27.** ________________

28. $\sqrt[3]{263}$ **28.** ________________

29. $\sqrt[6]{200}$ **29.** ________________

30. $-\sqrt[5]{990}$ **30.** ________________

Name: Date:
Instructor: Section:

Chapter 9 ROOTS, RADICALS, AND ROOT FUNCTIONS

9.2 Rational Exponents

Learning Objectives

1. Use exponential notation for *n*th roots.
2. Define and use expressions of the form $a^{m/n}$.
3. Convert between radicals and rational exponents.
4. Use the rules for exponents with rational exponents.

Key Terms

Use the vocabulary terms listed below to complete each statement in exercises 1–3.

product rule for exponents **quotient rule for exponents**

power rule for exponents

1. $\left(x^2y^3\right)^4 = x^8y^{12}$ is an example of the ______________________________.

2. $w^5w^3 = w^8$ is an example of the ______________________________.

3. $\frac{z^6}{z^4} = z^2$ is an example of the ______________________________.

Objective 1 Use exponential notation for *n*th roots.

Evaluate each exponential.

1. $(-8)^{1/3}$ **1.** ______________

2. $216^{1/3}$ **2.** ______________

3. $-81^{1/4}$ **3.** ______________

4. $1024^{1/10}$ **4.** ______________

Name: Date:
Instructor: Section:

5. $-256^{1/4}$ 5. ________

6. $16^{1/4}$ 6. ________

7. $3375^{1/3}$ 7. ________

Objective 2 Define and use expressions of the form $a^{m/n}$.

Evaluate each exponential.

8. $27^{2/3}$ 8. ________

9. $25^{-3/2}$ 9. ________

10. $16^{3/4}$ 10. ________

11. $-125^{-2/3}$ 11. ________

12. $729^{5/6}$ 12. ________

13. $36^{5/2}$ 13. ________

14. $(-8)^{-2/3}$ 14. ________

Name: Date:
Instructor: Section:

Objective 3 Convert between radicals and rational exponents.

Simplify each radical by rewriting it with a rational exponent. Write answers in radical form if necessary. Assume that variables represent positive real numbers.

15. $\sqrt[6]{4t^4}$ **15.** ______________

16. $\sqrt[8]{16x^{12}}$ **16.** ______________

17. $\sqrt[40]{y^{35}}$ **17.** ______________

18. $\dfrac{\sqrt{r}}{\sqrt[3]{r}}$ **18.** ______________

19. $\sqrt[3]{k^2} \cdot \sqrt[6]{k}$ **19.** ______________

20. $\sqrt[4]{x^3} \cdot \sqrt[3]{x}$ **20.** ______________

21. $\dfrac{\sqrt[5]{x^2}}{\sqrt[3]{x}}$ **21.** ______________

22. $\sqrt[15]{27t^6}$ **22.** ______________

Name: Date:
Instructor: Section:

Objective 4 Use the rules for exponents with rational exponents.

Use the rules of exponents to simplify each expression. Write all answers with positive exponents. Assume that variables represent positive real numbers.

23. $y^{7/3} \cdot y^{-4/3}$ **23.** ________

24. $\left(\frac{c^6}{x^3}\right)^{2/3}$ **24.** ________

25. $\frac{w^{7/4}w^{-1/2}}{w^{5/4}}$ **25.** ________

26. $\frac{a^{2/3} \cdot a^{-1/3}}{\left(a^{-1/6}\right)^3}$ **26.** ________

27. $\frac{8^{3/5} \cdot 8^{-8/5}}{8^{-2}}$ **27.** ________

28. $\left(a^{-1}\right)^{1/2}\left(a^{-3}\right)^{-1/2}$ **28.** ________

29. $\frac{\left(x^{-3}y^2\right)^{2/3}}{\left(x^2y^{-5}\right)^{2/5}}$ **29.** ________

30. $\frac{\left(x^{-1}y^{2/3}\right)^3}{\left(x^{1/3}y^{1/2}\right)^2}$ **30.** ________

Name: Date:
Instructor: Section:

Chapter 9 ROOTS, RADICALS, AND ROOT FUNCTIONS

9.3 Simplifying Radical Expressions

Learning Objectives

1. Use the product rule for radicals.
2. Use the quotient rule for radicals.
3. Simplify radicals.
4. Simplify products and quotients of radicals with different indexes.
5. Use the Pythagorean formula.
6. Use the distance formula.

Key Terms

Use the vocabulary terms listed below to complete each statement in exercises 1–3.

index **radicand** **hypotenuse** **legs**

1. In a right triangle, the side opposite the right angle is called the ____________________.

2. In the expression $\sqrt[4]{x^2}$, the "4" is the ____________________ and x^2 is the ____________________.

3. In a right triangle, the sides that form the right angle are called the ____________________.

Objective 1 Use the product rule for radicals.

Multiply. Assume that variables represent positive real numbers.

1. $\sqrt[4]{2} \cdot \sqrt[4]{2x}$ 1. ________________

2. $\sqrt[5]{6r^2t^3} \cdot \sqrt[5]{4r^2t}$ 2. ________________

3. $\sqrt[4]{8} \cdot \sqrt[4]{15}$ 3. ________________

4. $\sqrt[6]{4t} \cdot \sqrt[6]{5t^4}$ 4. ________________

Name: Date:
Instructor: Section:

5. $\sqrt[7]{8a^2t^3} \cdot \sqrt[7]{6at^3}$ 5. ________

Objective 2 Use the quotient rule for radicals.

Simplify each radical. Assume that variables represent positive real numbers.

6. $\sqrt[3]{\frac{27}{8}}$ 6. ________

7. $\sqrt{\frac{15}{169}}$ 7. ________

8. $\sqrt[5]{\frac{7x}{32}}$ 8. ________

9. $\sqrt[3]{-\frac{a^6}{125}}$ 9. ________

10. $\sqrt[3]{\frac{\ell^2}{27}}$ 10. ________

Objective 3 Simplify radicals.

Simplify each radical. Assume that variables represent positive real numbers.

11. $\sqrt[42]{x^{28}}$ 11. ________

Name: Date:
Instructor: Section:

12. $\sqrt[20]{x^{15}y^{10}}$ 12. ________

13. $\sqrt[12]{9^3x^6y^9}$ 13. ________

14. $\sqrt{8x^3y^6z^{11}}$ 14. ________

15. $\sqrt[3]{270b^4c^8}$ 15. ________

Objective 4 Simplify products and quotients of radicals with different indexes.

Simplify each radical. Assume that variables represent positive real numbers.

16. $\sqrt{r}\cdot\sqrt[3]{r}$ 16. ________

17. $\dfrac{\sqrt[4]{8}}{\sqrt[3]{8}}$ 17. ________

18. $\sqrt{3}\cdot\sqrt[5]{64}$ 18. ________

19. $\sqrt{6}\cdot\sqrt[3]{5}$ 19. ________

20. $\sqrt[3]{3}\cdot\sqrt[6]{7}$ 20. ________

Name: Date:
Instructor: Section:

Objective 5 Use the Pythagorean formula.

Find the unknown length in each right triangle. Simplify the answer if necessary.

21.

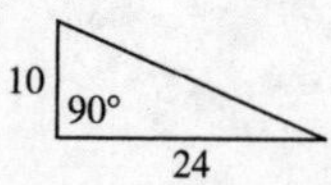

21. ______________

22.

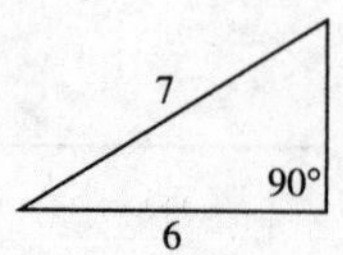

22. ______________

23.

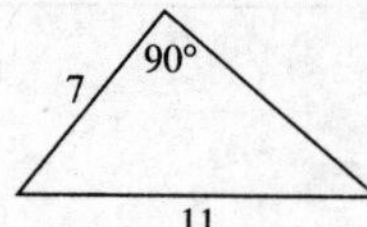

23. ______________

24.

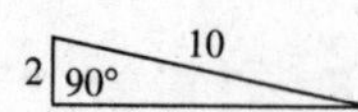

24. ______________

25.

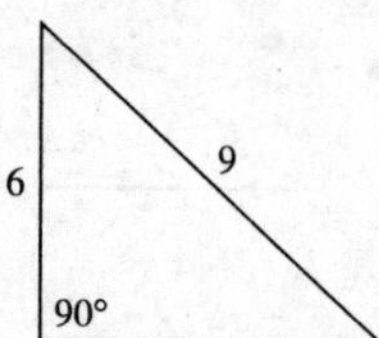

25. ______________

Objective 6 Use the distance formula.

Find the distance between each pair of points.

26. (3, 4) and (−1, −2)

26. ______________

Name: Date:
Instructor: Section:

27. $(-1, -2)$ and $(-4, 3)$ **27.** ________________

28. $\left(4\sqrt{3}, 2\sqrt{5}\right)$ and $\left(3\sqrt{3}, -\sqrt{5}\right)$ **28.** ________________

29. $(2x, x + y)$ and $(x, x - y)$ **29.** ________________

30. $(a - b, a)$ and $(a + b, b)$ **30.** ________________

Name: Date:
Instructor: Section:

Chapter 9 ROOTS, RADICALS, AND ROOT FUNCTIONS

9.4 Adding and Subtracting Radical Expressions

Learning Objectives
1 Simplify radical expressions involving addition and subtraction.

Key Terms

Use the vocabulary terms listed below to complete each statement in exercises 1–2.

like radicals **unlike radicals**

1. The expressions $2\sqrt{2}$ and $6\sqrt[3]{2}$ are ______________________.

2. The expressions $2\sqrt{2}$ and $7\sqrt{2}$ are ______________________.

Objective 1 Simplify radical expressions involving addition and subtraction.

Add or subtract. Assume that all variables represent positive real numbers.

1. $5\sqrt[3]{24}+4\sqrt[3]{81}$ 1. ______________

2. $2\sqrt{18}-5\sqrt{32}+7\sqrt{162}$ 2. ______________

3. $7\sqrt[4]{32}-9\sqrt[4]{2}$ 3. ______________

4. $\sqrt{48y}+\sqrt{12y}+\sqrt{27y}$ 4. ______________

5. $\sqrt[3]{81}+\sqrt[3]{24}+\sqrt[3]{192}$ 5. ______________

Name: Date:
Instructor: Section:

6. $\sqrt[3]{625}+\sqrt[3]{135}-\sqrt[3]{40}$ 6. ________

7. $\sqrt{100x}-\sqrt{9x}+\sqrt{25x}$ 7. ________

8. $-2\sqrt[3]{81}+\sqrt[3]{24}$ 8. ________

9. $3\sqrt{54}-5\sqrt{24}$ 9. ________

10. $7\sqrt[3]{54}-6\sqrt[3]{128}$ 10. ________

11. $6\sqrt[3]{135}+3\sqrt[3]{40}$ 11. ________

12. $2\sqrt[3]{16r}+\sqrt[3]{54r}-\sqrt[3]{16r}$ 12. ________

13. $3\sqrt{8z}+3\sqrt{2z}+\sqrt{32z}$ 13. ________

Name: Date:
Instructor: Section:

14. $3\sqrt{18z} + 2\sqrt{8z}$ 14. ____________

15. $6\sqrt[3]{54z^3} + 5\sqrt[3]{16z^3}$ 15. ____________

16. $4\sqrt[3]{r^5} + 3\sqrt[3]{27r^5}$ 16. ____________

17. $3\sqrt{125x} - \sqrt{80x} + 2\sqrt{45x}$ 17. ____________

18. $\dfrac{x\sqrt{48}}{5} + \dfrac{3\sqrt{75x^2}}{6}$ 18. ____________

19. $\sqrt[3]{\dfrac{216}{w^6}} + \sqrt{\dfrac{121}{w^4}}$ 19. ____________

20. $2\sqrt{\dfrac{k^3}{36}} - 3\sqrt{\dfrac{2k^3}{72}}$ 20. ____________

Name: Date:
Instructor: Section:

21. $4\sqrt{\frac{200}{r^2}}-7\sqrt{\frac{8}{r^2}}$ **21.** ______________

22. $\sqrt[3]{\frac{y^7}{125}}+y^2\sqrt[3]{\frac{y}{27}}$ **22.** ______________

23. $\sqrt{\frac{12y}{t^2}}-2\sqrt{\frac{75y}{t^4}}$ **23.** ______________

24. $\frac{1}{3}\sqrt[3]{27x^5r}+2x\sqrt[3]{x^2r}-\frac{1}{2}\sqrt[3]{8x^8r}$ **24.** ______________

Solve each problem. Give answers as simplified radical expressions.

25. Find the perimeter of a triangle with sides a, b, and c if $a=\sqrt{20}$ cm, $b=\sqrt{80}$ cm, and $c=\sqrt{125}$ cm. **25.** ______________

26. Find the perimeter and area of a rectangle whose length is $\sqrt{12}$ ft and whose width is $\sqrt{27}$ ft. **26.** perimeter______________

area ______________

Name: Date:
Instructor: Section:

27. Use $A = \frac{1}{2}h(B+b)$ to find A if $h = 10\sqrt{5}$ cm, $B = 6\sqrt{3}$ cm, and $b = \sqrt{3}$ cm.

27. 

28. Use $A = \frac{1}{2}h(B+b)$ to find A if $b = 2\sqrt{5}$ m, $h = 3\sqrt{2}$ m, and $B = 4\sqrt{5}$ m.

28. ______________

29. Find the width and area of a rectangle with length $3\sqrt{75}$ cm and perimeter $36\sqrt{3}$ cm.

29. width______________

area ______________

30. Use $A = \frac{1}{2}bh$ to find b if $A = 6\sqrt{105}$ in.^{2} and $h = 6\sqrt{3}$ in.

30. ______________

Name: Date:
Instructor: Section:

Chapter 9 ROOTS, RADICALS, AND ROOT FUNCTIONS

9.5 Multiplying and Dividing Radical Expressions

Learning Objectives	
1	Multiply radical expressions.
2	Rationalize denominators with one radical term.
3	Rationalize denominators with binomials involving radicals.
4	Write radical quotients in lowest terms.

Key Terms

Use the vocabulary terms listed below to complete each statement in exercises 1–2.

rationalizing the denominator **conjugate**

1. The ________________________ of $a + b$ is $a - b$.

2. The process of removing radicals from the denominator so that the denominator contains only rational quantities is called ________________________.

Objective 1 Multiply radical expressions.

Multiply each product, then simplify. Assume that variables represent positive real numbers.

1. $\left(4\sqrt{2}+3\sqrt{3}\right)\left(\sqrt{2}-7\sqrt{3}\right)$ **1.** ______________

2. $\left(\sqrt{6}-2\sqrt{5}\right)\left(4\sqrt{6}+\sqrt{10}\right)$ **2.** ______________

3. $\left(\sqrt{10}+\sqrt{3}\right)\left(\sqrt{6}-\sqrt{11}\right)$ **3.** ______________

Name: Date:
Instructor: Section:

4. $\left(\sqrt{5}+\sqrt{6}\right)\left(\sqrt{2}-4\right)$ 4. ____________

5. $\left(2\sqrt{x}-3\right)\left(3\sqrt{x}-2\right)$ 5. ____________

6. $\left(\sqrt{2}-\sqrt{12}\right)^2$ 6. ____________

7. $\left(2+\sqrt[3]{5}\right)\left(2-\sqrt[3]{5}\right)$ 7. ____________

Objective 2 Rationalize denominators with one radical term.

Simplify. Assume that variables represent positive real numbers.

8. $\sqrt{\frac{27}{98}}$ 8. ____________

Name: Date:
Instructor: Section:

9. $\sqrt{\dfrac{5a^2b^3}{6}}$

9. ______________

10. $\sqrt{\dfrac{7y^2}{12b}}$

10. ______________

11. $\dfrac{\sqrt{k^2m^4}}{\sqrt{k^5}}$

11. ______________

12. $\sqrt{\dfrac{20a^3b^4}{6a^2}}$

12. ______________

Name: Date:
Instructor: Section:

13. $\sqrt[3]{\dfrac{5}{49x}}$

13. ______________

14. $\sqrt[3]{\dfrac{7x^2}{81y^2}}$

14. ______________

Objective 3 Rationalize denominators with binomials involving radicals.

Rationalize each denominator. Write quotients in lowest terms. Assume that variables represent positive real numbers.

15. $\dfrac{4}{\sqrt{3}+2}$

15. ______________

16. $\dfrac{\sqrt{2}}{\sqrt{5}-2}$

16. ______________

Name: Date:
Instructor: Section:

17. $\dfrac{4}{\sqrt{5}+\sqrt{2}}$

17. ______________

18. $\dfrac{5}{\sqrt{3}-\sqrt{10}}$

18. ______________

19. $\dfrac{5\sqrt{5}}{4-\sqrt{15}}$

19. ______________

20. $\dfrac{\sqrt{6}+2}{\sqrt{2}-4}$

20. ______________

21. $\dfrac{\sqrt{5}-2}{\sqrt{3}+2}$

21. ______________

Name: Date:
Instructor: Section:

22. $\dfrac{\sqrt{x}+\sqrt{y}}{\sqrt{x}-\sqrt{y}}$

22. ______________

Objective 4 Write radical quotients in lowest terms.

Write each quotient in lowest terms. Assume that variables represent positive real numbers.

23. $\dfrac{7-\sqrt{98}}{14}$

23. ______________

24. $\dfrac{16-12\sqrt{72}}{24}$

24. ______________

25. $\dfrac{50+\sqrt{80x}}{10}$

25. ______________

Name: Date:
Instructor: Section:

26. $\dfrac{12-3\sqrt{8}}{9}$

26. ______________

27. $\dfrac{3+\sqrt{18x}}{3}$

27. ______________

28. $\dfrac{7y-\sqrt{98y^5}}{14y}$

28. ______________

29. $\dfrac{5\sqrt{5}-\sqrt{25}}{35}$

29. ______________

Name: Date:
Instructor: Section:

30. $\dfrac{2x-\sqrt{8x^2}}{4x}$

30. ________________

Name: Date:
Instructor: Section:

Chapter 9 ROOTS, RADICALS, AND ROOT FUNCTIONS

9.6 Solving Equations with Radicals

Learning Objectives
1 Solve radical equations using the power rule.
2 Solve radical equations that require additional steps.
3 Solve radical equations with indexes greater than 2.

Key Terms

Use the vocabulary terms listed below to complete each statement in exercises 1–2.

radical equation **extraneous solution**

1. A(n) ____________________ is a potential solution to an equation that does not satisfy the equation..

2. An equation with a variable in the radicand is a(n) ____________________.

Objective 1 Solve radical equations using the power rule.

Solve each equation.

1. $\sqrt{7x-6}=8$ 1. ______________

2. $\sqrt{12x+16}=16$ 2. ______________

3. $\sqrt{4x-19}=5$ 3. ______________

Name: Date:
Instructor: Section:

4. $\sqrt{3w+4}=7$ 4. ______________

5. $\sqrt{9c+1}=8$ 5. ______________

6. $3-\sqrt{7y-5}=0$ 6. ______________

7. $\sqrt{12p+1}+7=0$ 7. ______________

8. $\sqrt{5r-4}-9=0$ 8. ______________

9. $\sqrt{m}+7=-1$ 9. ______________

Name: Date:
Instructor: Section:

10. $\sqrt{x+13}-5=2$ **10.** ______________

Objective 2 Solve radical equations that require additional steps.

Solve each equation.

11. $\sqrt{27-18v}=2v-3$ **11.** ______________

12. $\sqrt{4x+3}=\sqrt{3x+5}$ **12.** ______________

13. $\sqrt{44-20x}=-8x$ **13.** ______________

Name: Date:
Instructor: Section:

14. $\sqrt{2p+5}+\sqrt{p+51}=8$ **14.** ______________

15. $\sqrt{2w+25}+\sqrt{2w+16}=9$ **15.** ______________

16. $\sqrt{7r+8}-\sqrt{r+1}=5$ **16.** ______________

17. $\sqrt{5y+4}-1=\sqrt{2y+2}$ **17.** ______________

18. $\sqrt{k+10}+\sqrt{2k+19}=2$ **18.** ______________

Name: Date:
Instructor: Section:

19. $\sqrt{t+2}-\sqrt{t-3}=1$ 19. ____________

20. $\sqrt{3k+7}+\sqrt{k+1}=2$ 20. ____________

Objective 3 Solve radical equations with indexes greater than 2.

Solve each equation.

21. $\sqrt[3]{2a-63}+5=0$ 21. ____________

22. $\sqrt[3]{r^2+3r+15}-\sqrt[3]{r^2}=0$ 22. ____________

Name: Date:
Instructor: Section:

23. $\sqrt[5]{5a+1}-\sqrt[5]{2a-11}=0$

23. ____________

24. $\sqrt[3]{7x-5}-\sqrt[3]{3x+7}=0$

24. ____________

25. $\sqrt[3]{t^2+5t+15}=\sqrt[3]{t^2}$

25. ____________

26. $\sqrt[4]{8x+5}=\sqrt[4]{7x+7}$

26. ____________

27. $\sqrt[5]{2w+5}=\sqrt[5]{7w}$

27. ____________

Name: Date:
Instructor: Section:

28. $\sqrt[4]{c^2+2c+18}=\sqrt[4]{c^2}$

28. ______________

29. $\sqrt[3]{3x-11}=\sqrt[3]{2x+10}$

29. ______________

30. $\sqrt[3]{x^3-125}+5=0$

30. ______________

Name: Date:
Instructor: Section:

Chapter 9 ROOTS, RADICALS, AND ROOT FUNCTIONS

9.7 Complex Numbers

Learning Objectives

1. Simplify numbers of the form $\sqrt{-b}$, where $b > 0$.
2. Recognize subsets of the complex numbers.
3. Add and subtract complex numbers.
4. Multiply complex numbers.
5. Divide complex numbers.
6. Find powers of i.

Key Terms

Use the vocabulary terms listed below to complete each statement in exercises 1–7.

complex number **real part** **imaginary part**

pure imaginary number **standard form (of a complex number)**

nonreal complex number **complex conjugate**

1. A ______________________________ is a number that can be written in the form $a + bi$, where a and b are real numbers.

2. The ______________________________ of $a + bi$ is $a - bi$.

3. The ______________________________ of $a + bi$ is bi.

4. The ______________________________ of $a + bi$ is b.

5. A complex number is in ______________________________ if it is written in the form $a + bi$.

6. A complex number $a + bi$ with $a = 0$ and $b \neq 0$ is called a ______________________________.

7. A complex number $a + bi$ $b \neq 0$ is called a ______________________________.

Objective 1 Simplify numbers of the form $\sqrt{-b}$, where $b > 0$.

Write each number as a product of a real number and i. Simplify all radical expressions.

1. $\sqrt{-60}$ 1. ______________

2. $\sqrt{-99}$ 2. ______________

Name: Date:
Instructor: Section:

3. $\sqrt{-1080}$ 3. ______________

4. $-\sqrt{-162}$ 4. ______________

Multiply or divide as indicated

5. $\sqrt{-5}\cdot\sqrt{-3}\cdot\sqrt{-7}$ 5. ______________

6. $\dfrac{\sqrt{-42}\cdot\sqrt{-6}}{\sqrt{-7}}$ 6. ______________

7. $\dfrac{\sqrt{-10}\cdot\sqrt{7}}{\sqrt{25}}$ 7. ______________

Objective 2 Recognize subsets of the complex numbers.

Classify each of the following complex numbers as real *or* imaginary.

8. $\sqrt{5}$ 8. ______________

9. $\sqrt{3}-i\sqrt{5}$ 9. ______________

10. $i\sqrt{7}$ 10. ______________

Name: Date:
Instructor: Section:

Objective 3 Add and subtract complex numbers.

Add or subtract as indicated. Write answers in standard form.

11. $(-7-2i)-(-3-3i)$ **11.** ______________

12. $4i-(9+5i)+(2+3i)$ **12.** ______________

13. $(7-9i)-(5-6i)$ **13.** ______________

14. $\left(\sqrt{3}-2i\sqrt{2}\right)+\left(2\sqrt{3}-2i\sqrt{2}\right)$ **14.** ______________

15. $[(8+4i)-(5-3i)]+(4-2i)$ **15.** ______________

Objective 4 Multiply complex numbers.

Multiply.

16. $(2-5i)(2+5i)$ **16.** ______________

17. $(1+3i)^2$ **17.** ______________

Name: Date:
Instructor: Section:

18. $(12+2i)(-1+i)$

18. ______________

19. $\left(\sqrt{2}-i\sqrt{3}\right)^2$

19. ______________

20. $(2+i)(3-i)$

20. ______________

Objective 5 Divide complex numbers.

Write each quotient in the form a + bi.

21. $\dfrac{3-2i}{2+i}$

21. ______________

22. $\dfrac{5+2i}{9-4i}$

22. ______________

Name: Date:
Instructor: Section:

23. $\dfrac{6-i}{2-3i}$ 23. ______________

24. $\dfrac{7+9i}{4-i}$ 24. ______________

25. $\dfrac{1+i}{(2-i)(2+i)}$ 25. ______________

Objective 6 Find powers of *i*.

Find each power of i.

26. i^{-13} 26. ______________

27. i^{14} 27. ______________

28. i^{-100} 28. ______________

Name: Date:
Instructor: Section:

29. i^{113}

29. ______________

30. i^{-21}

30. ______________

Name: Date:
Instructor: Section:

Chapter 10 QUADRATIC EQUATIONS, INEQUALITIES, AND FUNCTIONS

10.1 Solving Quadratic Equations by the Square Root Property

Learning Objectives

1 Review the zero-factor property.
2 Solve equations of the form $x^2 = k$, where $k > 0$.
3 Solve equations of the form $(ax+b)^2 = k$, where $k > 0$.
4 Solve quadratic equations with nonreal complex solutions.

Key Terms

Use the vocabulary terms listed below to complete each statement in exercises 1–2.

quadratic equation **zero-factor property**

1. An equation that can be written in the form $ax^2 + bx + c = 0$ is a ________________________.

2. The ________________________ states that if a product equals 0, then at least one of the factors of the product also equals zero.

Objective 1 Review the zero-factor property.

Solve each equation by factoring.

1. $3x^2 = 5x + 28$ 1. ____________

2. $15s^2 - 2 = s$ 2. ____________

3. $z^2 = 6z - 9$ 3. ____________

4. $16m^2 - 64 = 0$ 4. ____________

Name: Date:
Instructor: Section:

Objective 2 Solve equations of the form $x^2 = k$, where $k > 0$.

Solve each equation by using the square root property. Express all radicals in simplest form.

5. $r^2 = 900$ **5.** ______________

6. $d^2 - 250 = 0$ **6.** ______________

7. $c^2 + 36 = 0$ **7.** ______________

8. $t^2 - 12.25 = 0$ **8.** ______________

9. $121x^2 - 24 = 0$ **9.** ______________

10. $t^2 - 30.25 = 0$ **10.** ______________

11. $s^2 - 98 = 0$ **11.** ______________

12. $289h^2 - 90 = 0$ **12.** ______________

Name: Date:
Instructor: Section:

Objective 3 Solve equations of the form $(ax+b)^2 = k$, where $k > 0$.

Solve each equation by using the square root property. Express all radicals in simplest form.

13. $(y+2)^2 = 16$ **13.** ______________

14. $(7p-4)^2 = 289$ **14.** ______________

15. $(q-4)^2 - 7 = 0$ **15.** ______________

16. $(6p+9)^2 = 54$ **16.** ______________

17. $(10m-5)^2 - 9 = 0$ **17.** ______________

Name: Date:
Instructor: Section:

18. $(3f+4)^2 = 32$

18. ______________

19. $\left(\frac{1}{4}x-2\right)^2 = 49$

19. ______________

20. $\left(\frac{1}{2}z+4\right)^2 = 81$

20. ______________

21. $\left(\frac{1}{3}r-3\right)^2 = 50$

21. ______________

Objective 4 Solve quadratic equations with nonreal complex solutions.

Find the complex solutions of each equation.

22. $(10m-5)^2 + 9 = 0$

22. ______________

Name: Date:
Instructor: Section:

23. $(m+1)^2 = -36$

23. ______________

24. $9 = -(6q-7)^2$

24. ______________

25. $(4a+5)^2 = -12$

25. ______________

26. $x^2 - 2x + 3 = 0$

26. ______________

Name: Date:
Instructor: Section:

27. $0.03y^2 - 0.12y + 0.24 = 0$

27. ______________

28. $2a^2 + 8a + 9 = 0$

28. ______________

29. $49 + 16(2w + 4)^2 = 0$

29. ______________

30. $6x^2 - 2x = -1$

30. ______________

Name: Date:
Instructor: Section:

Chapter 10 QUADRATIC EQUATIONS, INEQUALITIES, AND FUNCTIONS

10.2 Solving Quadratic Equations by Completing the Square

Learning Objectives

1. Solve quadratic equations by completing the square when the coefficient of the second-degree term is 1.
2. Solve quadratic equations by completing the square when the coefficient of the second-degree term is not 1.
3. Simplify the terms of an equation before solving.

Key Terms

Use the vocabulary terms listed below to complete each statement in exercises 1–3.

completing the square **perfect square trinomial** **square root property**

1. A ______________________ can be written in the form $x^2 + 2kx + k^2$ or $x^2 - 2kx + k^2$.

2. The ______________________ says that, if k is positive and $a^2 = k$, then $a = \pm\sqrt{k}$.

3. Use the process called ______________________ in order to rewrite an equation so it can be solved using the square root property.

Objective 1 Solve quadratic equations by completing the square when the coefficient of the second-degree term is 1.

Solve each equation by completing the square.

1. $x^2 + 3x = 4$ **1.** ______________

2. $r^2 + 8r = -4$ **2.** ______________

Name: Date:
Instructor: Section:

3. $x^2 - 4x = 2$ 3. ______________

4. $w^2 - 3w - 4 = 0$ 4. ______________

5. $x^2 + 2x = 63$ 5. ______________

6. $d^2 + 10d - 11 = 0$ 6. ______________

7. $c^2 - c - \frac{5}{2} = 0$ 7. ______________

Name: Date:
Instructor: Section:

8. $z^2 - \frac{z}{2} - \frac{7}{4} = 0$

8. ______________

9. $x^2 - 9x + 8 = 0$

9. ______________

10. $x^2 - 12x + 27 = 0$

10. ______________

Objective 2 Solve quadratic equations by completing the square when the coefficient of the second-degree term is not 1.

Solve each equation by completing the square.

11. $3m^2 - 15m = 42$

11. ______________

12. $3r^2 = 6r + 2$

12. ______________

Name: Date:
Instructor: Section:

13. $2p^2 + 6p - 1 = 0$

13. ________________

14. $6x^2 - x = 15$

14. ________________

15. $3x^2 - 2x + 4 = 0$

15. ________________

16. $-y^2 - 2y + 8 = 0$

16. ________________

17. $3t^2 + t - 2 = 0$

17. ________________

Name: Date:
Instructor: Section:

18. $6q^2 + 4q = 1$ **18.** ______________

19. $2a^2 + 7a - 13 = 0$ **19.** ______________

20. $8x^2 - 4x = 21$ **20.** ______________

Objective 3 Simplify the terms of an equation before solving.

Simplify each of the following equations and then solve by completing the square.

21. $2x - 4 = x^2 - 2x$ **21.** ______________

22. $4y^2 + 6y = 2y + 3$ **22.** ______________

Name: Date:
Instructor: Section:

23. $6y^2 + 3y = 4y^2 + y - 5$ **23.** ______________

24. $2z^2 = 6z + 3 - 4z^2$ **24.** ______________

25. $(b-1)(b+7) = 9$ **25.** ______________

25. $(s+3)(s+1) = 1$ **26.** ______________

27. $(j+3)(j-2) = 5$ **27.** ______________

Name: Date:
Instructor: Section:

28. $(5m+2)^2 - 9(5m+2) - 36 = 0$ **28.** ________________

29. $4x^2 + 6x - 3 = 5x^2 - 6x$ **29.** ________________

30. $2a^2 + 3a + 1 = 5a^2 + 10a + 4$ **30.** ________________

Name: Date:
Instructor: Section:

Chapter 10 QUADRATIC EQUATIONS, INEQUALITIES, AND FUNCTIONS

10.3 The Quadratic Formula

Learning Objectives	
1	Derive the quadratic formula.
2	Solve quadratic equations by using the quadratic formula.
3	Use the discriminant to determine the number and type of solutions.

Key Terms

Use the vocabulary terms listed below to complete each statement in exercises 1–2.

quadratic formula **discriminant**

1. The expression under the radical in the quadratic formula is called the

__________________________________.

2. The formula $x = \dfrac{-b \pm \sqrt{b^2 - 4ac}}{2a}$ is called the ______________________________.

Objective 1 Derive the quadratic formula.

Objective 2 Solve quadratic equations by using the quadratic formula.

Use the quadratic formula to solve each equation. (All solutions for these equations are real numbers.)

1. $(z+2)^2 = 2(5z-2)$ **1.** ________________

2. $5m^2 + 5m - 1 = 0$ **2.** ________________

Name: Date:
Instructor: Section:

3. $4p(p+1)=1$ 3. ______________

4. $3x^2+1=6x$ 4. ______________

5. $x^2-x-3=0$ 5. ______________

6. $x^2-3x+1=8$ 6. ______________

7. $5k^2+4k-2=0$ 7. ______________

Name: Date:
Instructor: Section:

8. $(x-4)(x+3)=8$ **8.** ______________

9. $3p(p+2)=3p^2+7$ **9.** ______________

10. $-\frac{1}{4}x^2+4=\frac{1}{2}x$ **10.** ______________

Use the quadratic formula to solve each equation. (All solutions for these equations are nonreal complex numbers.)

11. $-7r^2=5r+3$ **11.** ______________

12. $\frac{1}{4}t^2-\frac{1}{3}t+\frac{5}{12}=0$ **12.** ______________

Name: Date:
Instructor: Section:

13. $t^2 - 2t + 3 = 0$ **13.** ______________

14. $2x^2 = 4x - 3$ **14.** ______________

15. $2x^2 + 16x + 34 = 0$ **15.** ______________

16. $3x^2 + 12x = -24$ **16.** ______________

17. $2x^2 - 4x + 1 = -4$ **17.** ______________

Name: Date:
Instructor: Section:

18. $34 - 10x = -x^2$ **18.** ______________

19. $2z^2 - 2z + 9 = 3 - 4z$ **19.** ______________

20. $x^2 - 14x = -52$ **20.** ______________

Objective 3 Use the discriminant to determine the number and type of solutions.

Use the discriminant to determine whether the solutions for each equation are

A. two rational numbers *B. one rational number*
C. two irrational numbers *D. two imaginary numbers*

Do not actually solve.

21. $n^2 + n = 2$ **21.** ______________

22. $2m^2 - 4m = 8$ **22.** ______________

Name: Date:
Instructor: Section:

23. $5y^2 - 5y + 2 = 0$ **23.** ________________

24. $m^2 - 4m + 4 = 0$ **24.** ________________

25. $z^2 + 6z + 3 = 0$ **25.** ________________

26. $2y^2 + 4y + 8 = 0$ **26.** ________________

27. $16x^2 - 40x + 25 = 0$ **27.** ________________

28. $16x^2 - 12x + 9 = 0$ **28.** ________________

29. $12r^2 - 40r + 25 = 0$ **29.** ________________

30. $6x^2 - 3x - 10 = 0$ **30.** ________________

Name:　　　　　　　　　　　　　　　　Date:
Instructor:　　　　　　　　　　　　　　Section:

Chapter 10 QUADRATIC EQUATIONS, INEQUALITIES, AND FUNCTIONS

10.4 Equations Quadratic in Form

Learning Objectives

1	Solve an equation with fractions by writing it in quadratic form.
2	Use quadratic equations to solve applied problems.
3	Solve an equation with radicals by writing it in quadratic form.
4	Solve an equation that is quadratic in form by substitution.

Key Terms

Use the vocabulary terms listed below to complete each statement in exercises 1–2.

quadratic in form　　　　**standard form**

1. A quadratic equation written in the form $ax^2 + bx + c = 0, a \neq 0$ is written in

____.

2. A nonquadratic equation that can be written as a quadratic equation is called

____.

Objective 1 Solve an equation with fractions by writing it in quadratic form.

Solve each equation. Check your solutions.

1. $\dfrac{x}{2x+15} = \dfrac{1}{3x-2}$　　　　**1.** ________________

2. $5 + \dfrac{6}{m+1} = \dfrac{14}{m}$　　　　**2.** ________________

Name: Date:
Instructor: Section:

3. $4-\frac{8}{x-1}=-\frac{35}{x}$ **3.** ______________

4. $9-\frac{12}{x}=-\frac{4}{x^2}$ **4.** ______________

5. $\frac{5}{x}+\frac{1}{2x+7}=-\frac{2}{3}$ **5.** ______________

6. $\frac{2m}{m-5}+\frac{7}{m+1}=0$ **6.** ______________

Name: Date:
Instructor: Section:

7. $1+\frac{49}{2x}=\frac{15}{x+1}$ **7.** ______

Objective 2 Use quadratic equations to solve applied problems.

Solve each problem. Round answers to the nearest tenth, if necessary.

8. Amy rows her boat 6 miles upstream and then returns in $2\frac{6}{7}$ hours. The speed of the current is 2 miles per hour. How fast can she row? **8.** ______

9. Two pipes together can fill a large tank in 10 hours. One of the pipes, used alone, takes 15 hours longer than the other to fill the tank. How long would each pipe used alone take to fill the tank? **9.** pipe 1 ______

pipe 2 ______

10. The distance from Appletown to Medina is 45 miles, as is the distance from Medina to Westmont. Karl drove from Westmont to Medina, stopped at Medina for a hamburger, and then drove on to Appletown at 10 miles per hour faster. Driving time for the entire trip was 99 minutes. Find Karl's speed from Westmont to Medina. **10.** ______

Name: Date:
Instructor: Section:

11. A jet plane traveling at a constant speed goes 1200 miles with the wind, then turns around and travels for 1000 miles against the wind. If the speed of the wind is 50 miles per hour and the total flight takes 4 hours, find the speed of the plane.

11. ______________

12. The sum of the reciprocal of a number and the reciprocal of 5 more than the number is $\frac{11}{24}$. What is the number?

12. ______________

13. A man rode a bicycle for 12 miles and then hiked an additional 8 miles. The total time for the trip was 5 hours. If his rate when he was riding the bicycle was 10 miles per hour faster than his rate walking, what was each rate?

13. bike ______________

hike ______________

14. The perimeter of a rectangle is 24 inches. The area is 32 square inches. What are the dimensions of the rectangle?

14. ______________

Name: Date:
Instructor: Section:

15. Andrew can do a job in 3 hours less time than Emily. If they can finish the job together in 10 hours, how long will it take Andrew working alone?

15. ______________

Objective 3 Solve an equation with radicals by writing it in quadratic form.

Solve each equation. Check your solutions.

16. $x = \sqrt{x+2}$

16. ______________

17. $\sqrt{7y-10} = y$

17. ______________

18. $\sqrt{2}y = \sqrt{6-y}$

18. ______________

Name: Date:
Instructor: Section:

19. $k - \sqrt{8k - 15} = 0$

19. ________________

20. $x = \sqrt{\frac{x+3}{2}}$

20. ________________

21. $y = \sqrt{\frac{1-2y}{8}}$

21. ________________

22. $\sqrt{3}y = \sqrt{28y - 49}$

22. ________________

Name: Date:
Instructor: Section:

23. $\sqrt{\dfrac{25r-1}{144}}=r$

23. ______________

Objective 4 Solve an equation that is quadratic in form by substitution.

Solve each equation. Check your solutions.

24. $c^4-13c^2+36=0$

24. ______________

25. $t^4=\dfrac{21t^2-5}{4}$

25. ______________

26. $p^{4/3}-12p^{2/3}+27=0$

26. ______________

Name: Date:
Instructor: Section:

27. $\sqrt{x} = x - 6$

27. ________________

28. $\dfrac{1}{(x+6)^2} - \dfrac{7}{2(x+6)} = -\dfrac{3}{2}$

28. ________________

29. $\left(t^2 - 3t\right)^2 = 14\left(t^2 - 3t\right) - 40$

29. ________________

30. $\left(\sqrt{x} + 3\right)^2 = 8\left(\sqrt{x} + 3\right) - 12$

30. ________________

Name: Date:
Instructor: Section:

Chapter 10 QUADRATIC EQUATIONS, INEQUALITIES, AND FUNCTIONS

10.5 Formulas and Further Applications

Learning Objectives

1. Solve formulas for variables involving squares and square roots.
2. Solve applied problems using the Pythagorean formula.
3. Solve applied problems using area formulas.
4. Solve applied problems using quadratic functions as models.

Key Terms

Use the vocabulary terms listed below to complete each statement in exercises 1–2.

quadratic function **Pythagorean theorem**

1. A function defined by $f(x) = ax^2 + bx + c$, for real numbers a, b, and c, with $a \neq 0$, is a ______________________.

2. The ______________________ states that the sum of the squares of the lengths of the legs of a right triangle equals the square of the length of the hypotenuse.

Objective 1 Solve formulas for variables involving squares and square roots.

Solve each equation for the indicated variable. (Leave ± in your answers.)

1. $F = \dfrac{kl}{\sqrt{d}}$ for d 1. ______________

2. $y = \dfrac{1}{2}gt^2$ for t 2. ______________

3. $p = \sqrt{\dfrac{kl}{g}}$ for k 3. ______________

Name: Date:
Instructor: Section:

4. $s = 30\sqrt{\frac{a}{p}}$ for p **4.** ________

5. $a = \sqrt{bc} + 1$ for c **5.** ________

6. $p^2q^2 + pkq = k^2$ for q **6.** ________

7. $b^2a^2 + 2bca = c^2$ for a **7.** ________

Objective 2 Solve applied problems using the Pythagorean formula.

Solve each problem.

8. A 13-foot ladder is leaning against a building. The distance from the bottom of the ladder to the building is 2 feet more than twice the distance from the top of the ladder to the ground. How far is the bottom of the ladder from the building? **8.** ________

Name: Date:
Instructor: Section:

9. Two cars left an intersection at the same time, one heading south, the other heading east. Some time later the car traveling south had gone 18 miles farther than the car headed east. At that time they were 90 miles apart. How far had each car traveled?

9. east ____________

south ____________

10. A child flying a kite has let out 45 feet of string to the kite. The distance from the kite to the ground is 9 feet more than the distance from the child to a point directly below the kite. How high up is the kite?

10. ____________

11. The longest side of a right triangle is 4 centimeters longer than the next longest side. The third side is 16 centimeters in length. Find the length of the longest side.

11. ____________

12. A ladder is leaning against a building so that the top is 8 feet above the ground. The length of the ladder is 2 feet less than twice the distance of the bottom of the ladder from the building. Find the length of the ladder.

12. ____________

Name: Date:
Instructor: Section:

13. The diagonal of a rectangle is 25 inches, and the length is 3 inches more than three times the width. What is the length of the rectangle?

13. ____________

14. Two cars left an intersection at the same time, one heading north, the other heading west. Later they were exactly 95 miles apart. The car headed west had gone 38 miles less than twice as far as the car headed north. How far had each car traveled?

14. north ____________

west ____________

Objective 3 Solve applied problems using area formulas.

Solve each problem.

15. A piece of plastic in the shape of a rectangle has a length 10 inches less than twice the width. A square 4 inches on a side is cut out of each corner and the sides turned up to form an open box with a volume of 160 cubic inches. Find the dimensions of the finished box.

15. ____________

Name: Date:
Instructor: Section:

16. A picture 9 inches by 12 inches is to be mounted on a piece of mat board so that there is an even width of mat all around the picture. How wide will the matted border be if the area of the mounted picture is 238 square inches?

16. ________________

17. An open box is to be made from a rectangular piece of tin by cutting 2-inch squares out of the corners and folding up the sides. The length of the finished box is to be twice the width. The volume of the box will be 100 cubic inches. Find the dimensions of the rectangular piece of tin.

17. ________________

18. A rug is to fit in a room so that a border of even width is left on all four sides. If the room is 16 feet by 20 feet and the area of the rug is 165 square feet, how wide to the nearest tenth of a foot will the border be?

18. ________________

Name: Date:
Instructor: Section:

19. A rectangular garden has an area of 12 feet by 5 feet. A gravel path of equal width is to be built around the garden. How wide can the path be if there is enough gravel for 138 square feet?

19. ______________

20. A doghouse 2 feet by 4 feet is to be built with a cement path around it of equal width on all sides. The area available for the doghouse and path is 120 square feet. How wide will the path be?

20. ______________

21. A rug is to fit in a room so that a border of even width is left on all four sides. If the room is 12 feet by 15 feet and the area of the rug is 108 square feet, how wide will the border be?

21. ______________

22. To make an open box, 3-centimeter squares are cut from the corners of a rectangular piece of cardboard and the sides folded up. The length of the cardboard is 6 centimeters more than its width. The volume of the finished box will be 120 cubic centimeters. Find the dimensions of the piece of cardboard.

22. ______________

Name: Date:
Instructor: Section:

Objective 4 Solve applied problems using quadratic functions as models.

Solve each problem. Round answers to the nearest tenth.

23. A population of microorganisms grows according to the function $p(x) = 100 + 0.2x + 0.5x^2$, where x is given in hours. How many hours does it take to reach a population of 250 microorganisms?

23. 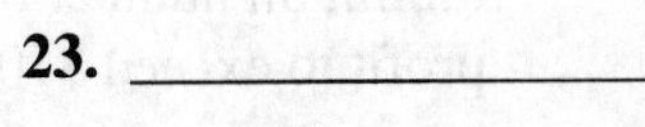

24. The charge potential for a certain experiment can be modeled by the function $c(x) = 5 - 0.1x - 0.2x^2$, where x is given in minutes. When does the potential equal 4.5?

24. 

25. The position of an object moving in a straight line is given by $s(t) = t^2 - 8t$, where s is in feet and t is in seconds. How long will it take the object to move 10 feet?

25. 

26. An object is thrown downward from a tower 280 feet high. The distance the object has fallen at time t in seconds is given by $s(t) = 16t^2 + 68t$. How long will it take the object to fall 100 feet?

26. 

Name: Date:
Instructor: Section:

27. The profit from the sale of x items is given by the function $P(x) = 2x^2 - 10x - 100$. What is the minimum number of items that must be sold for the profit to exceed \$1000?

27. 

28. A widget manufacturer estimates that her monthly revenue can be modeled by the function $R(x) = -0.006x^2 + 32x - 10{,}000$. What is the minimum number of items that must be sold for the revenue to equal \$30,000?

28. ______________

29. A baseball is thrown upward from a building 20 m high with a velocity of 15 m/sec. Its distance from the ground after t seconds is modeled by the function $f(t) = -4.9t^2 + 15t + 20$. When will the ball hit the ground?

29. ______________

30. A manufacturer finds that the number of items sold each month (n) is related to the price per item (p) by the equation $p = -2n + 200$. The monthly revenue is given by $R(n) = pn$. How many items where sold if the revenue was \$4800?

30. ______________

Name: Date:
Instructor: Section:

Chapter 10 QUADRATIC EQUATIONS, INEQUALITIES, AND FUNCTIONS

10.6 Graphs of Quadratic Functions

Learning Objectives

1. Graph a quadratic function.
2. Graph parabolas with horizontal and vertical shifts.
3. Use the coefficient of x^2 to predict the shape and direction in which a parabola opens.
4. Find a quadratic function to model data.

Key Terms

Use the vocabulary terms listed below to complete each statement in exercises 1–4.

parabola **vertex** **axis** **quadratic function**

1. The vertical (or horizontal) line through the vertex of a vertical (or horizontal) parabola is its ______________________.

2. The point on a parabola that has the least y-value (if the parabola opens up) or the greatest y-value (if the parabola opens down) is called the ________________ of the parabola.

3. A function defined by $f(x) = ax^2 + bx + c$, for real numbers a, b, and c, with $a \neq 0$, is a ______________________________.

4. The graph of a quadratic function is a ______________________________.

Objective 1 Graph a quadratic function.
Objective 2 Graph parabolas with horizontal and vertical shifts.

Sketch the graph of each parabola. Give the vertex, axis, domain, and range.

1. $f(x) = x^2 - 2$

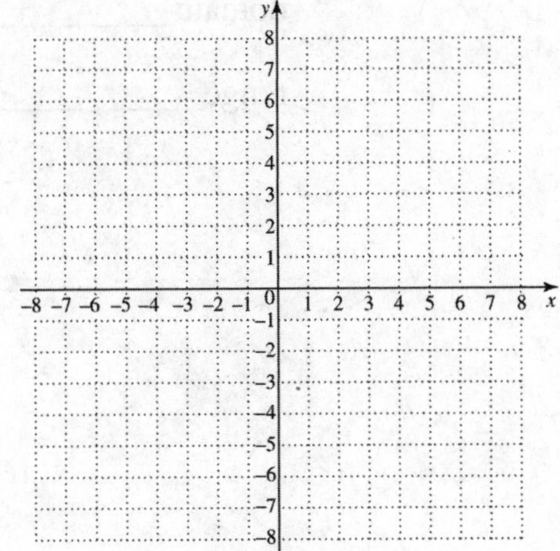

1. vertex ______________

axis ______________

domain ______________

range ______________

Name: Date:
Instructor: Section:

2. $f(x) = x^2 + 2$

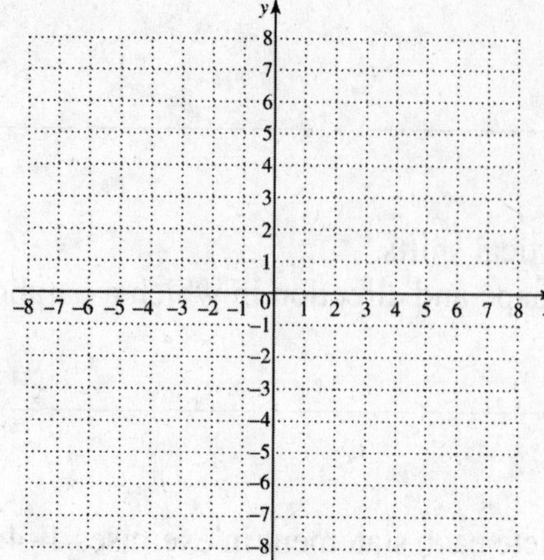

2. vertex ____________

axis ____________

domain ____________

range ____________

3. $f(x) = x^2 + 3$

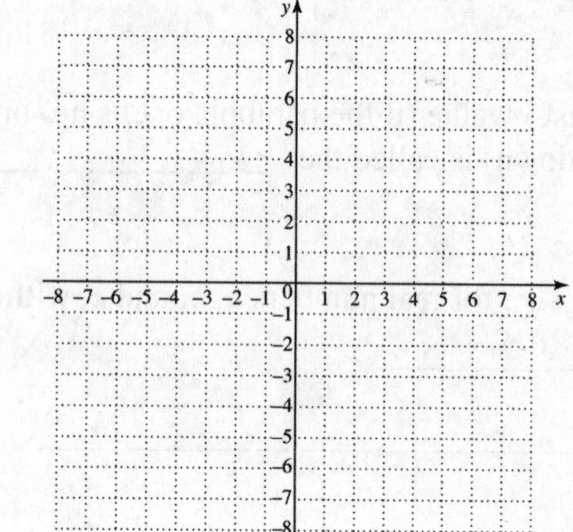

3. vertex ____________

axis ____________

domain ____________

range ____________

4. $f(x) = x^2 - 4$

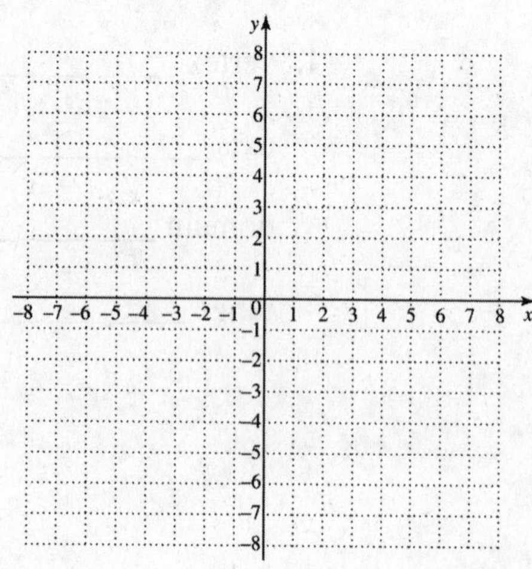

4. vertex ____________

axis ____________

domain ____________

range ____________

Name: Date:
Instructor: Section:

5. $f(x) = 2 - x^2$

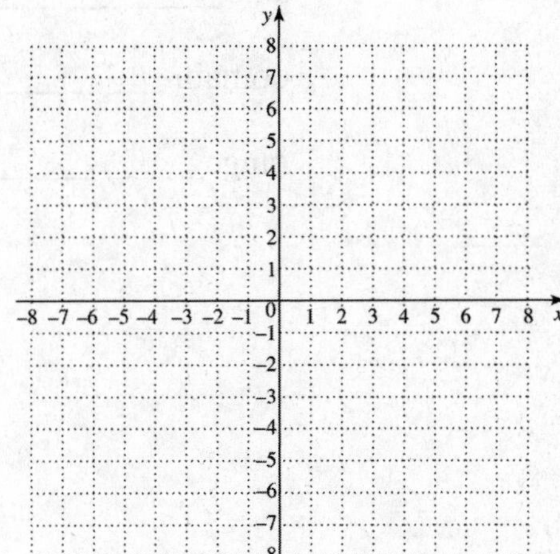

5. vertex ______________

axis ______________

domain ______________

range ______________

6. $f(x) = 5 - x^2$

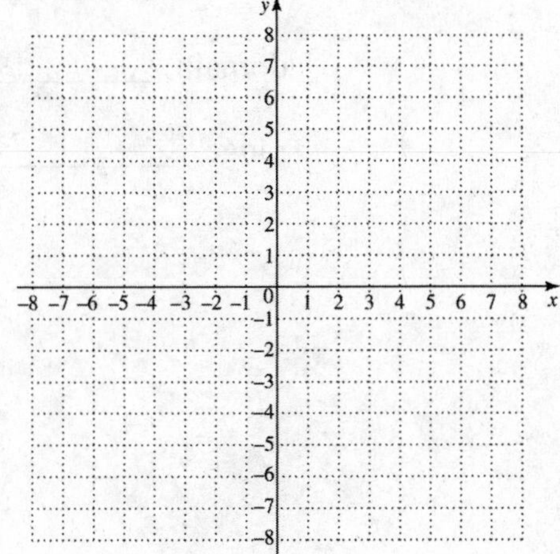

6. vertex ______________

axis ______________

domain ______________

range ______________

7. $f(x) = (x+2)^2$

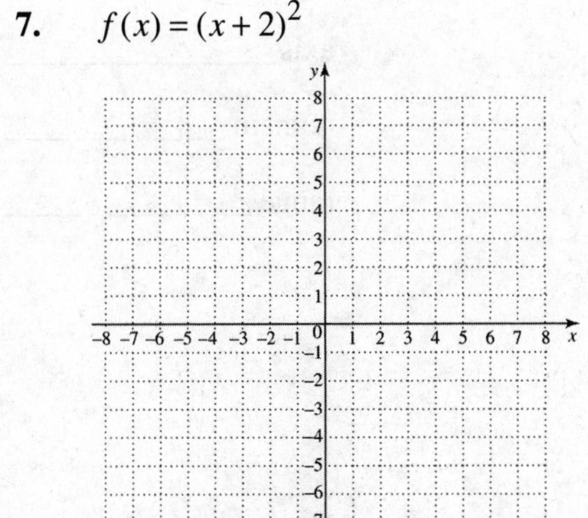

7. vertex ______________

axis ______________

domain ______________

range ______________

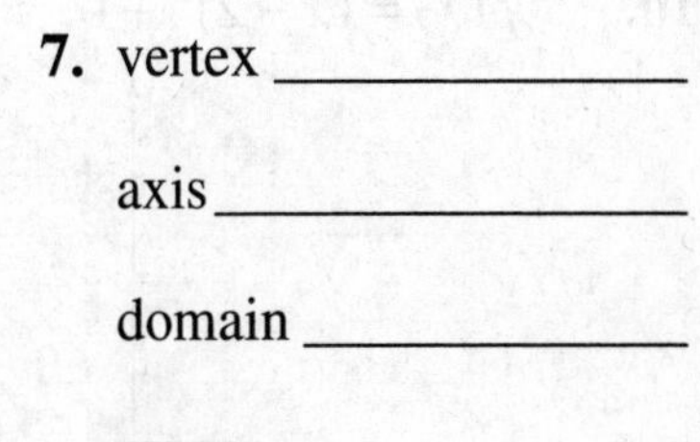

Name: Date:
Instructor: Section:

8. $f(x) = (x-3)^2$

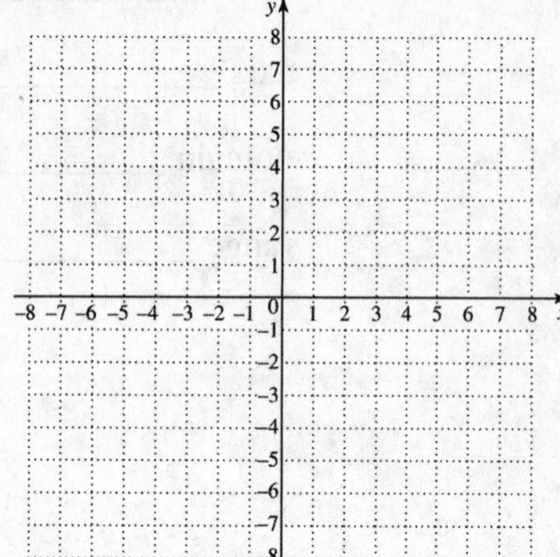

8. vertex ______________

axis ______________

domain ______________

range ______________

9. $f(x) = (x+3)^2 - 1$

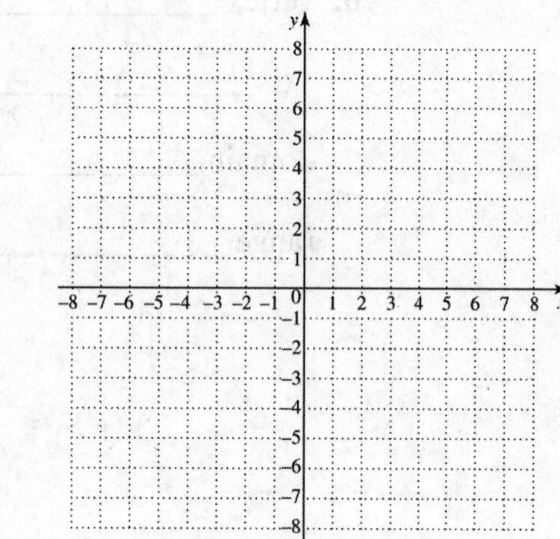

9. vertex ______________

axis ______________

domain ______________

range ______________

10. $f(x) = (x-2)^2 + 1$

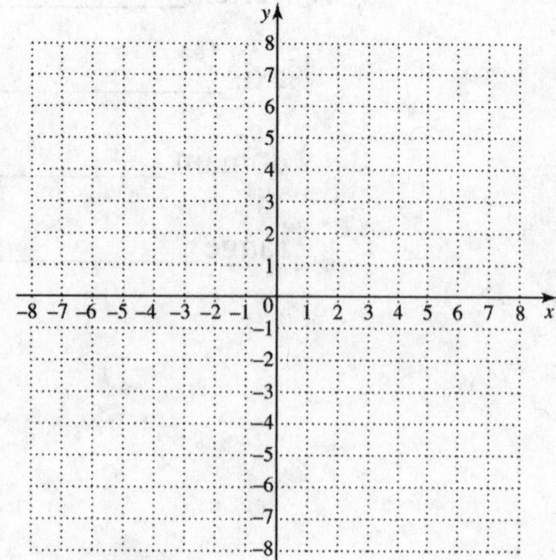

10. vertex ______________

axis ______________

domain ______________

range ______________

Name: Date:
Instructor: Section:

11. $f(x) = (x-1)^2$

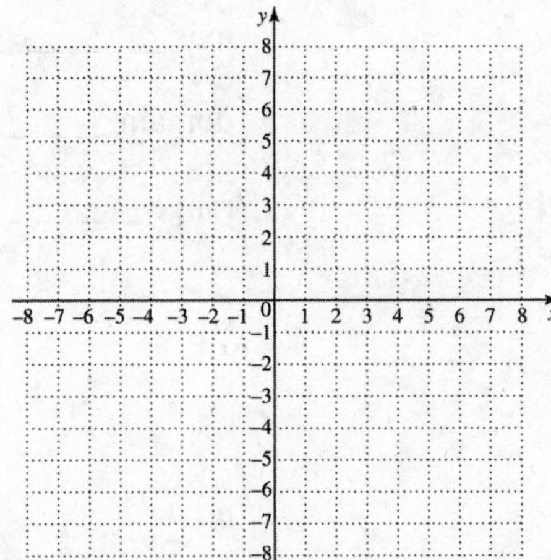

11. vertex ______________

axis______________

domain ______________

range______________

12. $f(x) = (x+2)^2 + 3$

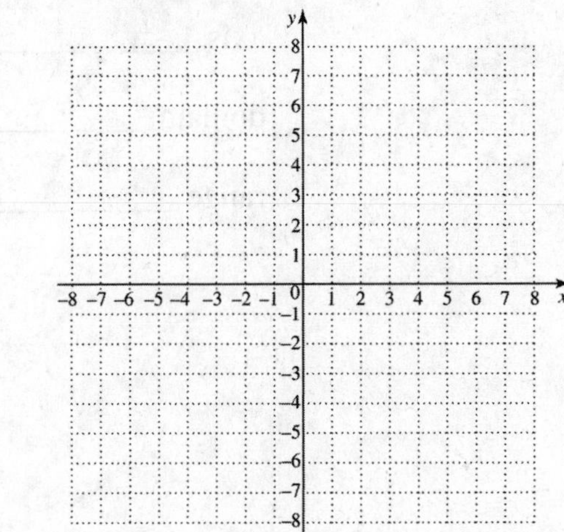

12. vertex ______________

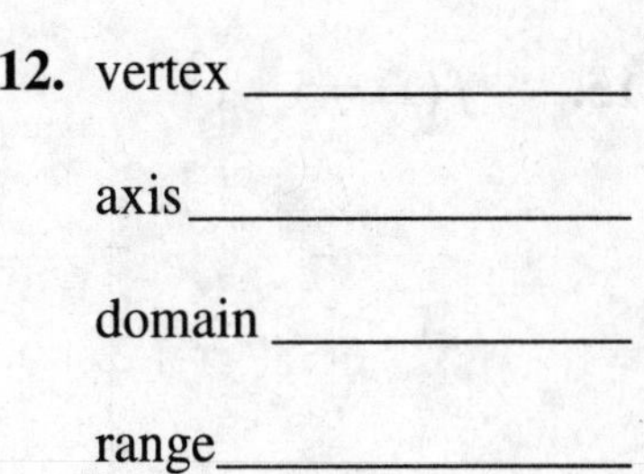

axis______________

domain ______________

range______________

13. $f(x) = (x-3)^2 - 1$

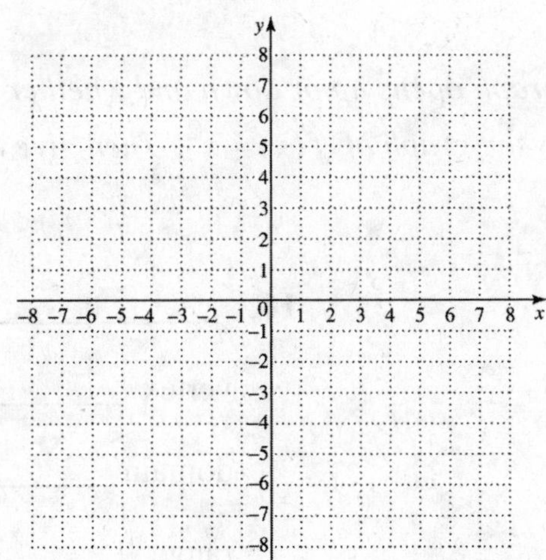

13. vertex ______________

axis______________

domain ______________

range______________

Name: Date:
Instructor: Section:

14. $f(x) = (x+3)^2$

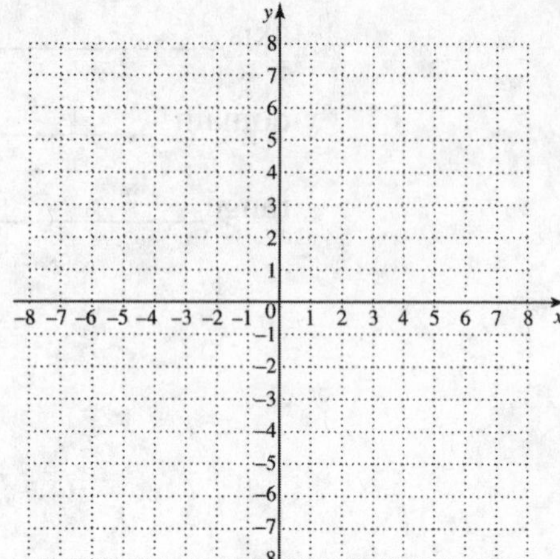

14. vertex ____________

axis ____________

domain ____________

range ____________

15. $f(x) = 5 - x^2$

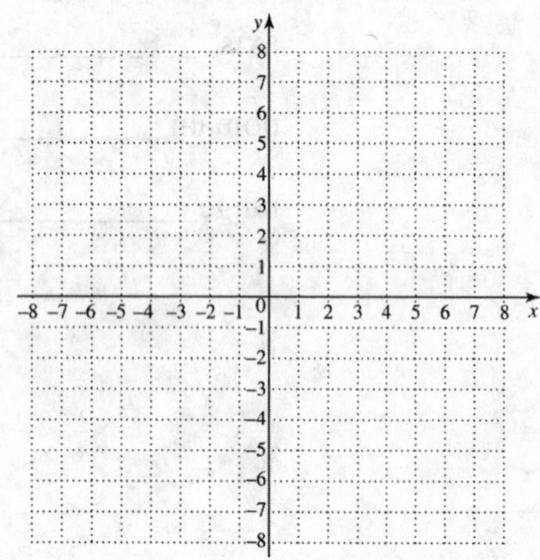

15. vertex ____________

axis ____________

domain ____________

range ____________

Objective 3 Use the coefficient of x^2 to predict the shape and direction in which a parabola opens.

For each quadratic function, tell whether the graph opens up or down and whether the graph is wider, narrower, or the same shape as the graph of $f(x) = x^2$. *Then give the vertex, domain, and range.*

16. $f(x) = \frac{1}{2}x^2$

16. ____________

vertex ____________

domain ____________

range ____________

Name: Date:
Instructor: Section:

17. $f(x) = -2x^2$

17. ______________

vertex ___________

domain ___________

range ___________

18. $f(x) = -\frac{4}{3}x^2 - 1$

18. ______________

vertex ___________

domain ___________

range ___________

19. $f(x) = \frac{3}{5}x^2 + 5$

19. ______________

vertex ___________

domain ___________

range ___________

20. $f(x) = 3(x-2)^2$

20. ______________

vertex ___________

domain ___________

range ___________

21. $f(x) = -2(x+1)^2$

21. ______________

vertex ___________

domain ___________

range ___________

22. $f(x) = -\frac{1}{3}(x+3)^2 - 4$

22. ______________

vertex ___________

domain ___________

range ___________

Name: Date:
Instructor: Section:

23. $f(x) = \frac{5}{4}(x-1)^2 + 7$

23. ______________

vertex ____________

domain __________

range ____________

24. $f(x) = 4 - x^2$

24. ______________

vertex ____________

domain __________

range ____________

Objective 4 Find a quadratic function to model data.

Tell whether a linear or quadratic function would be a more appropriate model for each set of graphed data. If linear, tell whether the slope should be positive or negative. If quadratic, tell whether the coefficient a of x^2 should be positive or negative.

25.

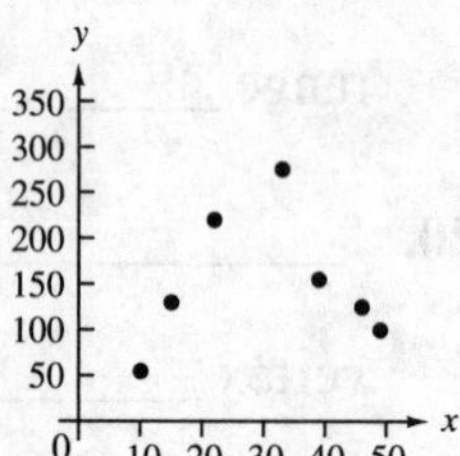

25. ______________

26.

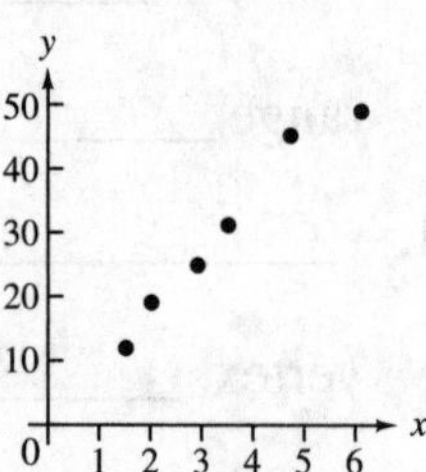

26. ______________

27.

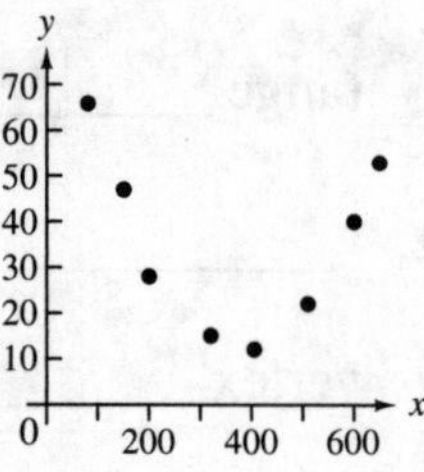

27. ______________

Name: Date:
Instructor: Section:

28.

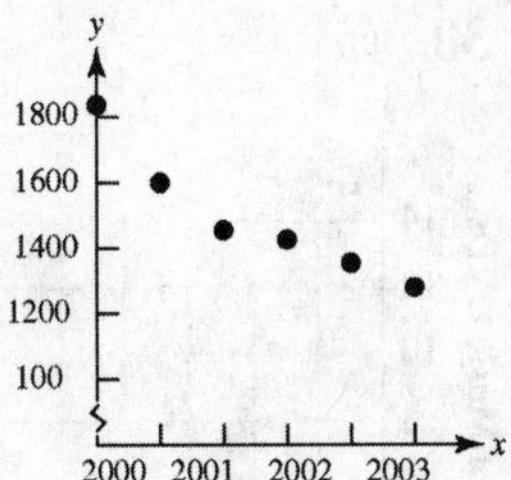

28. ____________

Solve each problem.

29. The number of publicly traded companies filing for bankruptcy for selected years between 1990 and 2000 are shown in the table, with 0 representing 1990, 2 representing 1992, etc.

Year	Number of Bankruptcies
0	115
2	91
4	70
6	84
8	120
10	176

(a) Use the ordered pairs to make a scatter diagram of the data.

(b) Use the scatter diagram to decide whether a linear or quadratic function would better model the data. If linear, is the slope positive or negative? If quadratic, should the coefficient a of x^2 be positive or negative?

(c) Use the ordered pairs (0, 115), (4, 70), and (8, 120) to find a function that models the data. Round the values of a, b, and c to three decimal places, if necessary.

(d) Use the model from part (c) to approximate the number of company bankruptcy filings in 2002. Round the answer to the nearest whole number.

29. (a)

Number of Filings: 20, 40, 60, 80, 100, 120, 140, 160, 180

0 2 4 6 8 10

Years Since 1990

(b) ____________

(c) ____________

(d) ____________

Source: Lial, Margaret L., John Hornsby, Terry McGinnis, *Intermediate Algebra* Eighth Edition. Boston: Pearson Education, 2006.

Name: Date:
Instructor: Section:

30. The number of music DVDs released from 2000 through 2007, with 0 representing 2000, 1 representing 2001, etc., is shown in the table below.

Year	Number of DVDs (thousands)
0	4.0
1	5.6
2	7.4
3	10.4
4	12.2
5	13.9
6	13.6
7	12.1

(a) Use the ordered pairs to make a scatter diagram of the data.

(b) Use the scatter diagram to decide whether a linear or quadratic function would better model the data. If linear, is the slope positive or negative? If quadratic, should the coefficient a of x^2 be positive or negative?

(c) Use the ordered pairs (0, 4.0), (5, 13.9), and (7, 12.1) to find a function that models the data. Round the values of a, b, and c to three decimal places, if necessary.

(d) Use the model from part (c) to predict the number of music DVDs in 2010.

Source: http://files.dvdnote.com/pdfs/press_kit.pdf.

30. (a)

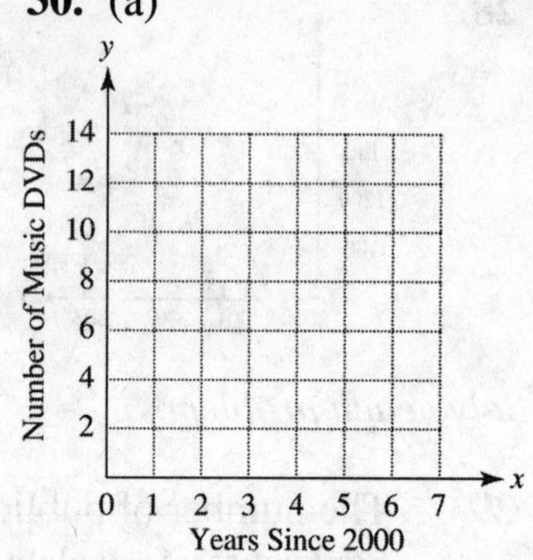

(b)__________

(c)__________

(d)__________

Name: Date:
Instructor: Section:

Chapter 10 QUADRATIC EQUATIONS, INEQUALITIES, AND FUNCTIONS

10.7 More about Parabolas and Their Applications

Learning Objectives

1. Find the vertex of a vertical parabola.
2. Graph a quadratic function.
3. Use the discriminant to find the number of x-intercepts of a parabola with a vertical axis.
4. Use quadratic functions to solve problems involving maximum or minimum value.
5. Graph parabolas with horizontal axes.

Key Terms

Use the vocabulary terms listed below to complete each statement in exercises 1–2.

discriminant **vertex**

1. The__________________________ of a quadratic function is found by using the formula $b^2 - 4ac$.

2. The maximum or minimum value of a quadratic function occurs at the __________________________ of its graph.

Objective 1 Find the vertex of a vertical parabola.
Objective 2 Graph a quadratic function.

Sketch the graph of each parabola. Give the vertex, axis, domain, and range.

1. $f(x) = x^2 + 6x + 10$

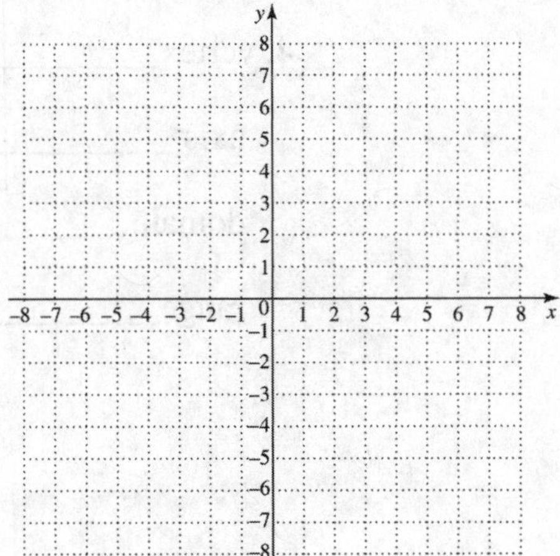

1. vertex ____________

axis____________

domain ____________

range____________

Name: Date:
Instructor: Section:

2. $f(x) = x^2 - 6x + 4$

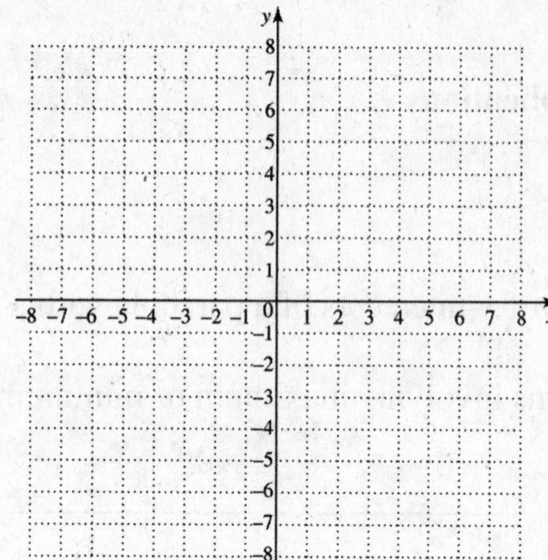

2. vertex___________

axis___________

domain___________

range___________

3. $f(x) = -x^2 + 8x - 10$

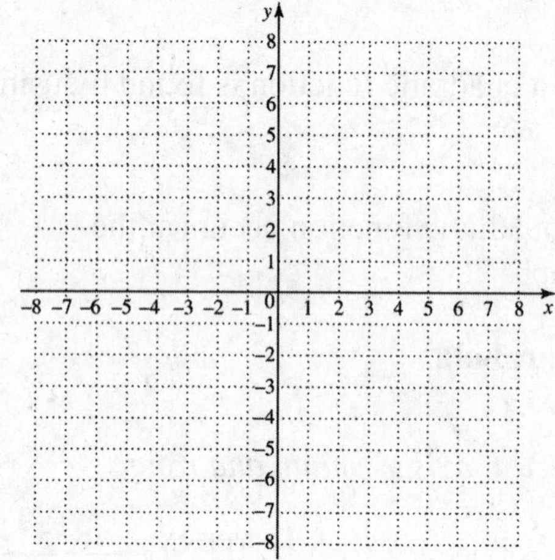

3. vertex___________

axis___________

domain___________

range___________

4. $f(x) = x^2 - 3x + 2$

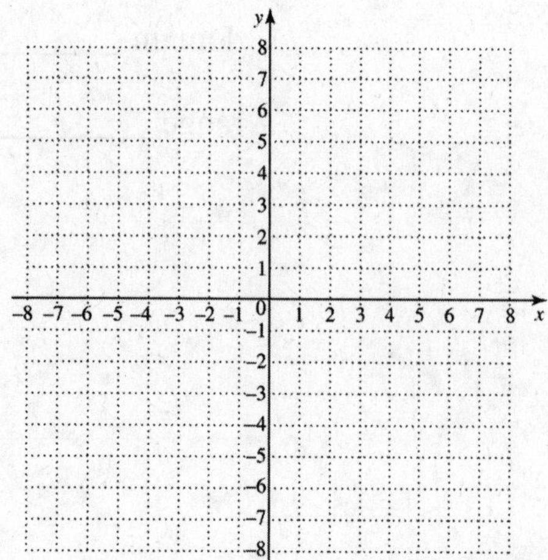

4. vertex___________

axis___________

domain___________

range___________

Name: Date:
Instructor: Section:

5. $f(x) = 3x^2 + 6x + 2$

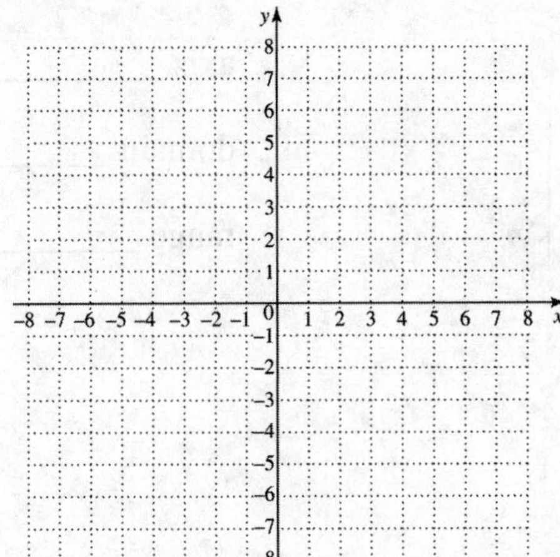

5. vertex ____________

axis ____________

domain ____________

range ____________

6. $f(x) = -2x^2 + 4x + 1$

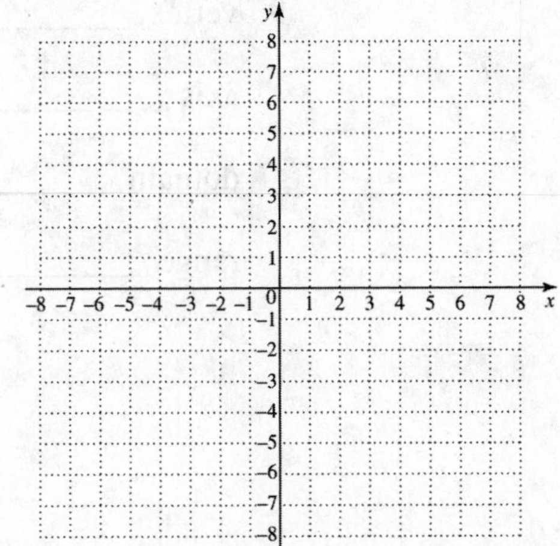

6. vertex ____________

axis ____________

domain ____________

range ____________

7. $f(x) = \frac{1}{2}x^2 + 2x + 3$

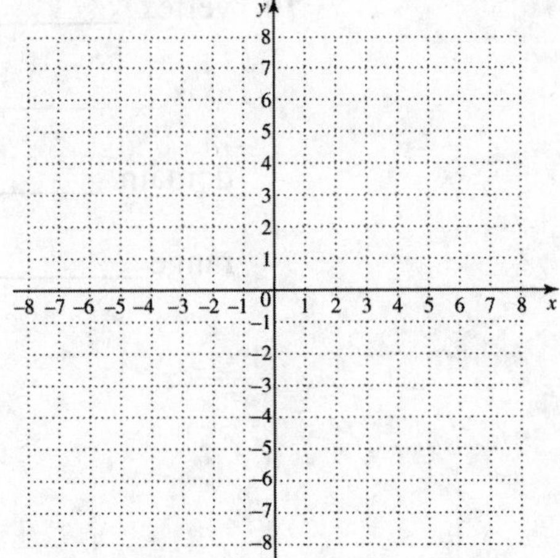

7. vertex ____________

axis ____________

domain ____________

range ____________

Name: Date:
Instructor: Section:

8. $f(x)=\frac{5}{4}x^2+5x+3$

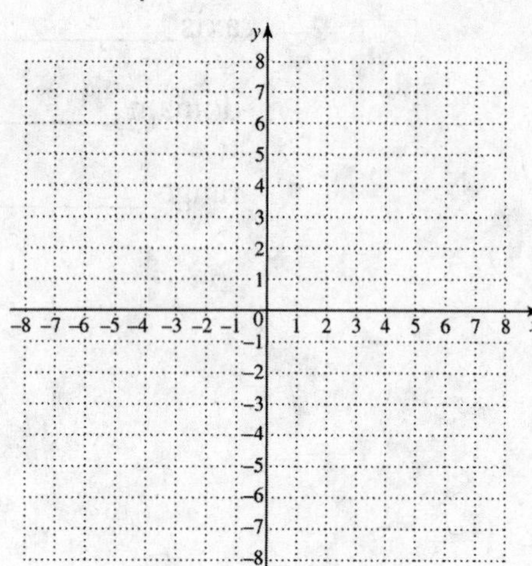

8. vertex ___________

axis ___________

domain ___________

range ___________

9. $y=-\frac{1}{3}x^2-2x-4$

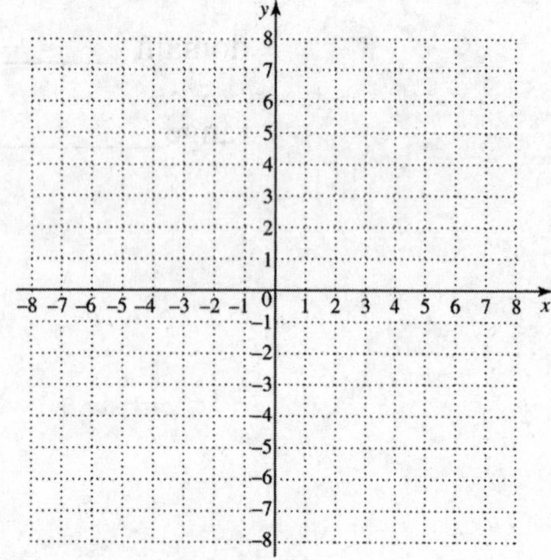

9. vertex ___________

axis ___________

domain ___________

range ___________

10. $y=-x^2-2x+8$

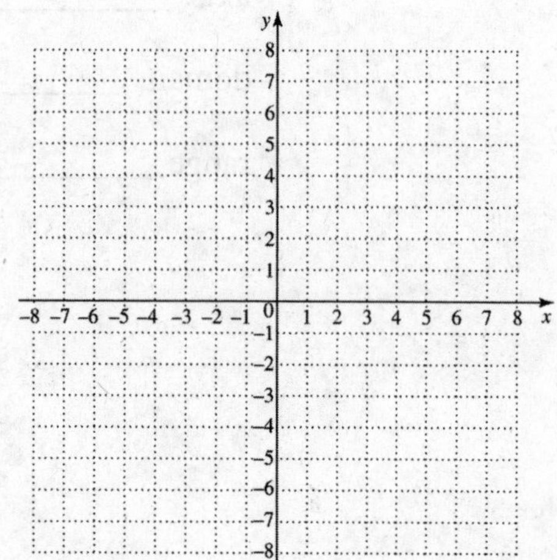

10. vertex ___________

axis ___________

domain ___________

range ___________

Name: Date:
Instructor: Section:

11. $y = 2x^2 + 4x - \dfrac{1}{2}$

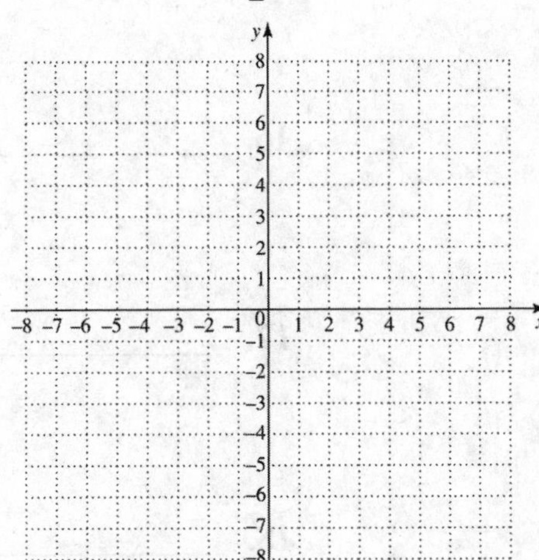

11. vertex ____________

axis ____________

domain ____________

range ____________

12. $y = \dfrac{1}{3}x^2 + x - 1$

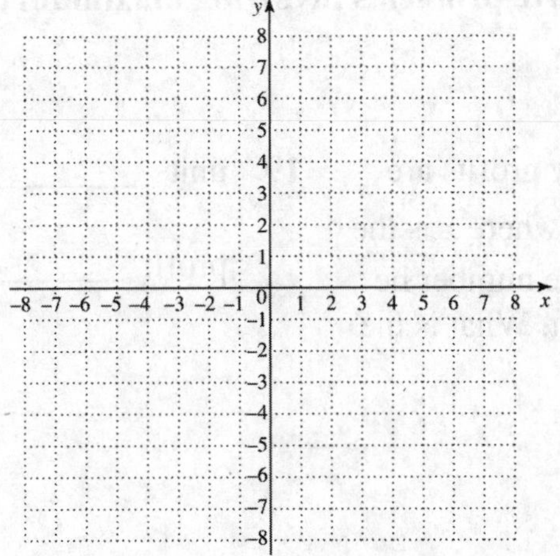

12. vertex ____________

axis ____________

domain ____________

range ____________

Objective 3 Use the discriminant to find the number of *x*-intercepts of a parabola with a vertical axis.

Use the discriminant to determine the number of x-intercepts of the graph of each function.

13. $f(x) = 6x^2 - 3x + 1$

13. ____________

14. $f(x) = x^2 + 5x + 4$

14. ____________

15. $f(x)=4x^2+12x+9$ **15.** ______________

16. $f(x)=5x^2-5x+2$ **16.** ______________

17. $f(x)=9x^2-24x+16$ **17.** ______________

18. $f(x)=2x^2-3x+2$ **18.** ______________

Objective 4 Use quadratic functions to solve problems involving maximum or minimum value.

Solve each problem.

19. A businessman has found that his daily profits are given by $P(x)=-2x^2+100x+2400$, where x is the number of units sold each day. Find the number he should sell daily to maximize his profit. What is the maximum profit?

19. units ______________

profit ______________

20. The same businessman has daily costs of $C(x)=x^2-50x+1625$, where x is the number of units sold each day. How many units must be sold to minimize his cost? What is the minimum cost?

20. units ______________

cost ______________

Name: Date:
Instructor: Section:

21. Jean sells ceramic pots. She has weekly costs of $C(x) = x^2 - 100x + 2700,$ where x is the number of pots she sells each week. How many pots should she sell to minimize her costs? What is the minimum cost?

21. units ___________

cost ___________

22. The length and width of a rectangle have a sum of 48. What width will produce the maximum area?

22. ___________

23. A projectile is fired upward so that its distance (in feet) above the ground t seconds after firing is given by $s(t) = -16t^2 + 80t + 156$. Find the maximum height it reaches and the number of seconds it takes to reach that height.

23. height ___________

time ___________

24. Of all pairs of numbers whose sum is 92, find the pair with the maximum product.

24. ___________

Name: Date:
Instructor: Section:

Objective 5 Graph parabolas with horizontal axes.

Sketch the graph of each parabola. Give the vertex, axis, domain, and range.

25. $x = -y^2 + 2$

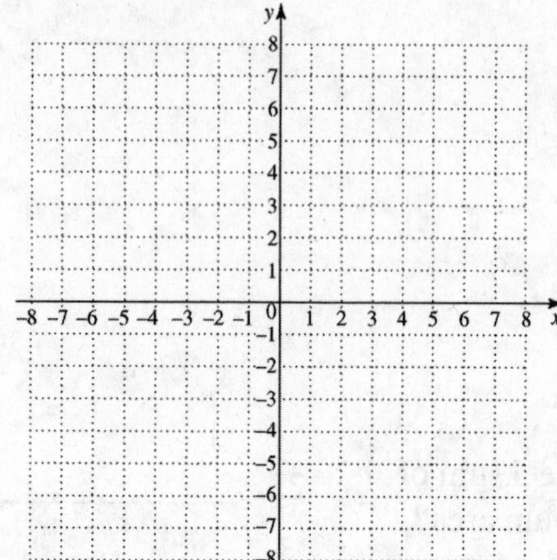

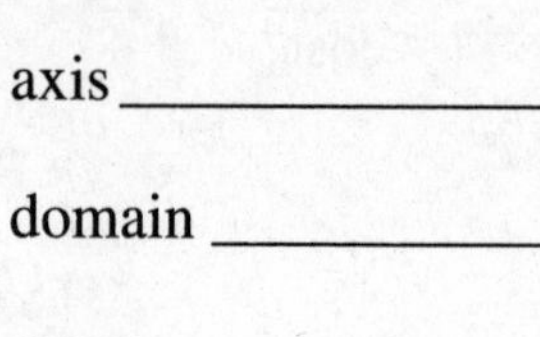

25. vertex ____________

axis ____________

domain ____________

range ____________

26. $x = y^2 - 3$

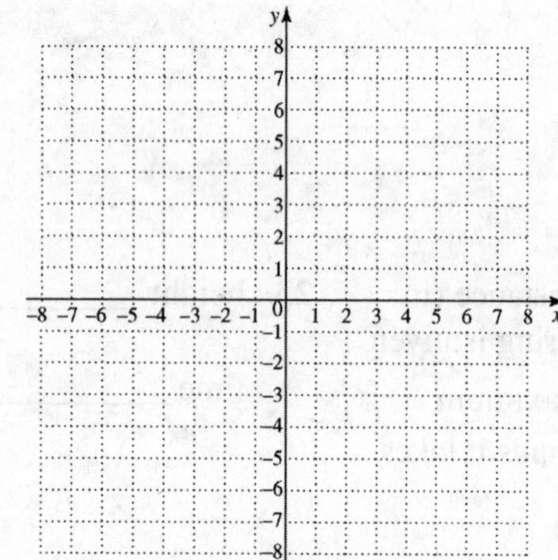

26. vertex ____________

axis ____________

domain ____________

range ____________

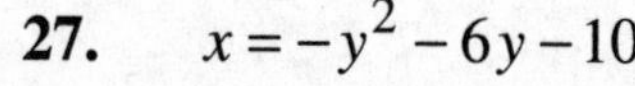

27. $x = -y^2 - 6y - 10$

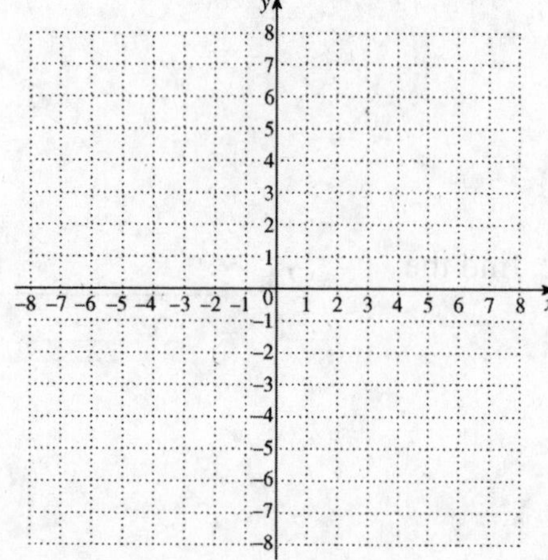

27. vertex ____________

axis ____________

domain ____________

range ____________

Name: Date:
Instructor: Section:

28. $x = -y^2 + 4y - 4$

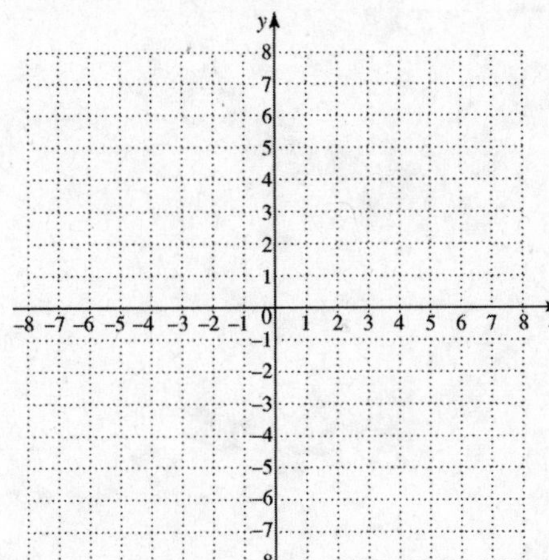

28. vertex ____________

axis ______________

domain ____________

range_____________

29. $x = y^2 - 4y + 7$

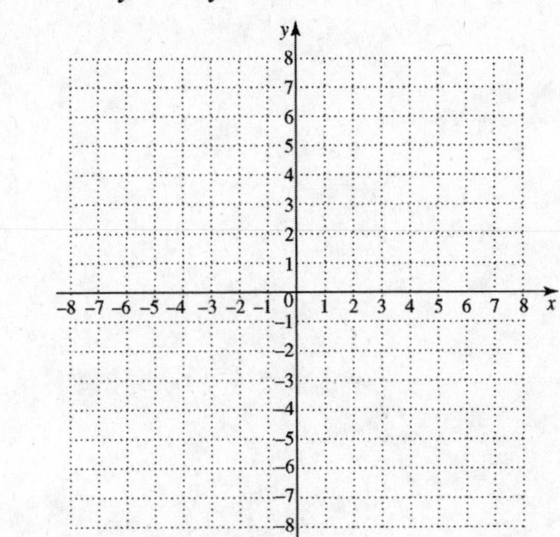

29. vertex ____________

axis ______________

domain ____________

range_____________

30. $3x = y^2 - 6y + 6$

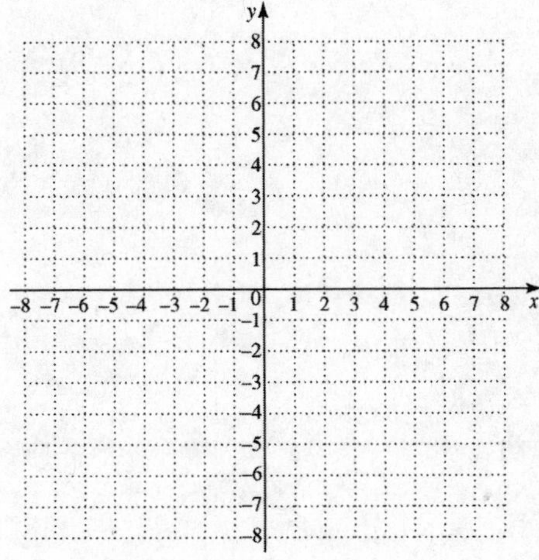

30. vertex ____________

axis ______________

domain ____________

range_____________

Name: Date:
Instructor: Section:

Chapter 10 QUADRATIC EQUATIONS, INEQUALITIES, AND FUNCTIONS

10.8 Polynomial and Rational Inequalities

Learning Objectives
1. Solve quadratic inequalities.
2. Solve polynomial inequalities of degree 3 or more.
3. Solve rational inequalities.

Key Terms

Use the vocabulary terms listed below to complete each statement in exercises 1–2.

quadratic inequality **rational inequality**

1. An inequality that involves a rational expression is a ____________________.

2. An inequality that can be written in the form $ax^2 + bx + c < 0$ or $ax^2 + bx + c > 0$, where a, b, and c are real numbers with $a \neq 0$ is called a ____________________.

Objective 1 Solve quadratic inequalities.

Solve each inequality, and graph the solution set.

1. $15k^2 + 2 \leq 11k$

1. ____________________

2. $k^2 + 7k + 12 > 0$

2. ____________________

3. $a^2 - a - 2 \leq 0$

3. ____________________

Name: Date:
Instructor: Section:

4. $2y^2 < y + 3$ **4.** ____________

5. $2m^2 - 5m > 12$ **5.** ____________

6. $6r^2 + 7r + 2 > 0$ **6.** ____________

7. $8k^2 + 10k > 3$ **7.** ____________

8. $(2k + 5)^2 \le -1$ **8.** ____________

9. $(3p + 2)^2 \ge -6$ **9.** ____________

Name: Date:
Instructor: Section:

10. $(3x-2)^2 < -1$

10. ____________________

Objective 2 Solve polynomial inequalities of degree 3 or more.

Solve each inequality, and graph the solution set.

11. $(x+1)(x-2)(x+4) \le 0$

11. ____________________

12. $(y+2)(y-1)(y-2) < 0$

12. ____________________

13. $(k+5)(k-1)(k+3) \le 0$

13. ____________________

14. $(x-1)(x-3)(x+2) \ge 0$

14. ____________________

15. $(y-4)(y+3)(y+1) \le 0$

15. ____________________

16. $(p-6)(p-4)(p-2) > 0$

16. ____________________

17. $(2x-1)(2x+3)(3x+1) \le 0$

17. ____________________

18. $(4b+1)(6b-1)(3b-7)>0$

18. ____________

19. $(4q-3)(2q-7)(3q-10)\geq 0$

19. ____________

20. $(z+1)(z-1)(3z-7)<0$

20. ____________

Objective 3 Solve rational inequalities.

Solve each inequality, and graph the solution set.

21. $\dfrac{7}{x-1}\leq 1$

21. ____________

22. $\dfrac{y}{y+1}\geq 3$

22. ____________

23. $\dfrac{4}{3q+5}\geq -3$

23. ____________

Name: Date:
Instructor: Section:

24. $\dfrac{-5}{2x-3} \le 2$

24. ______________

25. $\dfrac{-3}{4m-3} \ge 1$

25. ______________

26. $\dfrac{r}{r-2} \ge 4$

26. ______________

27. $\dfrac{2p-1}{3p+1} \le 1$

27. ______________

28. $\dfrac{z+2}{z-3} \le 2$

28. ______________

Name: Date:
Instructor: Section:

29. $\frac{5}{x-3} \leq -1$

29. ______________________

30. $\frac{4z}{3z-5} < -3$

30. ______________________

Name: Date:
Instructor: Section:

Chapter 11 INVERSE, EXPONENTIAL, AND LOGARITHMIC FUNCTIONS

11.1 Inverse Functions

Learning Objectives

1. Decide whether a function is one-to-one and, if it is, find its inverse.
2. Use the horizontal line test to determine whether a function is one-to-one.
3. Find the equation of the inverse of a function.
4. Graph f^{-1} from the graph of f.

Key Terms

Use the vocabulary terms listed below to complete each statement in exercises 1–2.

one-to-one function **inverse of a function *f***

1. A function in which each *x*-value corresponds to just one *y*-value and each *y*-value corresponds to just one *x*-value is a ______________________________.

2. If *f* is a one-to-one function, the ______________________________ is the set of all ordered pairs of the form (y, x) where (x, y) belongs to *f*.

Objective 1 Decide whether a function is one-to-one and, if it is, find its inverse.

If the function is one-to-one, find its inverse.

1. $\{(-3, -1), (-2, 0), (-1, 1), (0, 2)\}$ **1.** ______________

2. $\{(2, -1), (-2, 1), (1, 3), (-1, -3)\}$ **2.** ______________

3. $\{(4, 0), (2, 3), (0, 0), (3, 5)\}$ **3.** ______________

4. $\{(2, 4), (-1, 1), (0, 0), (1, 1), (2, 6)\}$ **4.** ______________

5. $\{(0, 0), (1, 1), (-1, -1), (2, 2), (-2, -2)\}$ **5.** ______________

6. $\{(3, 2), (-3, -2), (2, 3), (-2, -3)\}$ **6.** ______________

Name: Date:
Instructor: Section:

7. {(5, –1), (4, 7), (6, 3), (3, 3)} **7.** ______________

Objective 2 Use the horizontal line test to determine whether a function is one-to-one.

Use the horizontal line test to determine whether each function is one-to-one.

8. **8.** ______________

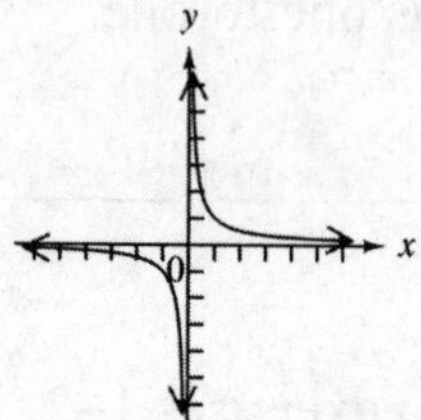

9. **9.** ______________

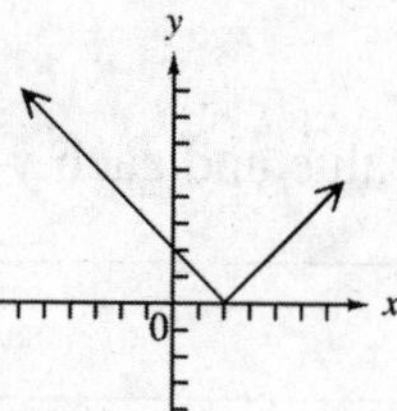

10. **10.** ______________

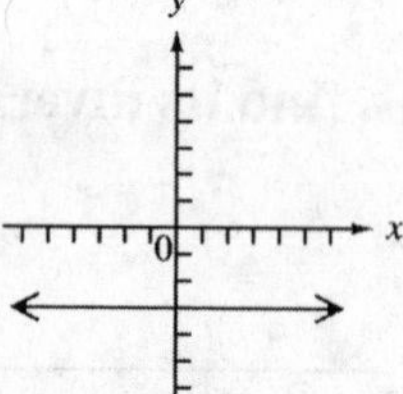

11. **11.** ______________

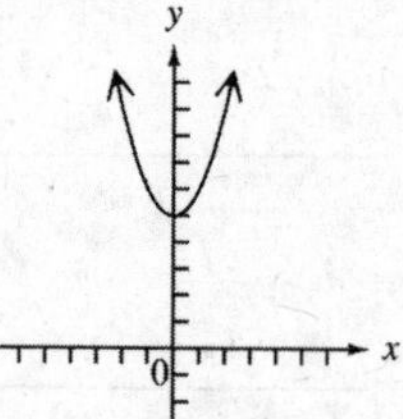

12. **12.** ______________

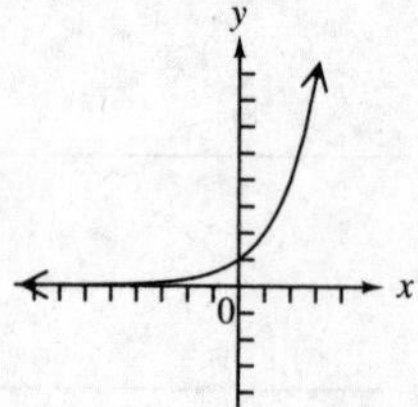

13. **13.** ______________

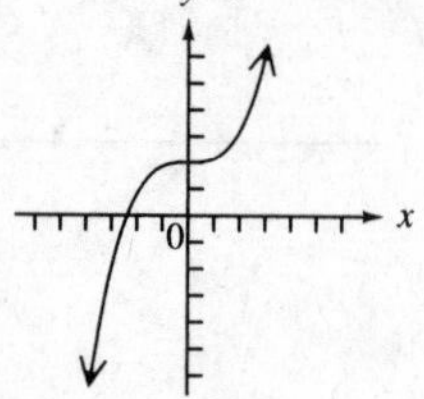

Name: Date:
Instructor: Section:

14. 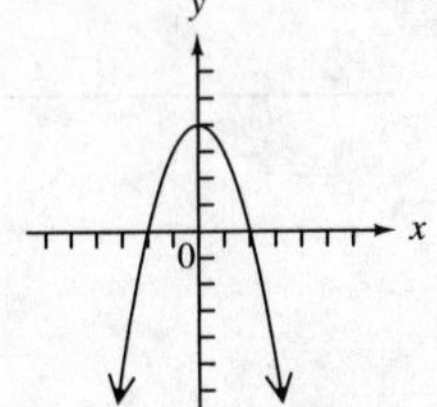

14. ______________

Objective 3 Find the equation of the inverse of a function.

If the function is one-to-one, find its inverse.

15. $f(x) = 3x - 5$

15. ______________

16. $f(x) = 2x^2 + 3$

16. ______________

17. $f(x) = 1 - 2x^2,\ x \leq 0$

17. ______________

18. $f(x) = \sqrt{x-1},\ x \geq 1$

18. ______________

19. $f(x) = 2\sqrt{3x},\ x \geq 0$

19. ______________

20. $f(x) = 2x^3 - 3$

20. ______________

21. $f(x) = \dfrac{x^2 + 3}{2}$

21. ______________

Name: Date:
Instructor: Section:

22. $f(x)=\dfrac{3}{x-1}$ 22. ________

Objective 4 Graph f^{-1} from the graph of f.

If the function is one-to-one, graph the function f and its inverse f^{-1} on the same set of axes.

23. 23. ________

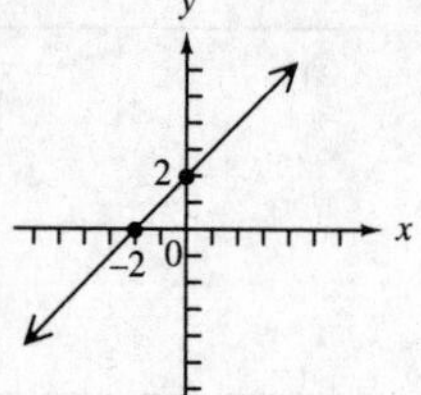

24. 24. ________

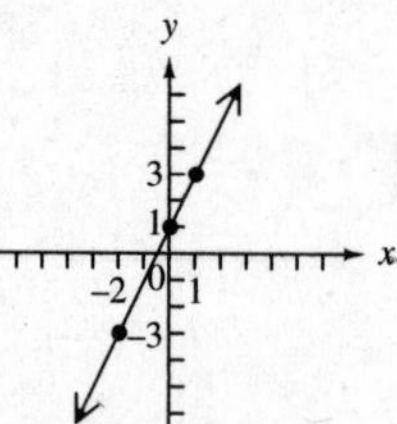

25. 25. ________

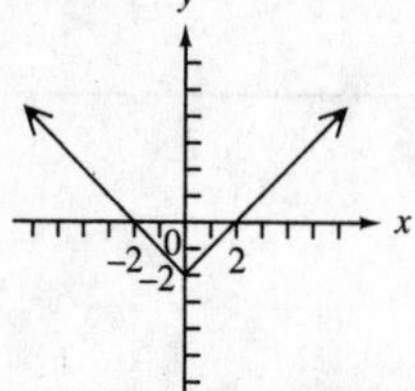

26. 26. ________

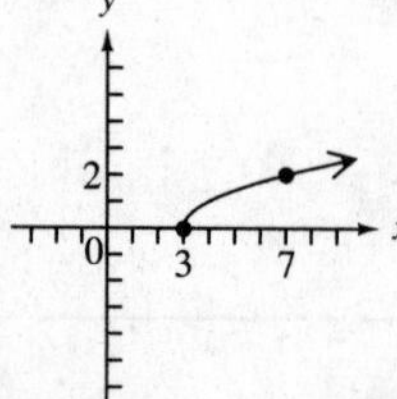

27. 27. ________

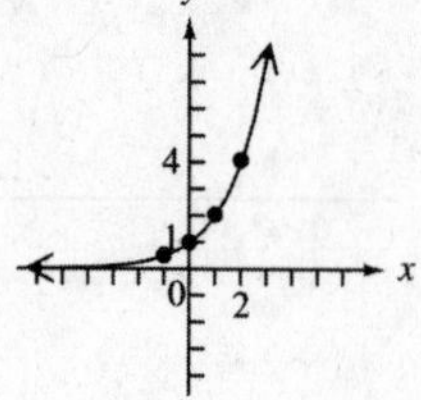

Name: Date:
Instructor: Section:

28.

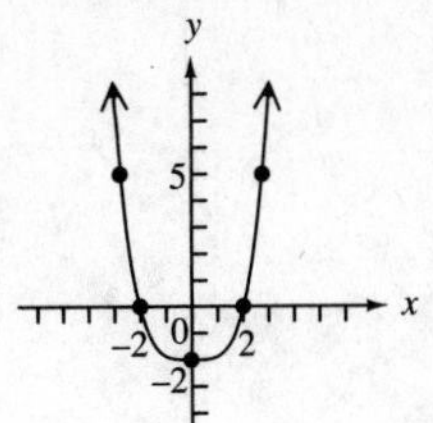

28. ______________

29.

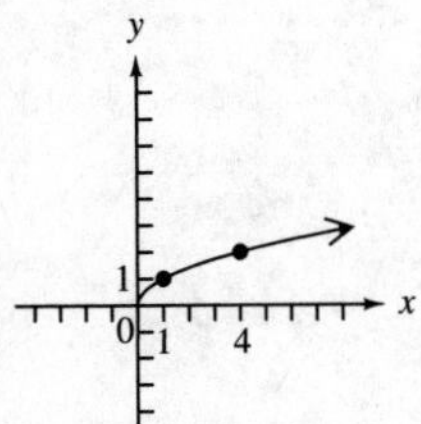

29. ______________

30.

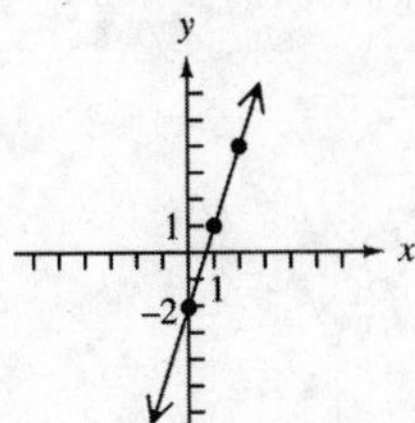

30. ______________

Name: Date:
Instructor: Section:

Chapter 11 INVERSE, EXPONENTIAL, AND LOGARITHMIC FUNCTIONS

11.2 Exponential Functions

Learning Objectives

1. Define exponential functions.
2. Graph exponential functions.
3. Solve exponential equations of the form $a^x = a^k$ for x.
4. Use exponential functions in applications involving growth or decay.

Key Terms

Use the vocabulary terms listed below to complete each statement in exercises 1–2.

exponential equation **inverse**

1. If f is a one-to-one function, then the ______________________ of f is the set of all ordered pairs formed by interchanging the coordinates of the ordered pairs of f.

2. An equation of the form $y = a^x$ for all real numbers x, is an ______________________.

Objective 1 Define exponential functions.

Decide whether or not each function defines an exponential function.

1. $f(x) = 2^x$ **1.** ______________

2. $f(x) = x^2$ **2.** ______________

3. $f(x) = x + 2$ **3.** ______________

4. $f(x) = 3^{2x}$ **4.** ______________

5. $f(x) = 2x^3$ **5.** ______________

Objective 2 Graph exponential functions.

Graph each exponential function.

6. $f(x) = 3^x$

6.

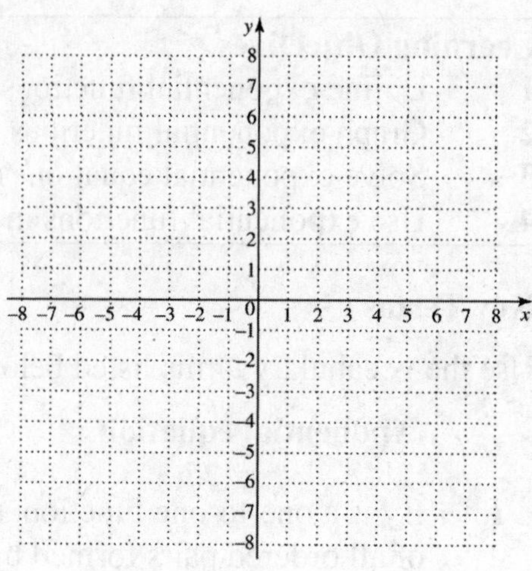

7. $f(x) = -4^x$

7.

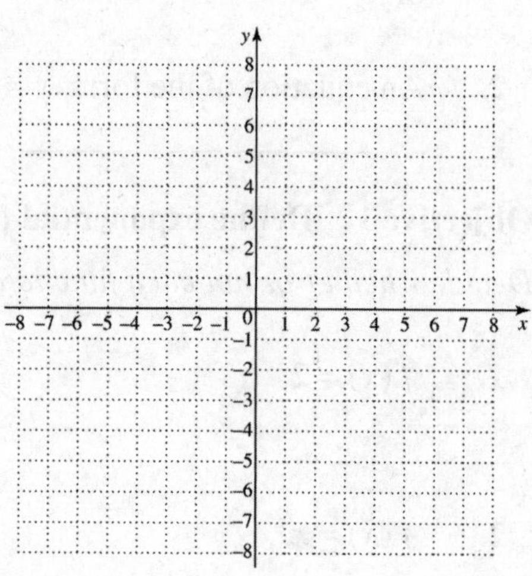

8. $f(x) = 2^{-x}$

8.

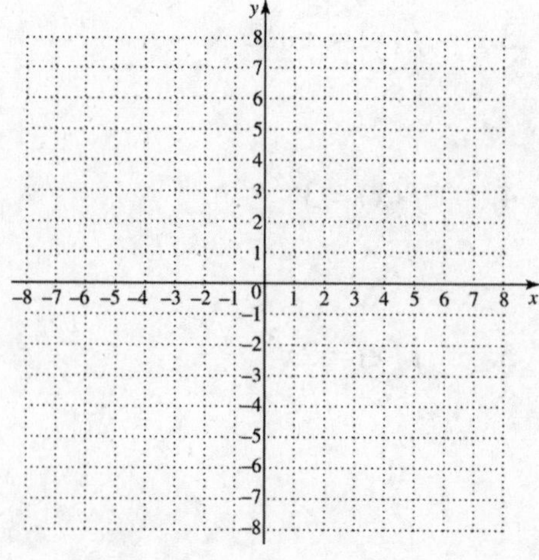

Name: Date:
Instructor: Section:

9. $f(x) = \left(\frac{1}{8}\right)^x$

9.

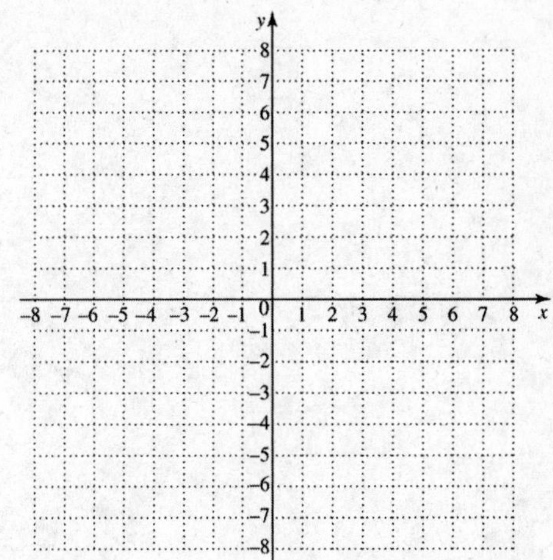

10. $f(x) = 2^{1-x}$

10.

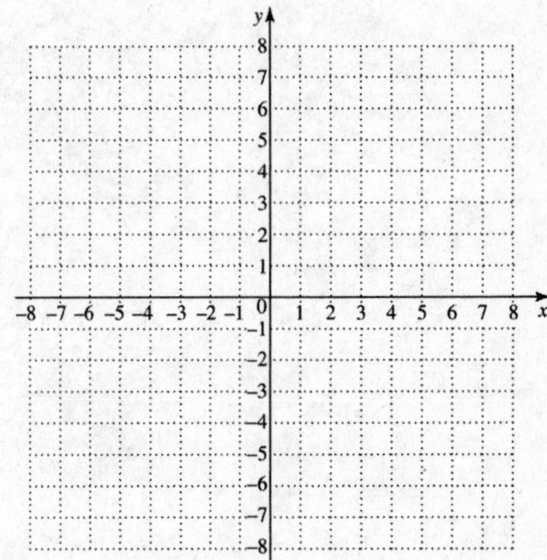

11. $f(x) = -2^{x-2}$

11.

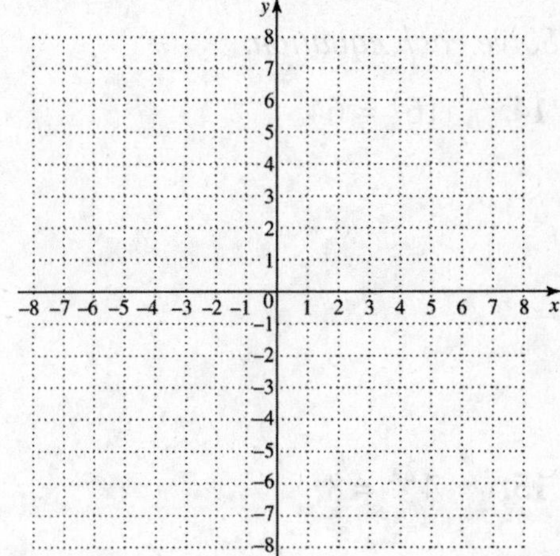

Name: Date:
Instructor: Section:

12. $f(x) = 4^{2x-3}$

12.

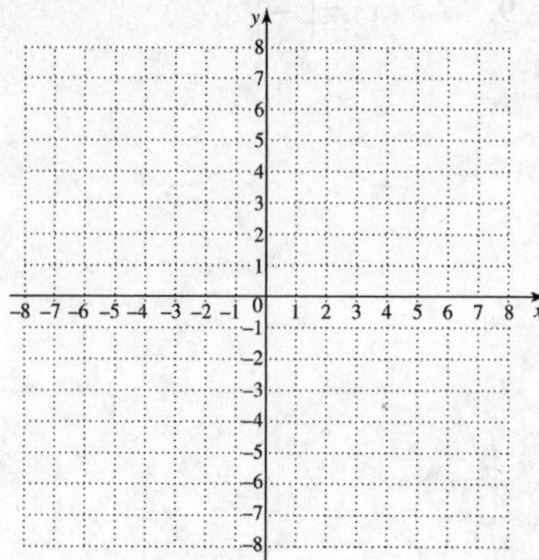

13. $f(x) = -3^{-x}$

13.

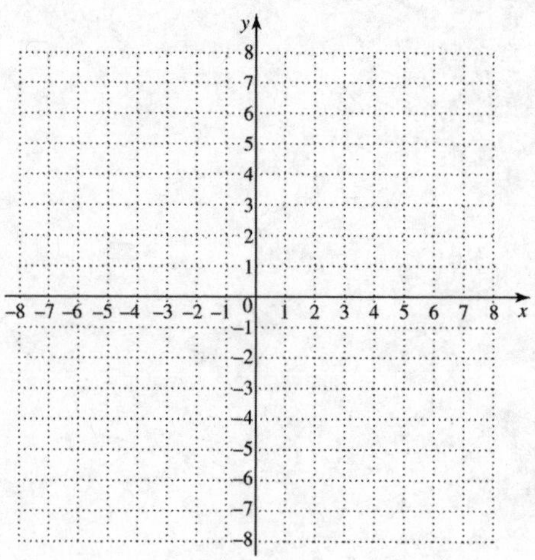

Objective 3 Solve exponential equations of the form $a^x = a^k$ for x.

Solve each equation.

14. $16^x = 64$

14. ______________

15. $4^{2x} = 8$

15. ______________

Name: Date:
Instructor: Section:

16. $4^{k+2} = 32$ **16.** ________

17. $25^{2x-1} = 5$ **17.** ________

18. $25^{1-t} = 5$ **18.** ________

19. $100^{2+t} = 1000$ **19.** ________

20. $16^{-x+1} = 8$ **20.** ________

21. $\left(\frac{3}{4}\right)^x = \frac{16}{9}$ **21.** ________

Name: Date:
Instructor: Section:

Objective 4 Use exponential functions in applications involving growth or decay.

Solve each problem.

22. The population of Canadian geese that spend the summer at Gemini Lake each year has been growing according to the function $f(x) = 56(2)^{.2x}$, where x is the time in years from 1998. Find the number of geese in 2004.

22. 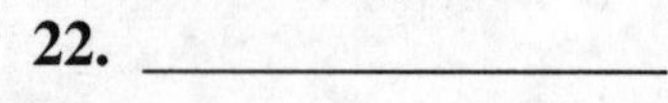

23. A culture of a certain kind of bacteria grows according to the function $f(x) = 3650(2)^{.8x}$, where x is the time in hours after 12 noon. Find the number of bacteria in the culture at 12 noon.

23. 

24. When a bactericide is placed in a certain culture of bacteria, the number of bacteria decreases according to the function $f(x) = 3200(4)^{-.1x}$, where x is the time in hours. Find the number of bacteria in the culture after 20 hours.

24. 

25. Corinna's savings grows according to the function $A(t) = P(1.01)^{4t}$, where P is the amount of her original deposit and t is the number of years since the deposit was made. If $P = \$10{,}000$, how much will she have in 25 years?

25. 

Name: Date:
Instructor: Section:

26. The diameter in inches of a tree during a certain period grew according to the function $f(x) = 2.5(9)^{.05x}$, where x was the number of years after the start of this growth period. Find the diameter of the tree after 10 years.

26. ______________

27. An industrial city in Pennsylvania has found that its population is declining according to the function $f(x) = 70{,}000(2)^{-.01x}$, where x is the time in years from 1900. What is the city's anticipated population in the year 2000?

27. ______________

28. A sample of a radioactive substance with mass in grams decays according to the function $f(x) = 100(10)^{-.2x}$, where x is the time in hours after the original measurement. Find the mass of the substance after 10 hours.

28. ______________

29. Suppose the number of bacteria present in a certain culture after t minutes is given by the function $Q(t) = 500(2)^{.5t}$. Find the number of bacteria present after 2 minutes.

29. ______________

Name: Date:
Instructor: Section:

30. The population of Evergreen Park is now 16,000. The population t years from now is given by the function $P(t) = 16{,}000(2)^{t/10}$. What will the population be 40 years from now?

30. ______________

Name: Date:
Instructor: Section:

Chapter 11 INVERSE, EXPONENTIAL, AND LOGARITHMIC FUNCTIONS

11.3 Logarithmic Functions

Learning Objectives

1. Define a logarithm.
2. Convert between exponential and logarithmic forms.
3. Solve logarithmic equations of the form $\log_a b = k$ for a, b, or k.
4. Define and graph logarithmic functions.
5. Use logarithmic functions in applications involving growth or decay.

Key Terms

Use the vocabulary terms listed below to complete each statement in exercises 1–2.

logarithm **logarithmic equation**

1. The ________________________ of a positive number is the exponent indicating the power to which it is necessary to raise a given number (the base) to give the original number.

2. An equation with a logarithm in at least one term is a ___________________________________.

Objective 1 Define a logarithm.

Simplify. (*Example:* $\log_3 9 = 2$).

1. $\log_8 64$ **1.** ______________

2. $\log_3 \frac{1}{3}$ **2.** ______________

3. $\log_{10} .0001$ **3.** ______________

4. $\log_{1/2} 4$ **4.** ______________

5. $\log_7 \sqrt{7}$ **5.** ______________

Name: Date:
Instructor: Section:

6. $\log_{81} 27$ **6.** ________

Objective 2 Convert between exponential and logarithmic forms.

Write in exponential form.

7. $\log_{10} .001 = -3$ **7.** ________

8. $\log_{16} 2 = \frac{1}{4}$ **8.** ________

9. $\log_4 \frac{1}{16} = -2$ **9.** ________

Write in logarithmic form.

10. $5^{1/3} = \sqrt[3]{5}$ **10.** ________

11. $3^2 = 9$ **11.** ________

12. $2^{-7} = \frac{1}{128}$ **12.** ________

Objective 3 Solve logarithmic equations of the form $\log_a b = k$ for a, b, or k.

Solve each equation.

13. $\log_2 64 = p$ **13.** ________

Name: Date:
Instructor: Section:

14. $x = \log_{32} 8$ **14.** ______________

15. $\log_{1/3} r = -4$ **15.** ______________

16. $\log_5 1 = x$ **16.** ______________

17. $\log_a 4 = \frac{1}{2}$ **17.** ______________

18. $\log_n .01 = -2$ **18.** ______________

Objective 4 Define and graph logarithmic functions.

Graph each logarithmic function.

19. $y = \log_9 x$ **19.** 

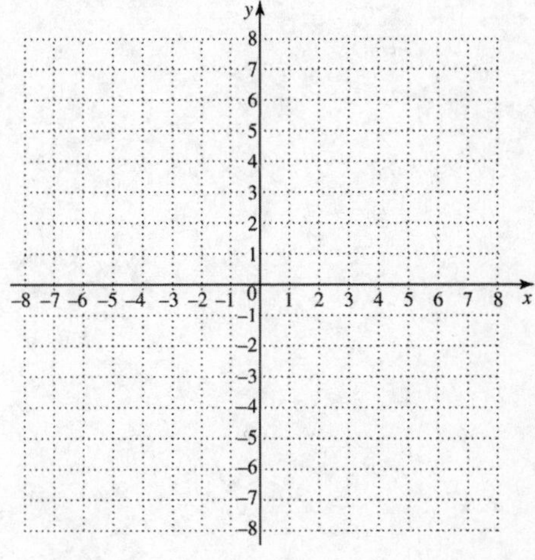

Name: Date:
Instructor: Section:

20. $y = -\log_4 x$

20.

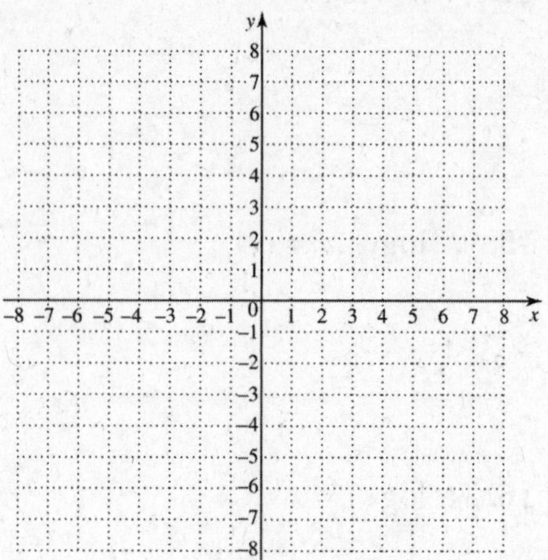

21. $y = \log_{1/2} x$

21.

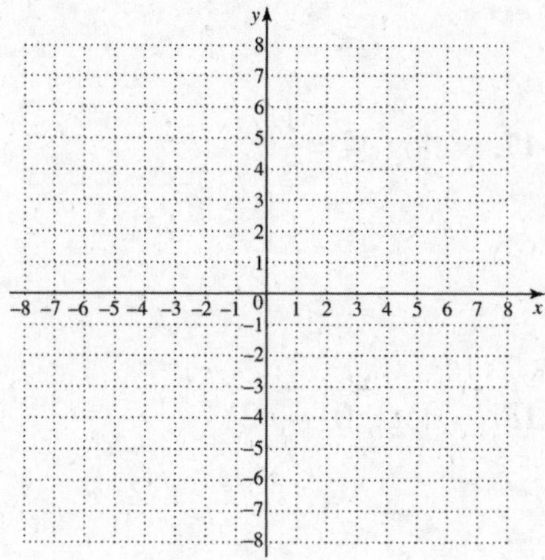

22. $y = -\log_{1/4} x$

22.

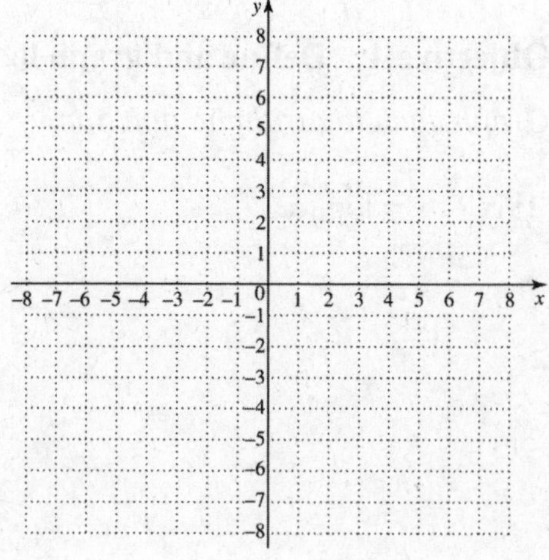

Name: Date:
Instructor: Section:

23. $y = \log_2(-x)$

23.

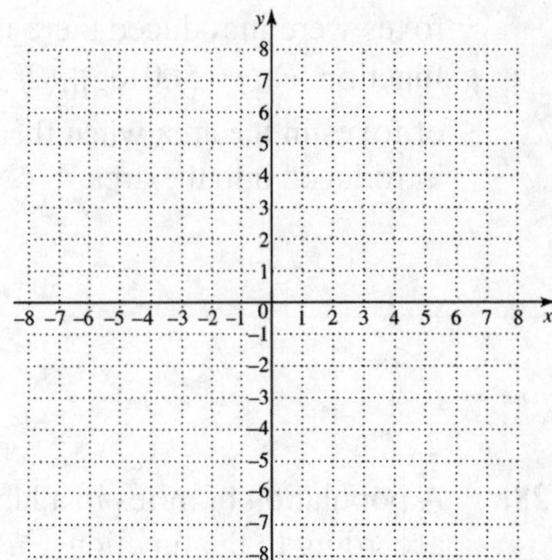

24. $y = \log_3 3x$

24.

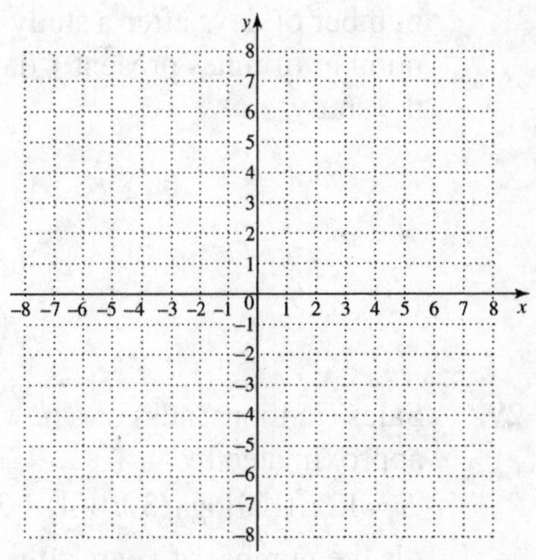

Objective 5 Use logarithmic functions in applications involving growth or decay.

Solve each problem.

25. A company analyst has found that total sales in thousands of dollars after a major advertising campaign are given by $S(x) = 100\log_2(x+2)$, where x is time in weeks after the campaign was introduced. Find the sales when the campaign was introduced.

25. ________________

26. The number of fish in an aquarium is given by the function $f(t) = 8\log_5(2t+5)$, where t is time in months. Find the number of fish present when $t = 10$.

26. ________________

Name: Date:
Instructor: Section:

27. The population of foxes in an area t months after the foxes were introduced there is approximated by the function $F(t) = 500\log_{10}(2t + 3)$. Find the number of foxes in the area when the foxes were first introduced into the area.

27. ______________

28. A population of mites in a laboratory is growing according to the function $p = 50\log_3(20t + 7) - 25\log_9(80t + 1)$, where t is the number of days after a study is begun. Find the number of mites present 1 day after the beginning of the study.

28. ______________

29. Sales (in thousands) of a new product are approximated by $S = 125 + 20\log_2(30t + 4) + 30\log_4(35t - 6)$, where t is the number of years after the product is introduced. Find the total sales 2 years after the product is introduced.

29. ______________

30. A decibel is a measure of the loudness of a sound. A very faint sound is assigned an intensity of I_0, then another sound is given an intensity I found in terms of I_0, the faint sound. The decibel rating of the sound is given in decibels by $d = 10\log_{10}\frac{I}{I_0}$. Find the decibel rating of rock music that has intensity $I = 100{,}000{,}000{,}000I_0$.

30. ______________

Name: Date:
Instructor: Section:

Chapter 11 INVERSE, EXPONENTIAL, AND LOGARITHMIC FUNCTIONS

11.4 Properties of Logarithms

Learning Objectives	
1	Use the product rule for logarithms.
2	Use the quotient rule for logarithms.
3	Use the power rule for logarithms.
4	Use properties to write alternative forms of logarithmic expressions.

Key Terms

Use the vocabulary terms listed below to complete each statement in exercises 1–2.

base **exponent** **logarithmic function**

1. A function defined by $f(x) = \log_a x$, for real numbers a and x positive and $a \neq 0$, is a ______________________.

2. In the expression $\log_2 8 = 3$, "2" is the ______________ and "3" is the ______________.

Objective 1 Use the product rule for logarithms.

Use the product rule to express each logarithm as a sum of logarithms, or as a single number if possible.

1. $\log_3 (6)(5)$ 1. ______________

2. $\log_7 5m$ 2. ______________

3. $\log_2 6xy$ 3. ______________

Use the product rule to express each sum as a single logarithm.

4. $\log_4 7 + \log_4 3$ 4. ______________

5. $\log 4 + \log 3$ 5. ______________

6. $\log_7 11y + \log_7 2y + \log_7 3y$ 6. ______________

7. $\log_9 8r^2 + \log_9 5r^2 + \log_9 3r$ 7. ____________

Objective 2 Use the quotient rule for logarithms.

Use the quotient rule for logarithms to express each logarithm as a difference of logarithms, or as a single number if possible.

8. $\log_2 \frac{8}{m}$ 8. ____________

9. $\log_6 \frac{k}{3}$ 9. ____________

10. $\log_4 \frac{5}{8}$ 10. ____________

Use the quotient rule for logarithms to express each difference as a single logarithm

11. $\log_2 7q^4 - \log_2 5q^2$ 11. ____________

12. $\log 9x^3 - \log 3x^2$ 12. ____________

13. $\log_7 60r^3 - \log_7 100r^7$ 13. ____________

14. $\log_9 40y^5 - \log_9 20y^7$ 14. ____________

Objective 3 Use the power rule for logarithms.

Use the power rule for logarithms to rewrite each logarithm or as a single number if possible.

15. $\log_3 4^3$ 15. ____________

16. $\log_m 2^7$ 16. ____________

Name: Date:
Instructor: Section:

17. $\log_b \sqrt{5}$ 17. ______________

18. $\log_3 \sqrt[3]{7}$ 18. ______________

19. $\log_2 32^{2/5}$ 19. ______________

20. $\log_3 \sqrt{27}$ 20. ______________

21. $\log_5 125^{1/3}$ 21. ______________

Objective 4 Use properties to write alternative forms of logarithmic expressions.

Use the properties of logarithms to express each logarithm as a sum or difference of logarithms, or as a single number if possible.

22. $\log_2 4p^3$ 22. ______________

23. $\log_5 \sqrt[4]{3p}$ 23. ______________

24. $\log_7 \frac{8r^7}{3a^3}$ 24. ______________

25. $\log_4 \frac{3m}{m+2}$ 25. ______________

Name: Date:
Instructor: Section:

Use the properties of logarithms to express each sum or difference of logarithms as a single logarithm, or as a single number if possible.

26. $\log_4 10y + \log_4 3y - \log_4 6y^3$

26. ______________

27. $2\log_5 m + 3\log_5 m^2 - 4\log_5 m^3$

27. ______________

28. $\log_6 14m + \log_6 7m^2 - \log_6 14m^4$

28. ______________

29. $\log_2(x-1) + \log_2(x+1) - \log_2(x^2-1)$

29. ______________

30. $2\log_2 y^2 + \log_2 y - 2\log_2 y^3$

30. ______________

Name: Date:
Instructor: Section:

Chapter 11 INVERSE, EXPONENTIAL, AND LOGARITHMIC FUNCTIONS

11.5 Common and Natural Logarithms

Learning Objectives
1. Evaluate common logarithms using a calculator.
2. Use common logarithms in applications.
3. Evaluate natural logarithms using a calculator.
4. Use natural logarithms in applications.

Key Terms

Use the vocabulary terms listed below to complete each statement in exercises 1–2.

common logarithm **natural logarithm**

1. A logarithm to the base e is a ________________________________.

2. A logarithm to the base 10 is a ________________________________.

Objective 1 Evaluate common logarithms using a calculator.

Use a calculator to find each logarithm. Give an approximation to four decimal places.

1. log 57.23 1. ______________

2. log 8 2. ______________

3. log 0.091419 3. ______________

4. $\log\left(3.184\times10^{-5}\right)$ 4. ______________

5. log 87,123 5. ______________

6. $\log\left(5.14\times10^{3}\right)$ 6. ______________

7. log 4.0014 7. ______________

Name: Date:
Instructor: Section:

Objective 2 Use common logarithms in applications.

Find the pH of solutions with the given hydronium ion concentrations using the formula $\mathrm{pH} = -\log\left[\mathrm{H_3O^+}\right]$, *where* $\left[\mathrm{H_3O^+}\right]$ *is the hydronium ion concentration in moles per liter. Round answers to the nearest tenth.*

8. 5.6×10^{-8} **8.** ________

9. 1.7×10^{-9} **9.** ________

10. 7.4×10^{-11} **10.** ________

Find the hydronium ion concentration of solutions with the given pH values.

11. 10.2 **11.** ________

12. 6.5 **12.** ________

13. 1.3 **13.** ________

Use the formula $D = 10\log\left(\frac{I}{I_0}\right)$ *to find the loudness of sound in decibels given intensity I. Round to the nearest whole number.*

14. $I = 5.012\times10^{10} I_0$ **14.** ________

15. $I = 3.16\times10^{8} I_0$ **15.** ________

Name: Date:
Instructor: Section:

Objective 3 Evaluate natural logarithms using a calculator.

Find each natural logarithm. Give an approximation to four decimal places.

16. ln 100 **16.** ______________

17. ln 76.3 **17.** ______________

18. ln 4.102 **18.** ______________

19. ln 874 **19.** ______________

20. ln 0.000806 **20.** ______________

21. ln 50 **21.** ______________

22. ln 0.00214 **22.** ______________

Objective 4 Use natural logarithms in applications.

The time t in years for an amount increasing at a rate of r (in decimal form) to double (the doubling time) is given by $t = \frac{\ln 2}{\ln(1+r)}$. *Find the doubling time for an investment at each interest rate. Round to the nearest whole number.*

23. 3% **23.** ______________

24. 4.5% **24.** ______________

25. 6.25% **25.** ______________

Name: Date:
Instructor: Section:

The half-life of a radioactive substance is the time it takes for half of the material to decay. The amount A in pounds of substance remaining after t years is given by $\ln\frac{A}{C} = -\frac{t}{h}\ln 2$, *where C is the initial amount in pounds, and h is its half-life in years. Use the formula to solve the following problems. Round to the nearest whole number.*

26. The half-life of radium-226 is 1620 years. How long, to the nearest year, will it take for 100 pounds to decay to 25 pounds?

26. ______________

27. The half-life of carbon-14 is 5730 years. How long, to the nearest year, will it take for 157 pounds to decay to 123 pounds?

27. ______________

28. The half-life of strontium-90 is 28.1 years. How long, to the nearest year, will it take for 200 pounds to decay to 150 pounds?

28. ______________

Name: Date:
Instructor: Section:

Newton's Law of Cooling describes the cooling of a warmer object to the cooler temperature of the surrounding environment. The formula can be given as $t = \frac{1}{k}\ln\frac{T_s - T_1}{T_s - T_2}$, *where t is the elapsed time,* T_1 *is the initial temperature measurement of the object,* T_2 *is the second temperature measurement of the object, and* T_s *is the temperature of the surrounding environment. Use this formula to solve the following problems. Round to the nearest tenth.*

29. A corpse was discovered in a motel room at midnight and its temperature was 80°F. The temperature in the room was 60°F. Assuming that the person's temperature at the time of death was 98.6° F and using $k = 0.1438$, determine t and the time of death.

29. t __________

time __________

30. A corpse was found lying on a street at 3:00 A.M. Its temperature was 85.7°F. The air temperature was 55°F. Assuming that the person's temperature at the time of death was 98.6° F and using $k = 0.0596$, determine t and the time of death.

30. t __________

time __________

Name: Date:
Instructor: Section:

Chapter 11 INVERSE, EXPONENTIAL, AND LOGARITHMIC FUNCTIONS

11.6 Exponential and Logarithmic Equations; Further Applications

Learning Objectives

1 Solve equations involving variables in the exponents.
2 Solve equations involving logarithms.
3 Solve applications of compound interest.
4 Solve applications involving base *e* exponential growth and decay.
5 Use the change-of-base rule.

Key Terms

Use the vocabulary terms listed below to complete each statement in exercises 1–3.

product rule for logarithms **quotient rule for logarithms**

power rule for logarithms

1. An example of the ______________________________
is $\log_4 8 - \log_4 5 = \log_4 \frac{8}{5}$.

2. An example of the ______________________________
is $\log_4 8 + \log_4 5 = \log_4 (8 \cdot 5)$.

3. An example of the ______________________________
is $2\log_4 8 = \log_4 8^2$.

Objective 1 Solve equations involving variables in the exponents.

Solve each equation. Give solutions to three decimal places.

1. $3^{-q} = 2$ **1.** ______________

2. $5^{2k-1} = 17$ **2.** ______________

3. $25^{x+2} = 125^{3-x}$ **3.** ______________

Name: Date:
Instructor: Section:

4. $4^{x-1} = 3^{2x}$ 4. ________

5. $e^{2x} = e^{5x-3}$ 5. ________

6. $e^{0.005x} = 9$ 6. ________

Objective 2 Solve equations involving logarithms.

Solve each equation. Give the exact solution.

7. $\log_t 100 = \frac{2}{3}$ 7. ________

8. $\log_2 x + \log_2(3x - 1) = 1$ 8. ________

9. $\log_3(x - 1) + \log_3 x = \log_3 6$ 9. ________

10. $\log(-y) + \log 4 = \log(2y + 5)$ 10. ________

Name: Date:
Instructor: Section:

11. $\log_3 a = \log_3(a-1)+2$ **11.** ________

12. $\log_4 u = 1 - \log_4(u+3)$ **12.** ________

Objective 3 Solve applications of compound interest.

Find how long each investment will take to double if interest is compounded annually. Use $A = P\left(1+\frac{r}{n}\right)^{nt}$ *, where A is the doubled amount, P is the initial amount, r is the interest rate, n is the number of compounding periods per year, and t is the time in years. Round to the nearest tenth.*

13. \$25,000 at 10%, compounded annually **13.** ________

14. \$5600 at 8%, compounded annually **14.** ________

Find how long each investment will take to double if interest is compounded continuously. Use $A = Pe^{rt}$ *, where A is the doubled amount, P is the initial amount, r is the interest rate, and t is the time in years. Round to the nearest tenth.*

15. \$2700 at 9% **15.** ________

Name: Date:
Instructor: Section:

16. $3950 at 7% **16.** ______________

Use the compound interest formulas given on the previous page to solve the following problems.

17. How much money will be in an account if $32,800 is invested for 7 years at 11% compounded annually? **17.** ______________

18. How much money will be in an account if $32,800 is invested for 7 years at 11% compounded continuously? **18.** ______________

Objective 4 Solve applications involving base *e* exponential growth and decay.

Solve each problem.

19. Radioactive strontium decays according to the function $y = y_0 e^{-0.0239t}$, where *t* is the time in years. If an initial sample contains $y_0 = 15$ g of radioactive strontium, how many grams will be present after 25 years? Round to the nearest hundredth of a gram. **19.** ______________

Name: Date:
Instructor: Section:

20. How long will it take the initial sample of strontium in exercise 19 to decay to half of its original amount? **20.** ____________

21. Lead-210 decays according to the function $y = y_0 e^{-0.032t}$, where t is the time in years. If an initial sample contains $y_0 = 500$ g of lead-210, how much lead will be left in the sample after 20 years? Round to the nearest gram. **21.** ____________

22. How long will it take the initial sample of lead in exercise 21 to decay to half of its original amount? **22.** ____________

23. Cesium-137, a radioactive isotope used in radiation therapy, decays according to the function $y = y_0 e^{-0.0231t}$, where t is the time in years. If an initial sample contains $y_0 = 36$ mg of cesium-137, how much cesium will be left in the sample after 50 years? Round to the nearest mg. **23.** ____________

Name: Date:
Instructor: Section:

24. How long will it take the initial sample of cesium in exercise 23 to decay to half of its original amount?

24. ______________

Objective 5 Use the change-of-base rule.

Use the change-of-base rule to find each logarithm. Give approximations to four decimal places.

25. $\log_4 8$

25. ______________

26. $\log_4 0.25$

26. ______________

27. $\log_{.5} 75$

27. ______________

28. $\log_{\sqrt{3}} 9$

28. ______________

29. $\log_{2/3} 5$

29. ______________

30. $\log_{1/3} \frac{1}{6}$

30. ______________

Name: Date:
Instructor: Section:

Chapter 12 NONLINEAR FUNCTIONS, CONIC SECTIONS, AND NONLINEAR SYSTEMS

12.1 Additional Graphs of Functions; Operations and Composition

Learning Objectives

1 Recognize the graphs of the elementary functions defined by $|x|, \frac{1}{x}$, and $\sqrt{x}$, and graph their translations.
2 Recognize and graph step functions.
3 Perform operations on functions.
4 Find the composition of functions.

Key Terms

Use the vocabulary terms listed below to complete each statement in exercises 1–4.

asymptotes **greatest integer function** **step function**

composition (composite function)

1. A ______________________________ is a function that looks like a series of steps.

2. If *f* and *g* are functions, then the ______________________________ of *g* and *f* is defined by $(g \circ f)(x) = g(f(x))$ for all *x* in the domain of *f* such that $f(x)$ is in the domain of *g*. The function $g(f(x))$ is a ______________________________.

3. The function defined by $f(x) = [\![x]\!]$ is called the ______________________________.

4. Lines that a graph approaches without actually touching are called ____________________.

Name:
Instructor:

Date:
Section:

Objective 1 Recognize the graphs of the elementary functions defined by $|x|$, $\frac{1}{x}$, and $\sqrt{x}$, and graph their translations.

Graph each function. Give the domain and range.

1. $f(x) = |x-2| + 3$

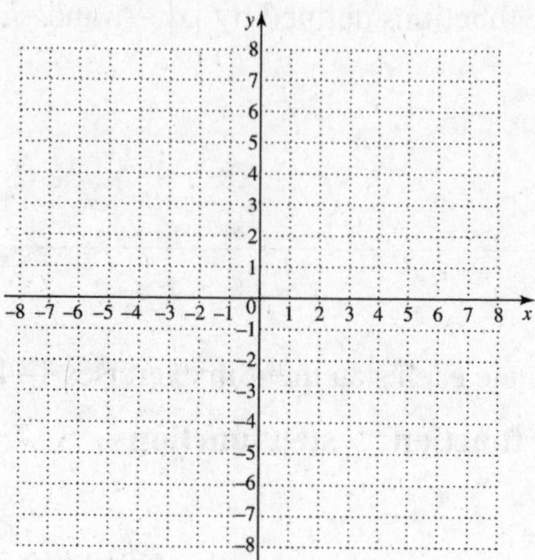

1. domain ________

range ________

2. $f(x) = \sqrt{x+3}$

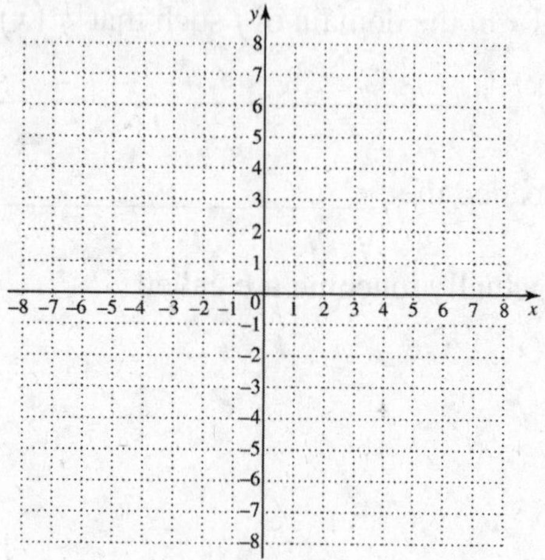

2. domain ________

range ________

Name: Date:
Instructor: Section:

3. $f(x) = \dfrac{1}{x-1}$

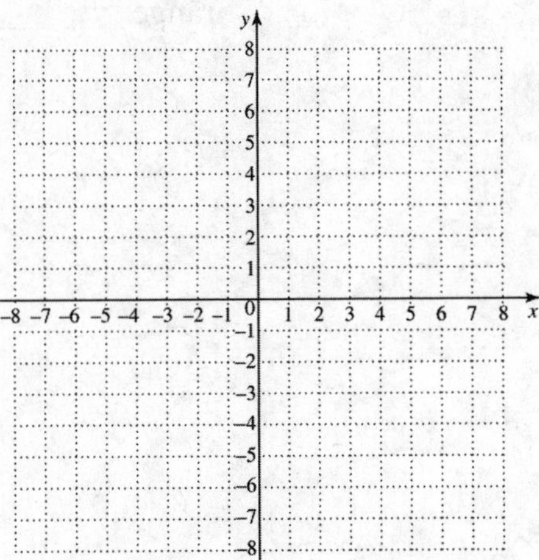

3. domain ____________

range ____________

4. $f(x) = -|x+3| - 2$

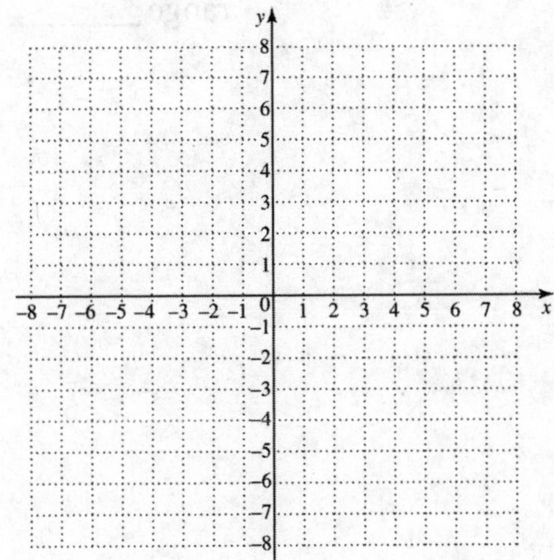

4. domain ____________

range ____________

5. $f(x) = \dfrac{1}{x} + 3$

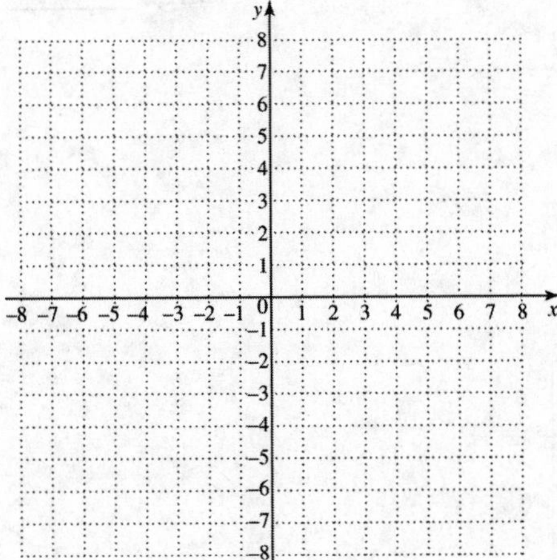

5. domain ____________

range ____________

Name: Date:
Instructor: Section:

6. $f(x) = \sqrt{x} + 3$

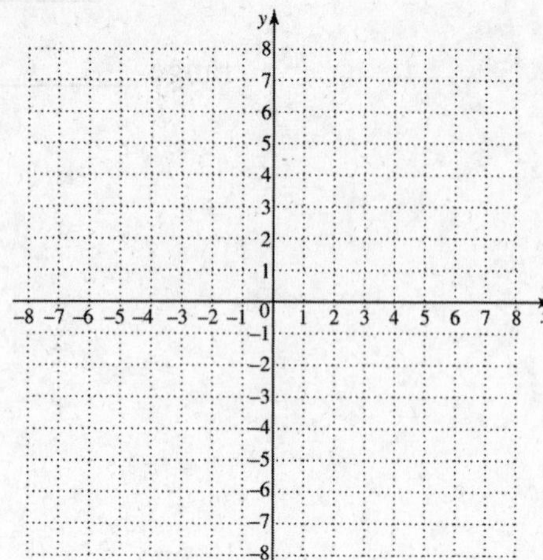

6. domain __________

range __________

7. $f(x) = \dfrac{1}{x-3}$

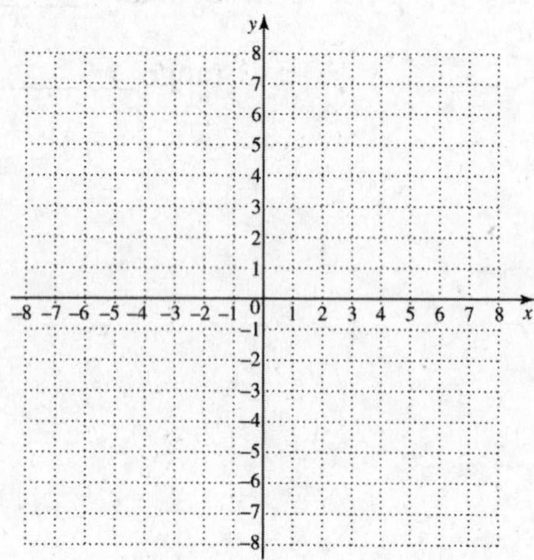

7. domain __________

range __________

8. $f(x) = -\sqrt{1-x^2}$

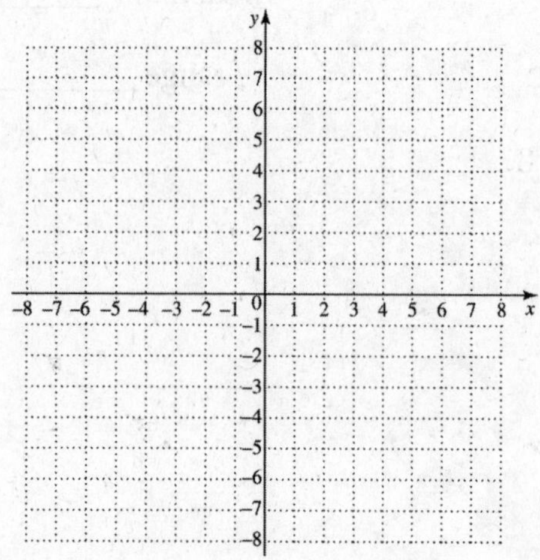

8. domain __________

range __________

Name: Date:
Instructor: Section:

Objective 2 Recognize and graph step functions.

Graph each function.

9. $f(x) = -[\![x]\!]$

9.

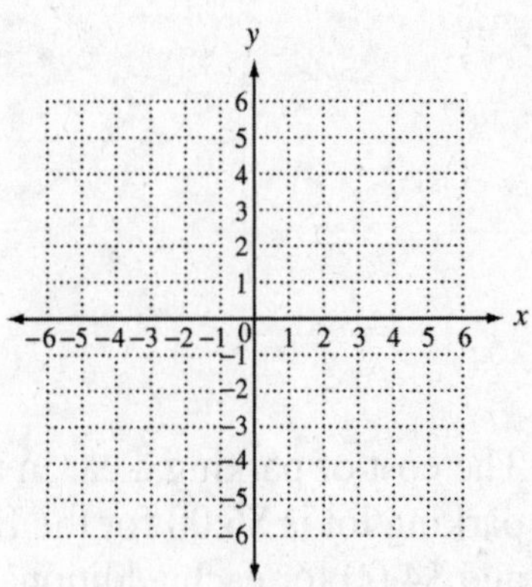

10. $f(x) = -[\![x]\!] + 2$

10.

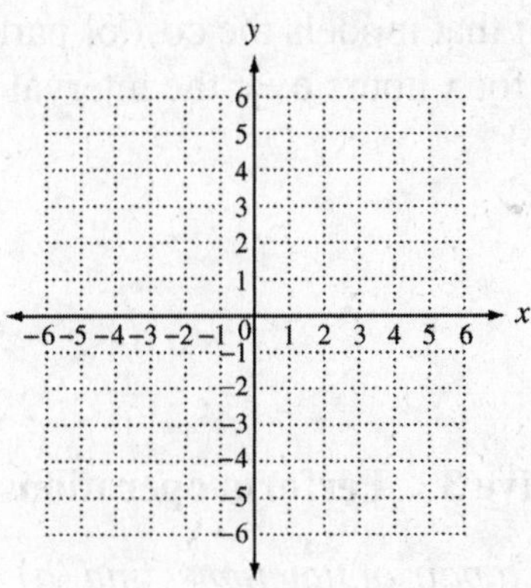

11. $f(x) = -[\![x+2]\!]$

11.

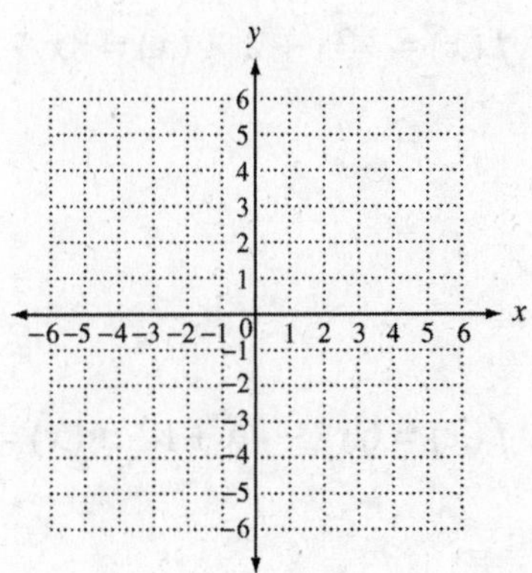

Name: Date:
Instructor: Section:

12. $f(x) = [\![x+2]\!] - 3$

12.

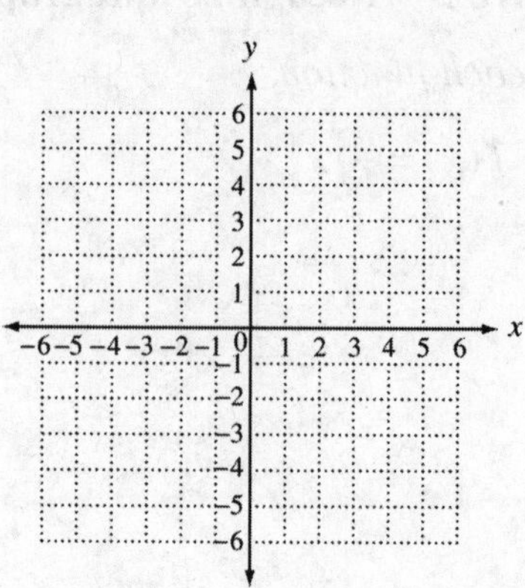

13. The cost of parking a car in an hourly parking lot is \$6.00 for the first hour and \$4.00 for each additional hour or fraction of an hour. Graph the function *f* that models the cost of parking a car for *x* hours over the interval (0, 6].

13.

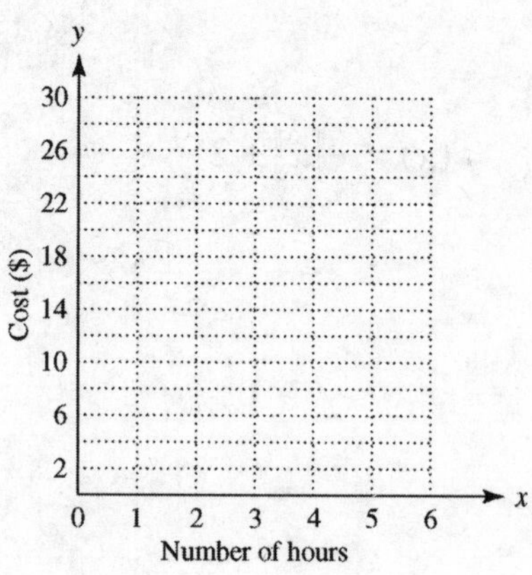

Objective 3 Perform operations on functions.

For each pair of functions, find (a) $(f+g)(x)$ *and* $(f-g)(x)$.

14. $f(x) = -3x+2,\ g(x) = 8x+1$

14. a.________________

b.________________

15. $f(x) = 6x^2 - 7x + 12,\ g(x) = -3x^2 + x + 9$

15. a.________________

b.________________

Name: Date:
Instructor: Section:

Let $f(x) = 3x^2 + 2, g(x) = -5x,$ *and* $h(x) = x + 2$. *Find each of the following.*

16. $(f - g)(-1)$ **16.** ______________

17. $(g - h)\left(\frac{1}{2}\right)$ **17.** ______________

18. $(fg)(-1)$ **18.** ______________

19. $(fh)(2)$ **19.** ______________

Let $f(x) = 4x^2 - 81, g(x) = 3x,$ *and* $h(x) = 2x + 9$. *Find each of the following. Give any x-values that are not in the domain of the quotient function.*

20. $\left(\frac{f}{h}\right)(x)$ **20.** ______________

Name: Date:
Instructor: Section:

21. $\left(\frac{f}{g}\right)(0)$ 21. ____________

Objective 4 Find the composition of functions.

Let $f(x) = x^2 + 3$, $g(x) = 3x + 2$, *and* $h(x) = x + 4$. *Find each composite function.*

22. $(f \circ g)(1)$ 22. ____________

23. $(g \circ f)(3)$ 23. ____________

24. $(h \circ f)(4)$ 24. ____________

25. $(f \circ g)(x)$ 25. ____________

Name: Date:
Instructor: Section:

26. $(g \circ h)(x)$ **26.** ______________

27. $(f \circ h)(x)$ **27.** ______________

Let $f(x) = x - 3$ *and* $g(x) = x^2 + 6$. *Find each composite function.*

28. $(g \circ f)(x)$ **28.** ______________

29. $(f \circ g)(2)$ **29.** ______________

30. $(f \circ g)\left(x^2\right)$ **30.** ______________

Name: Date:
Instructor: Section:

Chapter 12 NONLINEAR FUNCTIONS, CONIC SECTIONS, AND NONLINEAR SYSTEMS

12.2 The Circle and the Ellipse

Learning Objectives

1. Find an equation of a circle given the center and radius.
2. Determine the center and radius of a circle given its equation.
3. Recognize the equation of an ellipse.
4. Graph ellipses.

Key Terms

Use the vocabulary terms listed below to complete each statement in exercises 1–5.

conic sections **circle** **center** **radius** **ellipse**

1. A(n) ______________________ is the set of all points in a plane that lie a fixed distance from a fixed point.

2. A(n) ______________________ is the set of all points in a plane the sum of whose distances from two fixed points is constant.

3. Figures that result from the intersection of an infinite cone with a plane are called ______________________.

4. A fixed point such that every point on a circle is a fixed distance from it is the ______________________.

5. The distance from the center of a circle to a point on the circle is called the ______________________.

Objective 1 Find an equation of a circle given the center and radius.

Find the equation of a circle satisfying the given conditions.

1. center: (–3, 2); radius: 5 — 1. ______________

2. center: (1, 4); radius: 2 — 2. ______________

3. center: (3, –4); radius: 5 — 3. ______________

Name: Date:
Instructor: Section:

4. center: (–2, –2); radius: 3 **4.** ____________

5. center: (0, 3); radius: $\sqrt{2}$ **5.** ____________

6. center: (–2, –4); radius: $2\sqrt{5}$ **6.** ____________

7. center: (7, 1); radius: 2 **7.** ____________

Objective 2 Determine the center and radius of a circle given its equation.

Find the center and radius of each circle.

8. $x^2 + y^2 + 4x + 6y - 3 = 0$ **8.** center: ____________

radius: ____________

9. $x^2 + y^2 - 10x + 12y + 52 = 0$ **9.** center: ____________

radius: ____________

10. $x^2 + y^2 - 8x - 2y + 15 = 0$ **10.** center: ____________

radius: ____________

11. $4x^2 + 4y^2 - 24x + 16y + 43 = 0$ **11.** center: ____________

radius: ____________

Name: Date:
Instructor: Section:

12. $3x^2 + 3y^2 + 12y + 30x = 21$

12. center: ___________

radius: ___________

13. $x^2 + y^2 + 8x + 4y - 29 = 0$

13. center: ___________

radius: ___________

Find the center and radius of each circle. Then graph each circle.

14. $x^2 + y^2 - 4x + 8y + 11 = 0$

14. center: ___________

radius: ___________

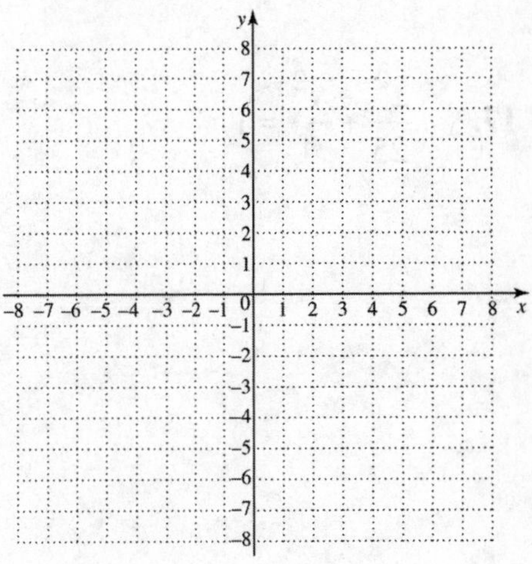

15. $x^2 + y^2 + 6x - 4y + 12 = 0$

15. center: ___________

radius: ___________

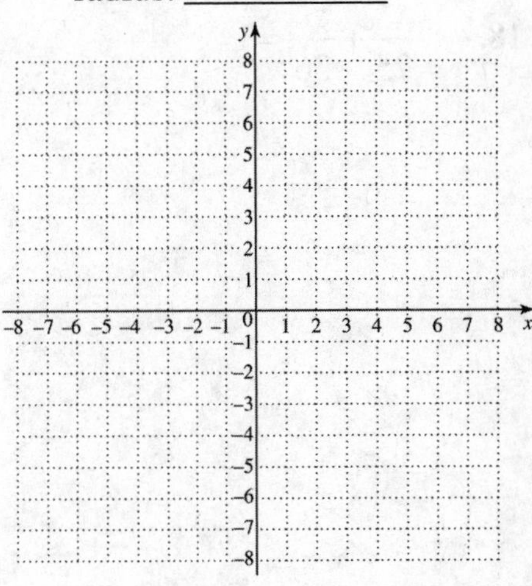

Name: Date:
Instructor: Section:

Objective 3 Recognize the equation of an ellipse.
Objective 4 Graph ellipses.

Graph each ellipse.

16. $\frac{x^2}{9}+\frac{y^2}{49}=1$

16.

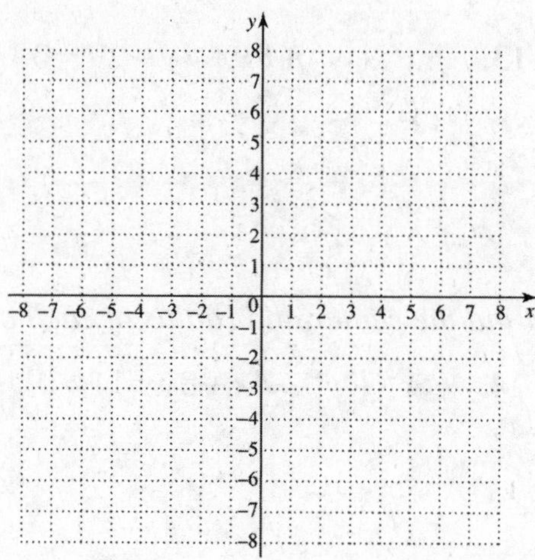

17. $\frac{x^2}{25}+\frac{y^2}{4}=1$

17.

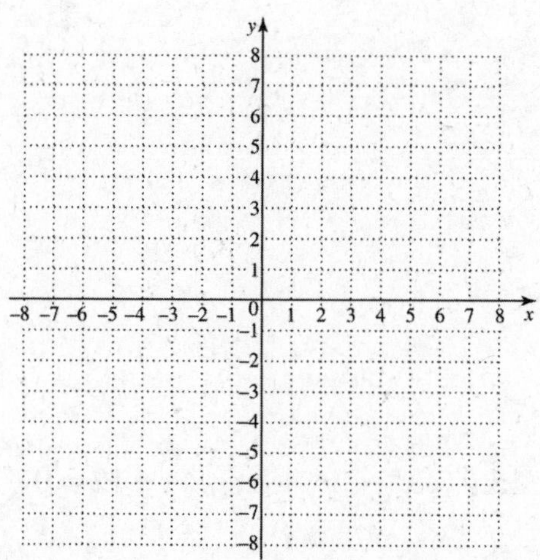

18. $\frac{x^2}{25}+\frac{y^2}{36}=1$

18.

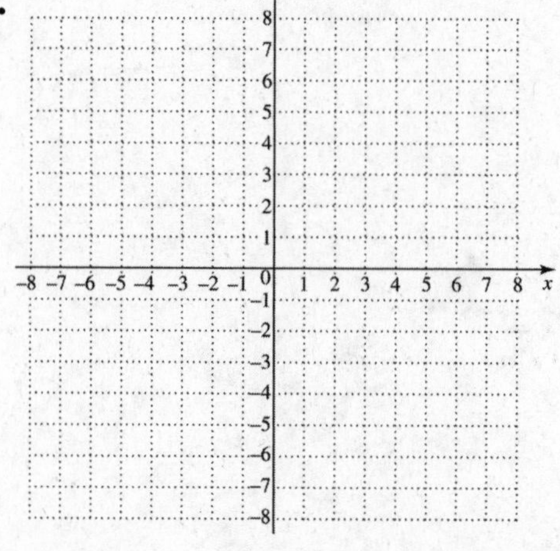

Name: Date:
Instructor: Section:

19. $\dfrac{x^2}{4}+\dfrac{y^2}{9}=1$

19.

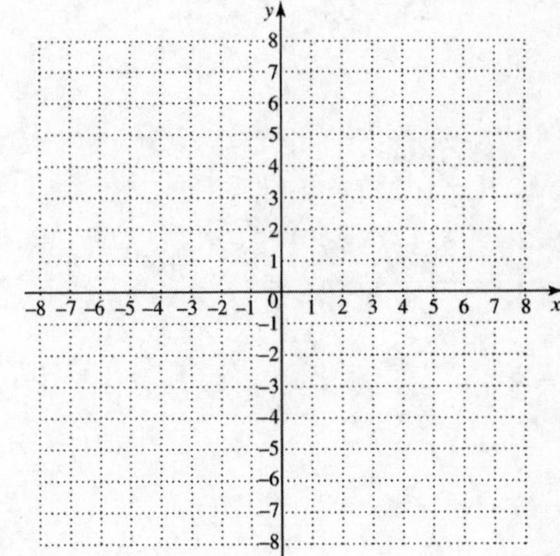

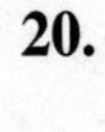

20. $\dfrac{x^2}{16}+\dfrac{y^2}{25}=1$

20.

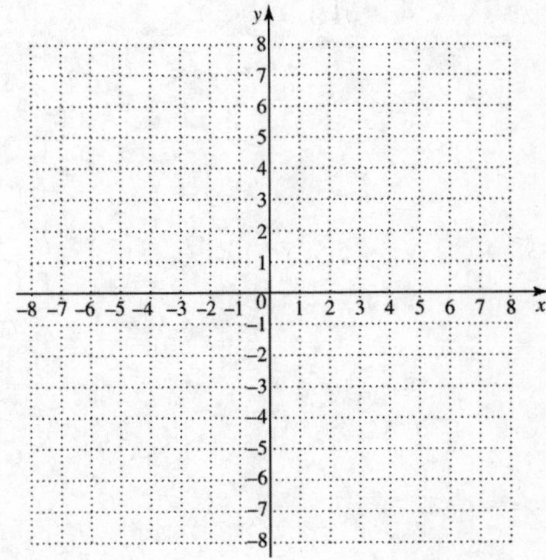

21. $\dfrac{x^2}{36}+\dfrac{y^2}{9}=1$

21.

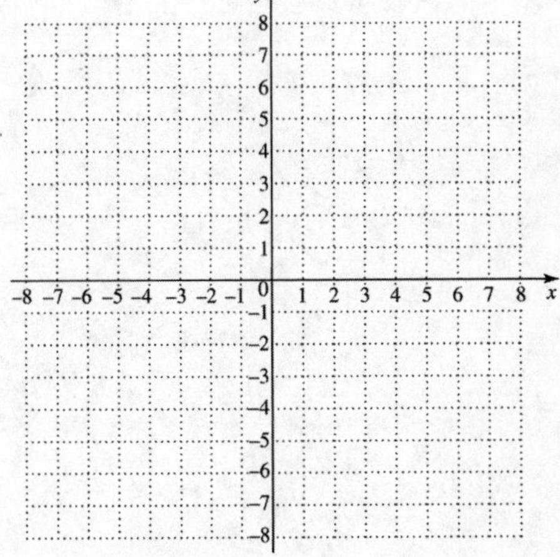

Name: Date:
Instructor: Section:

22. $\frac{x^2}{25}+\frac{y^2}{64}=1$

22.

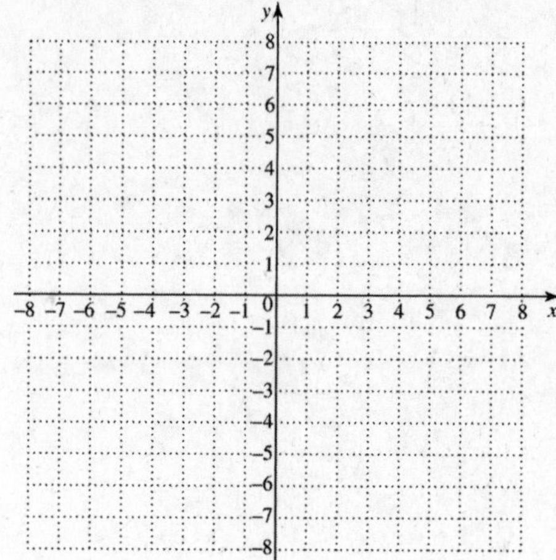

23. $\frac{x^2}{4}+\frac{y^2}{16}=1$

23.

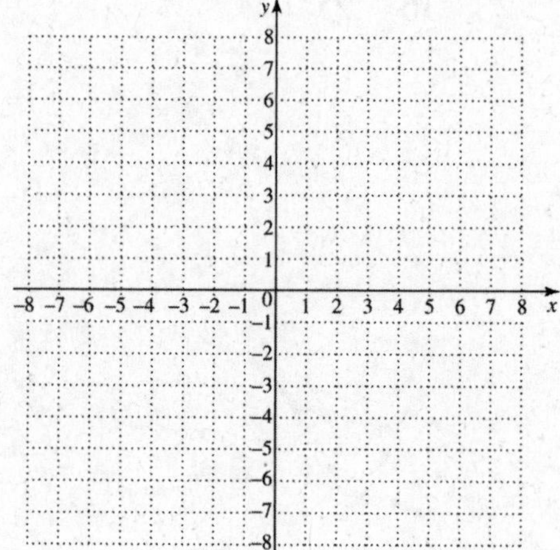

24. $\frac{x^2}{16}+\frac{y^2}{49}=1$

24.

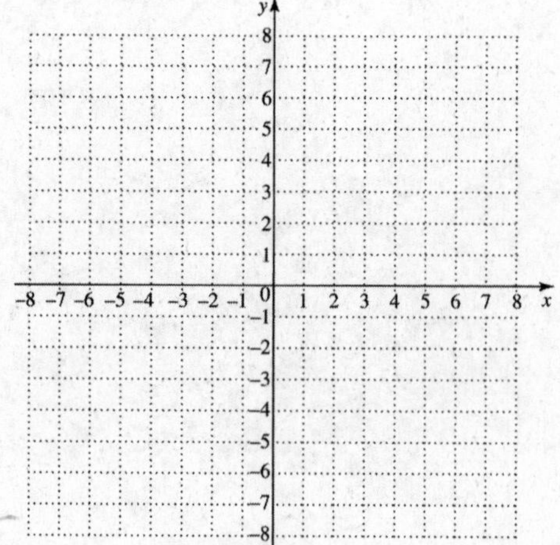

Name: Date:
Instructor: Section:

25. $\frac{x^2}{25}+\frac{y^2}{81}=1$

25.

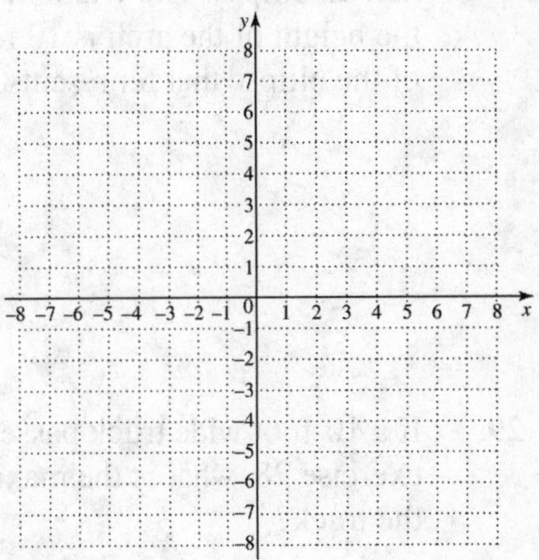

Solve.

26. An elliptical arch under a bridge has the shape of half an ellipse. The equation of the ellipse is $625x^2 + 900y^2 = 562{,}500$, where x and y are in feet.

(a) How high is the center of the arch?

(b) How wide is the arch across the bottom?

26. (a) ____________

(b) ____________

27. The playing field in Australian Rules Football is an ellipse that must be between 135 m and 185 m long and between 110 m and 155 m wide. An equation of the ellipse the represents a team's playing field is $\frac{x^2}{8464}+\frac{y^2}{4900}=1$, where x and y are in meters.

(a) What are the dimensions of the field?

(b) Does the team's field meet the regulations?

27. (a) ____________

(b) ____________

Name: Date:
Instructor: Section:

28. An elliptical arch under a bridge has the shape of half an ellipse. The width of the arch is 16 feet and the height of the arch is 10 feet. What is the equation of the ellipse that represents the arch?

28. 

29. If a 10-foot wide truck passes under the arch in exercise 28, what is the maximum possible height of the truck?

29. 

30. The orbit of Mercury is an ellipse with the sun at one focus. For x and y in millions of miles, the equation of the orbit is $\frac{x^2}{36^2}+\frac{y^2}{35.3^2}=1$. The minimum distance between Mercury and the sun is about 29 million miles. What is the maximum distance between Mercury and the sun? Use the diagram below to help you solve the problem. A represents the position of Mercury farthest from the sun, and B represents the position of Mercury closest to the sun.

30. ____________

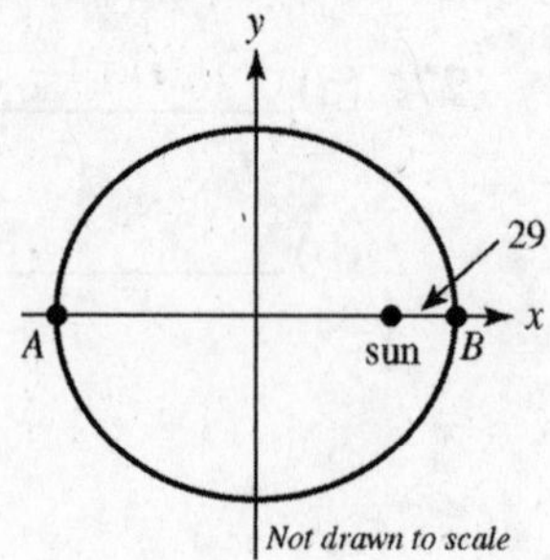

Name: Date:
Instructor: Section:

Chapter 12 NONLINEAR FUNCTIONS, CONIC SECTIONS, AND NONLINEAR SYSTEMS

12.3 The Hyperbola and Other Functions Defined by Radicals

Learning Objectives

1. Recognize the equation of a hyperbola.
2. Graph hyperbolas by using asymptotes.
3. Identify conic sections by their equations.
4. Graph certain square root functions.

Key Terms

Use the vocabulary terms listed below to complete each statement in exercises 1–3.

hyperbola **asymptotes of a hyperbola**

fundamental rectangle

1. A(n) __________________________ is the set of all points in a plane such that the absolute value of the difference of the distances from two fixed points is constant.

2. The two intersecting lines that the branches of a hyperbola approach are the____________________________________.

3. The asympotes of a hyperbola are the extended diagonals of its _________________________________.

Objective 1 Recognize the equation of a hyperbola.
Objective 2 Graph hyperbolas by using asymptotes.

Graph each hyperbola.

1. $\frac{x^2}{9} - \frac{y^2}{16} = 1$

1.

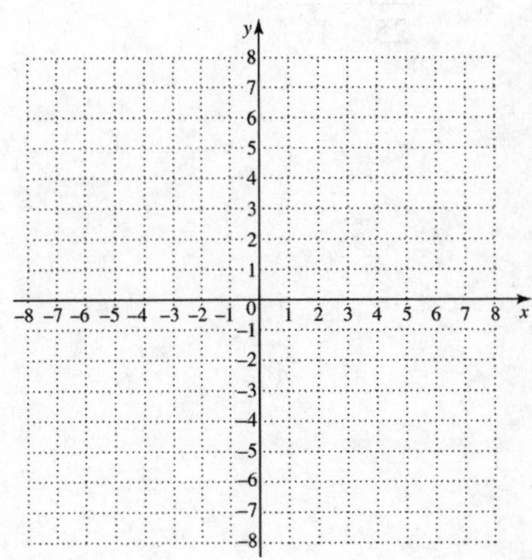

Name: Date:
Instructor: Section:

2. $\dfrac{x^2}{25} - \dfrac{y^2}{9} = 1$

2.

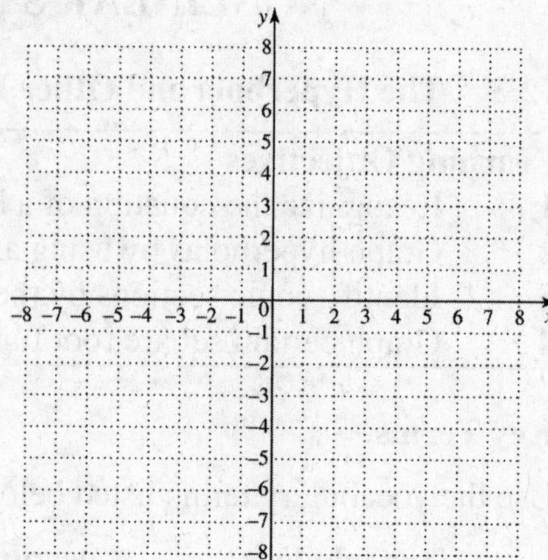

3. $\dfrac{y^2}{4} - \dfrac{x^2}{9} = 1$

3.

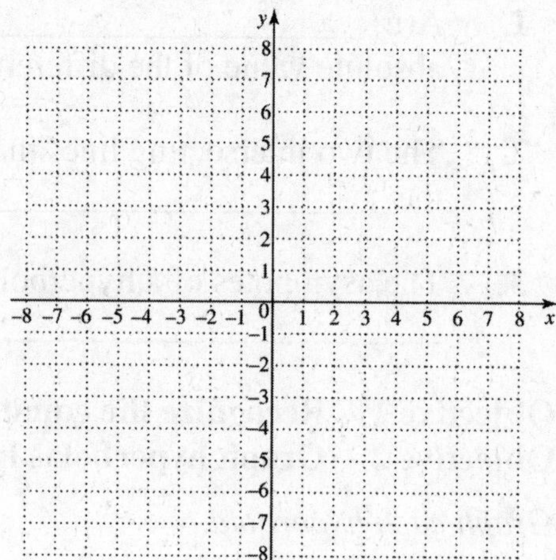

4. $\dfrac{y^2}{25} - \dfrac{x^2}{16} = 1$

4.

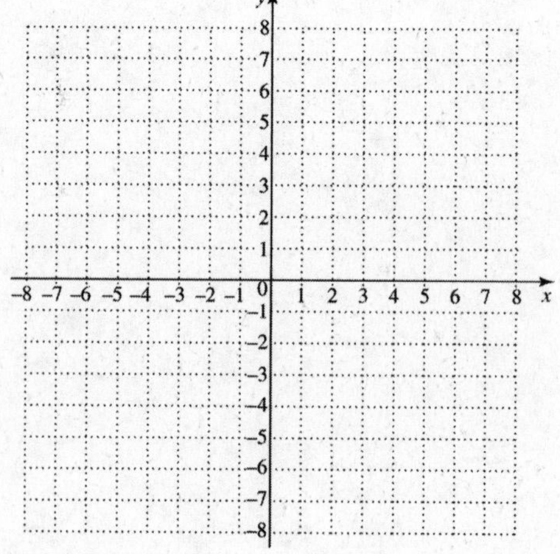

Name: Date:
Instructor: Section:

5. $\frac{x^2}{36} - \frac{y^2}{49} = 1$

5.

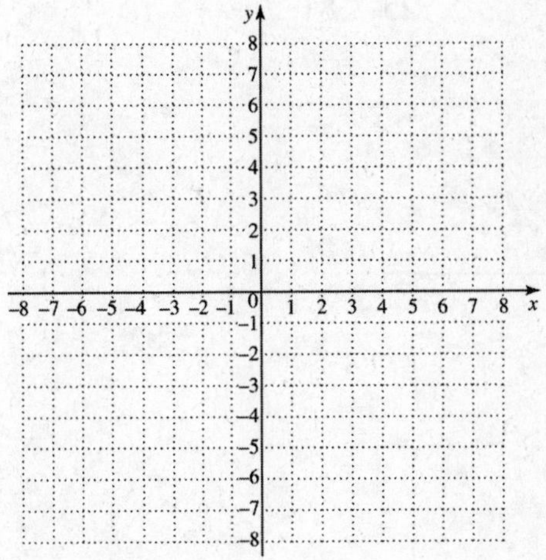

6. $\frac{y^2}{4} - \frac{x^2}{4} = 1$

6.

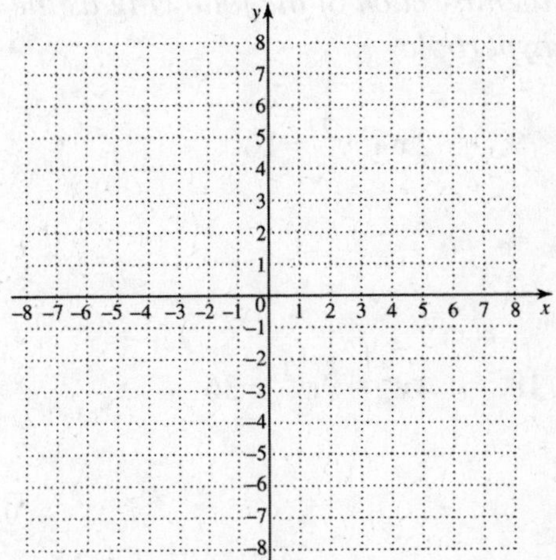

7. $\frac{x^2}{25} - \frac{y^2}{4} = 1$

7.

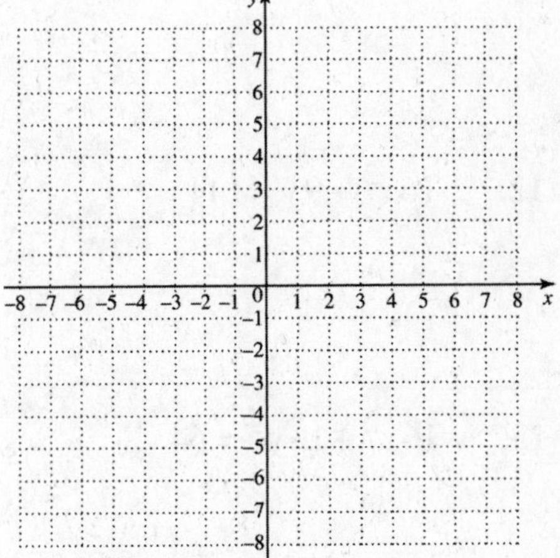

Name: Date:
Instructor: Section:

8. $\dfrac{x^2}{25}-\dfrac{y^2}{81}=1$

8.

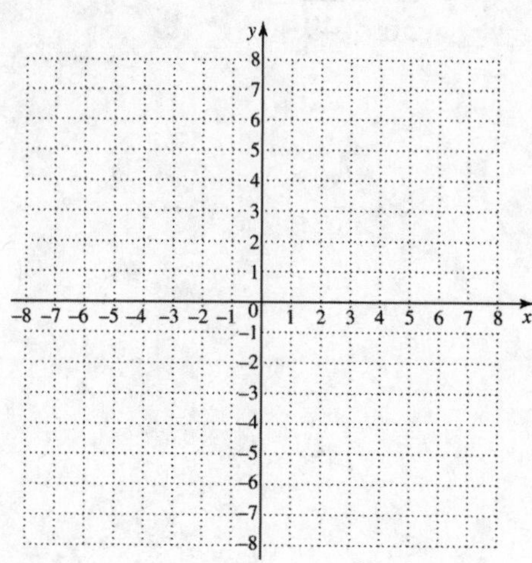

Objective 3 Identify conic sections by their equations.

Identify each of the following as the equation of a parabola, a circle, an ellipse, or a hyperbola.

9. $2x+y^2=16$ **9.** ______________

10. $4x^2-9y^2=36$ **10.** ______________

11. $25y^2+100=4x^2$ **11.** ______________

12. $16x^2+9y=144$ **12.** ______________

13. $16x^2+16y^2=64$ **13.** ______________

Name: Date:
Instructor: Section:

14. $2x^2 + 4y^2 = 8$ **14.** ______________

15. $5x^2 = 25 - 5y^2$ **15.** ______________

16. $y^2 = 36 - 36x^2$ **16.** ______________

17. $3x^2 - 3y = 9$ **17.** ______________

18. $3x^2 + 3y^2 = 1$ **18.** ______________

19. $x^2 = 49 - y^2$ **19.** ______________

20. $x^2 = 16 - y$ **20.** ______________

21. $3x^2 = 3y^2 + 1$ **21.** ______________

22. $x^2 = y^2 + 16$ **22.** ______________

Name: Date:
Instructor: Section:

Objective 4 Graph certain square root functions.

Graph each function. Give the domain and range.

23. $f(x)=\sqrt{36-x^2}$

23.

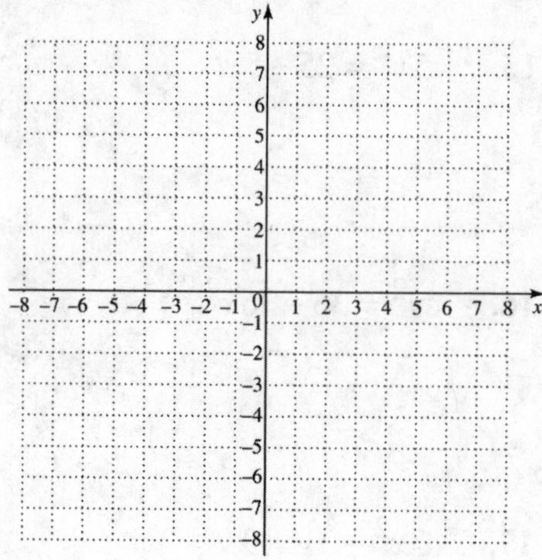

24. $f(x)=\sqrt{25-x^2}$

24.

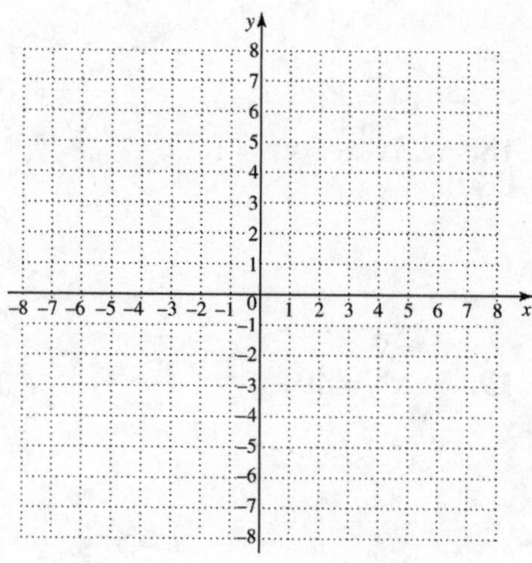

25. $f(x)=-\sqrt{4-x^2}$

25.

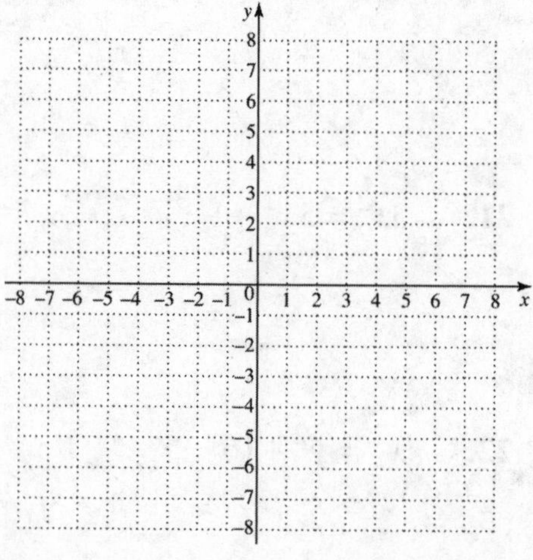

Name: Date:
Instructor: Section:

26. $f(x) = -\sqrt{9 - x^2}$

26.

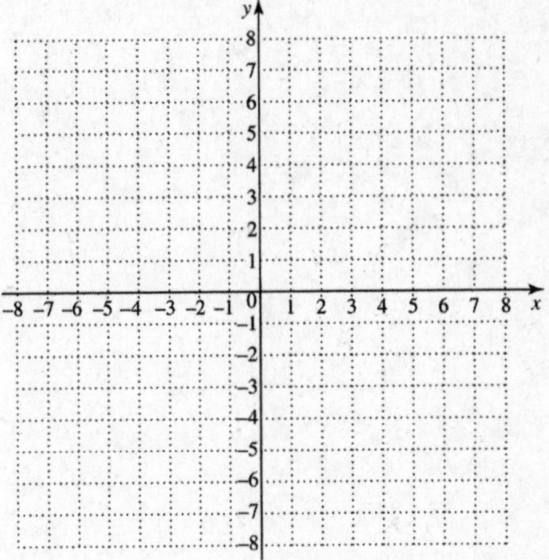

27. $f(x) = \sqrt{1 + \frac{x^2}{4}}$

27.

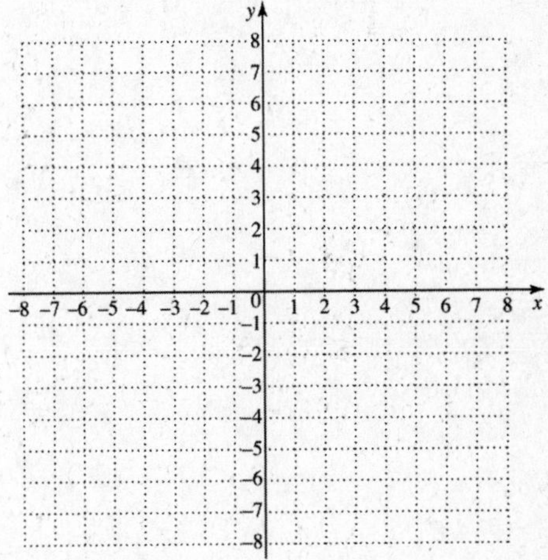

28. $f(x) = -3\sqrt{1 + \frac{x^2}{25}}$

28.

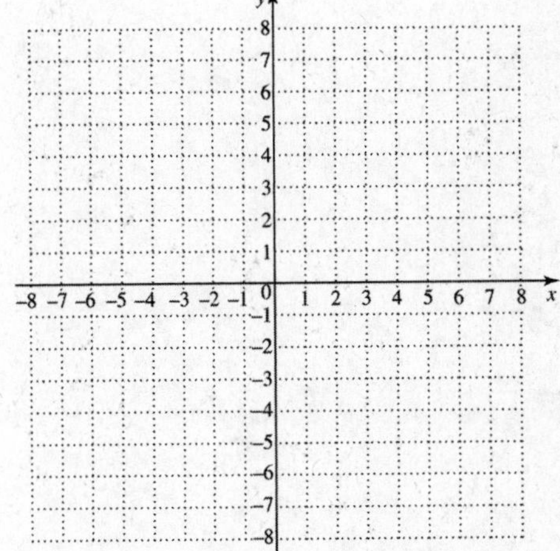

Name: Date:
Instructor: Section:

29. $f(x) = \sqrt{9 - 9x^2}$

29.

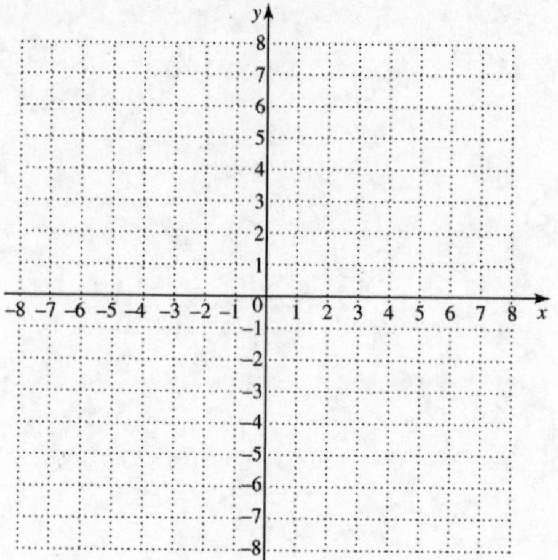

30. $f(x) = -5\sqrt{1 - \frac{x^2}{9}}$

30.

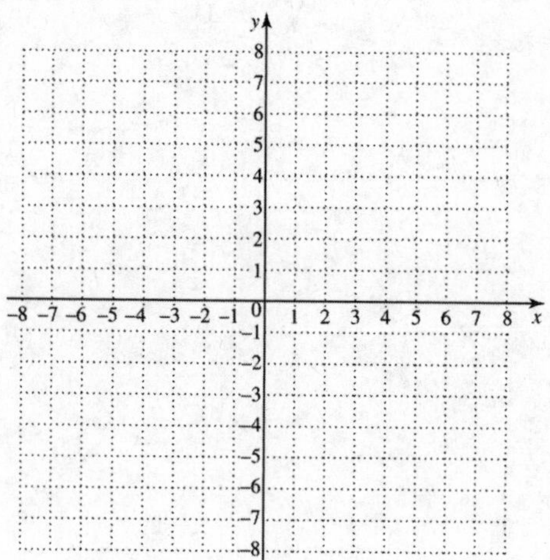

Name: Date:
Instructor: Section:

Chapter 12 NONLINEAR FUNCTIONS, CONIC SECTIONS, AND NONLINEAR SYSTEMS

12.4 Nonlinear Systems of Equations

Learning Objectives

1 Solve a nonlinear system by substitution.
2 Use the elimination method to solve a system with two second-degree equations.
3 Solve a system that requires a combination of methods.

Key Terms

Use the vocabulary terms listed below to complete each statement in exercises 1–2.

nonlinear equation **nonlinear system of equations**

1. An equation in which some terms have more than one variable or a variable of degree 2 or greater is a ______________________________.

2. A system with at least one nonlinear equation is a ______________________________.

Objective 1 Solve a nonlinear system by substitution.

Solve each system by the substitution method.

1. $x^2 + y^2 = 17$
   $2x = y + 9$

   1. ______________

2. $2x^2 - y^2 = -1$
   $2x + y = 7$

   2. ______________

3. $4x^2 + 3y^2 = 7$
   $2x - 5y = -7$

   3. ______________

Name:
Instructor:

Date:
Section:

4. $x^2 = 2y^2 + 2$
$y = 3x + 7$

4. ______________

5. $y = x^2 - 3x - 8$
$x = y + 3$

5. ______________

6. $x = y^2 + 5y$
$3y = x$

6. ______________

7. $xy = -6$
$x + y = 1$

7. ______________

8. $x^2 - xy + y^2 = 6$
$x + y = 0$

8. ______________

9. $xy = -10$
$2x - y = 9$

9. ______________

Name: Date:
Instructor: Section:

10. $x^2 + 3y^2 = 3$
$x = 3y$

10. ____________

Objective 2 Use the elimination method to solve a system with two second-degree equations.

Solve each system using the elimination method

11. $x^2 + y^2 = 10$
$2x^2 - y^2 = -7$

11. ____________

12. $2x^2 + y^2 = 54$
$x^2 - 3y^2 = 13$

12. ____________

13. $2x^2 - 3y^2 = -19$
$4x^2 + y^2 = 25$

13. ____________

14. $x^2 - y^2 = 3$
$2x^2 + y^2 = 9$

14. ____________

Name: Date:
Instructor: Section:

15. $x^2 + 2y^2 = 11$
$2x^2 - y^2 = 17$

15. ________________

16. $3x^2 + 2y^2 = 30$
$2x^2 + y^2 = 17$

16. ________________

17. $5x^2 + y^2 = 6$
$2x^2 - 3y^2 = -1$

17. ________________

18. $4x^2 - 3y^2 = -8$
$2x^2 + y^2 = 5$

18. ________________

19. $3x^2 - 3y^2 = 9$
$4x^2 + y^2 = 17$

19. ________________

20. $3x^2 - 2y^2 = 12$
$x^2 + 3y^2 = 4$

20. ________________

Name: Date:
Instructor: Section:

Objective 3 Solve a system that requires a combination of methods.

Solve each system.

21. $x^2 + xy + y^2 = 43$
$x^2 + 2xy + y^2 = 49$

21. ________________

22. $x^2 + xy - y^2 = 5$
$-x^2 + xy + y^2 = -1$

22. ________________

23. $x^2 + 2xy + 3y^2 = 6$
$x^2 + 4xy + 3y^2 = 8$

23. ________________

24. $2x^2 + 3xy - 2y^2 = 50$
$x^2 - 4xy - y^2 = -41$

24. ________________

Name: Date:
Instructor: Section:

25. $$\begin{aligned} 4x^2 - 2xy + 4y^2 &= 64 \\ x^2 + y^2 &= 13 \end{aligned}$$

25. ______________

26. $$\begin{aligned} 5x^2 - xy + 5y^2 &= 89 \\ x^2 + y^2 &= 17 \end{aligned}$$

26. ______________

27. $$\begin{aligned} 3x^2 - 4xy + 2y^2 &= 59 \\ -3x^2 + 5xy - 2y^2 &= -65 \end{aligned}$$

27. ______________

28. $$\begin{aligned} x^2 + 3xy + 2y^2 &= 12 \\ -x^2 + 8xy - 2y^2 &= 10 \end{aligned}$$

28. ______________

Name: Date:
Instructor: Section:

29. $x^2 + 5xy - y^2 = 20$

$x^2 - 2xy - y^2 = -8$

29. ______________

30. $2x^2 + xy + y^2 = 16$

$-2x^2 + 3xy - y^2 = -28$

30. ______________

Name: Date:
Instructor: Section:

Chapter 12 NONLINEAR FUNCTIONS, CONIC SECTIONS, AND NONLINEAR SYSTEMS

12.5 Second-Degree Inequalities and Systems of Inequalities

Learning Objectives
1 Graph second-degree inequalities.
2 Graph the solution set of a system of inequalities.

Key Terms

Use the vocabulary terms listed below to complete each statement in exercises 1–2.

second-degree inequality **system of inequalities**

1. A ________________________________ consists of two or more inequalities to be solved at the same time.

2. A(n)____ is an inequality with at least one variable of degree 2 and no variable with degree greater than 2.

Objective 1 Graph second-degree inequalities.

Graph each inequality.

1. $x \geq y^2$

1.

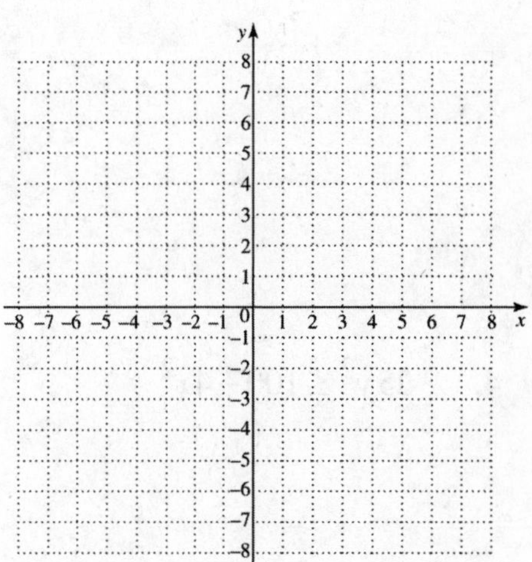

Name: Date:
Instructor: Section:

2. $y^2 \geq 9 - x^2$

2.

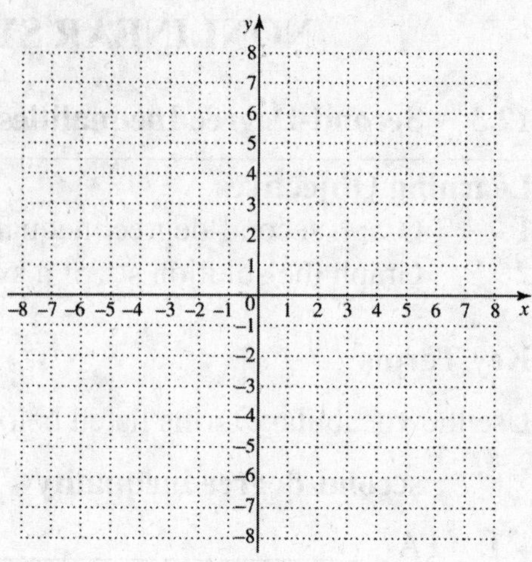

3. $16x^2 < 9y^2 + 144$

3.

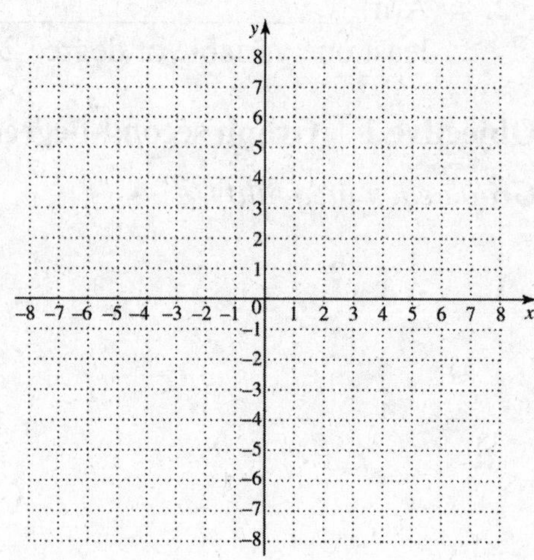

4. $25y^2 \leq 100 - 4x^2$

4.

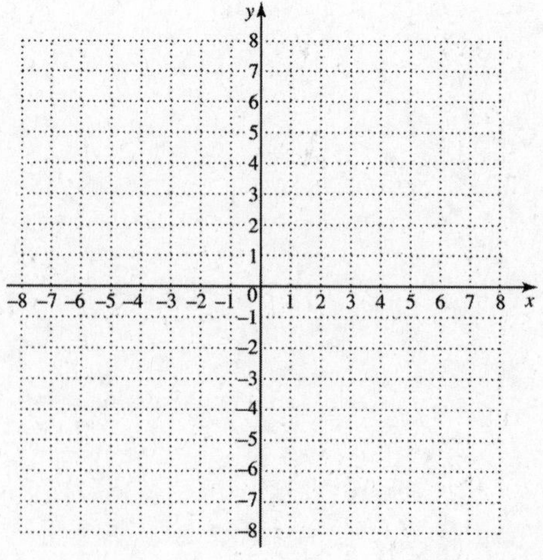

Name: Date:
Instructor: Section:

5. $x^2 + 4y^2 > 4$

5.

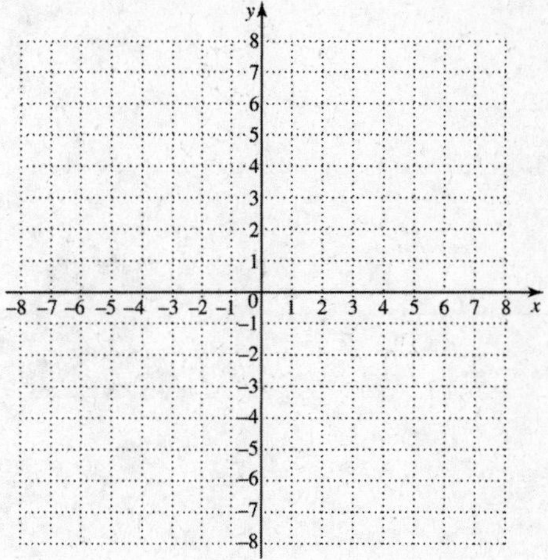

6. $y \geq x^2 - 4$

6.

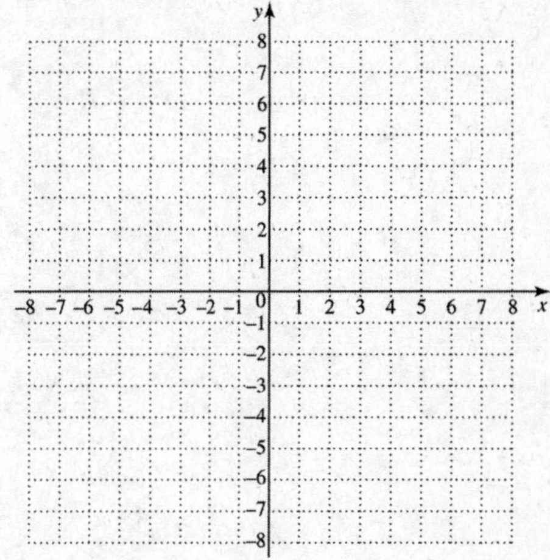

7. $x \leq 2y^2 + 8y + 9$

7.

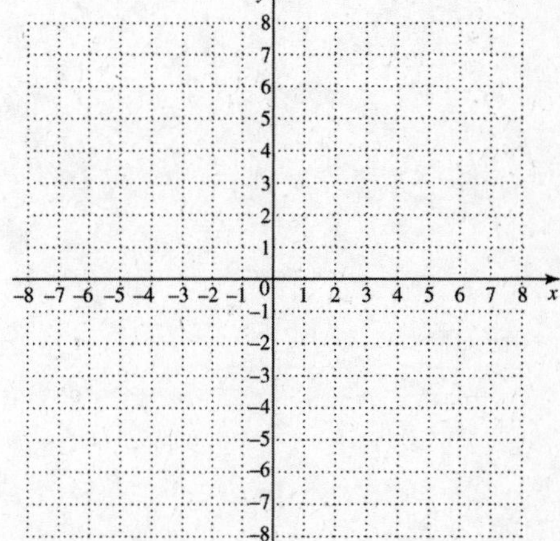

Name: Date:
Instructor: Section:

8. $4y^2 \geq 196 + 49x^2$

8.

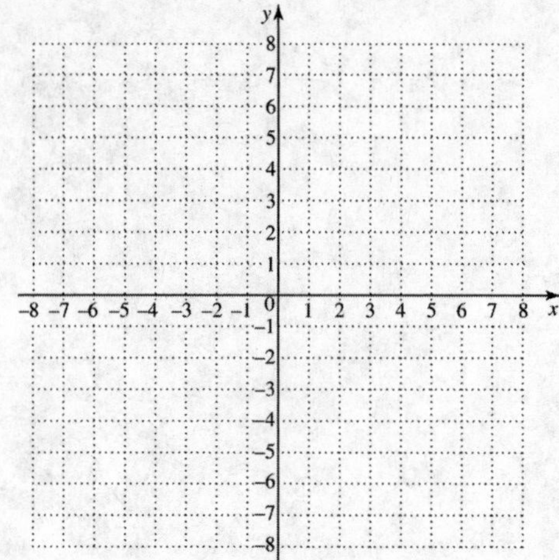

9. $7x^2 \leq 42 - 6y^2$

9.

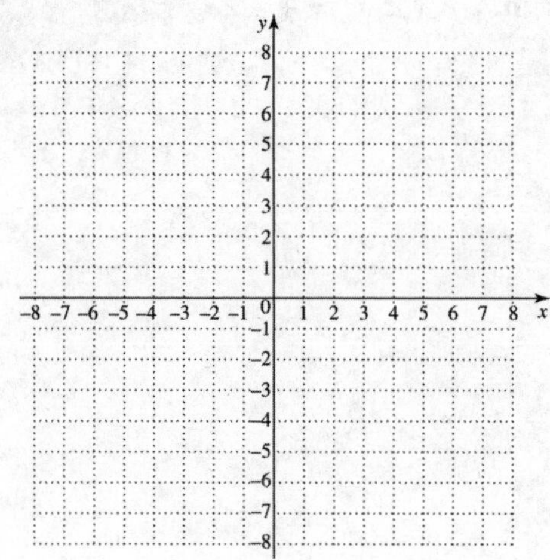

10. $x^2 + 9y^2 > 36$

10.

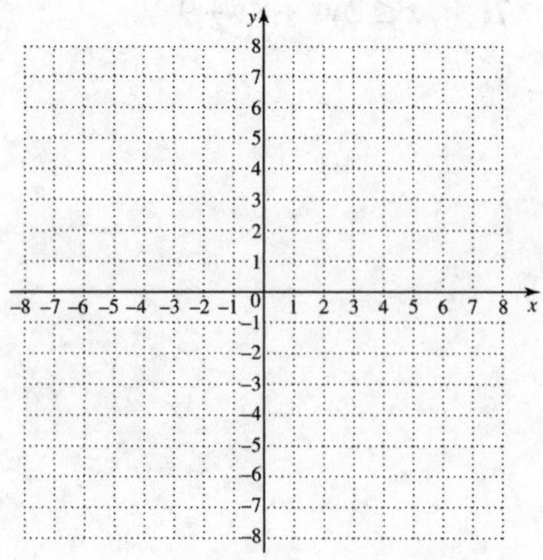

Name: Date:
Instructor: Section:

11. $y < -x^2 + 3$

11.

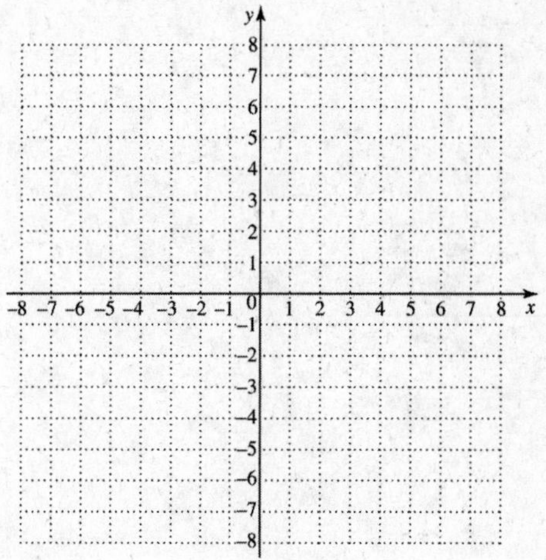

12. $9y^2 - 36x^2 > 144$

12.

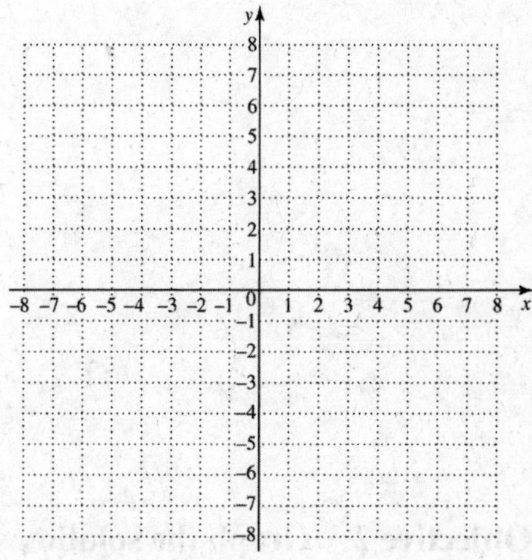

13. $(x+3)^2 + y^2 \le 6$

13.

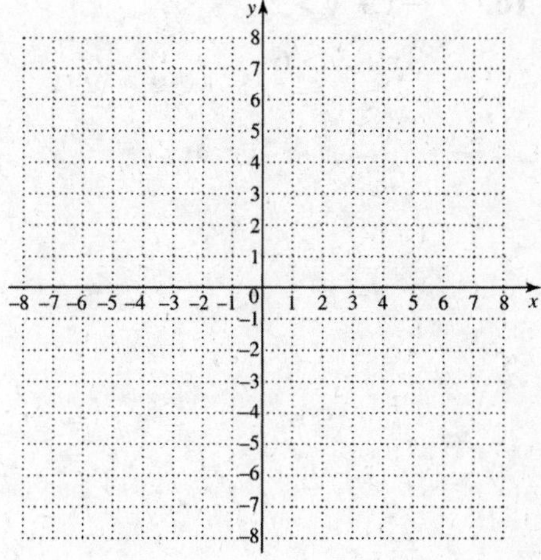

Name: Date:
Instructor: Section:

14. $8x^2 + 2y^2 < 32$

14.

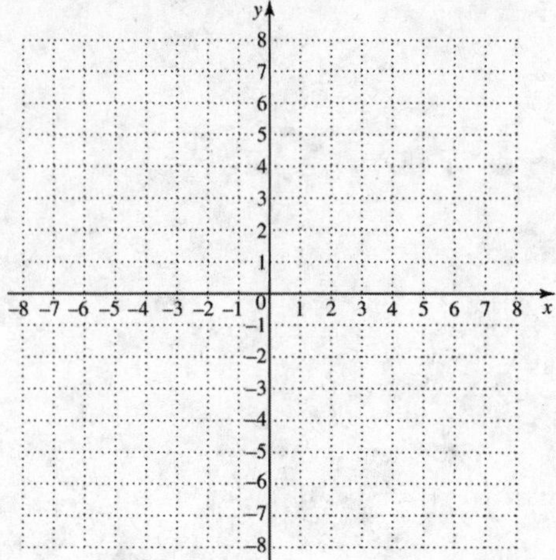

15. $9x^2 - y^2 < 36$

15.

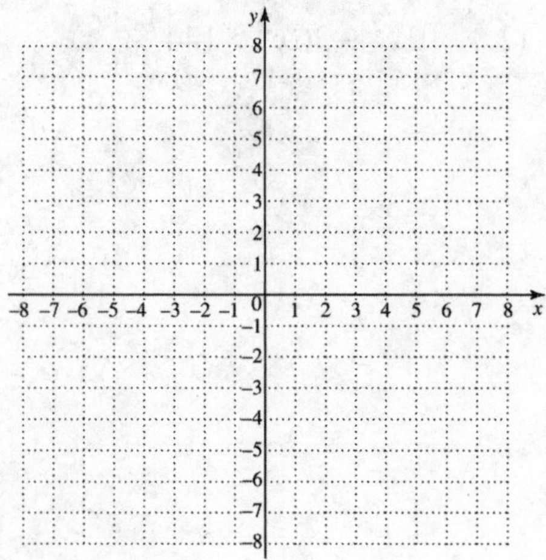

Objective 2 Graph the solution set of a system of inequalities.

Graph each system of inequalities.

16. $-x + y > 2$
$3x + y > 6$

16.

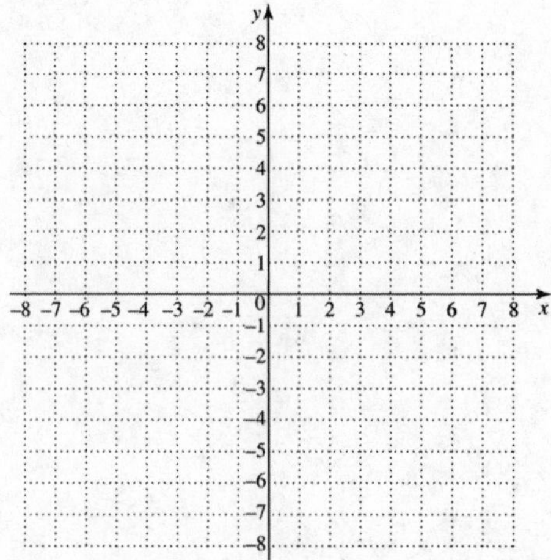

Name: Date:
Instructor: Section:

17. $x + y > -2$
$2x - y \leq -4$

17.

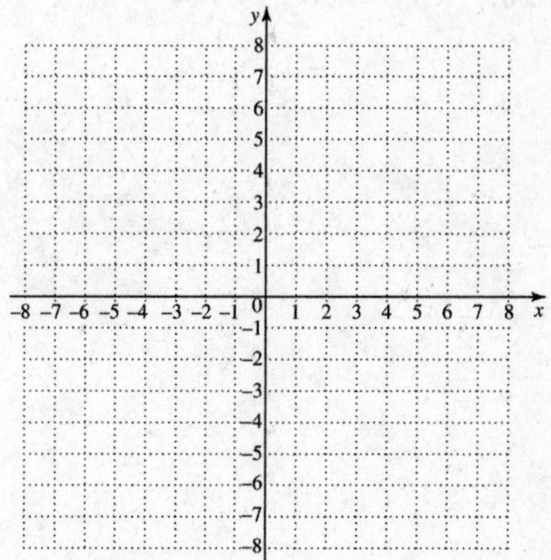

18. $x - 2y \geq -6$
$x + 4y \geq 12$

18.

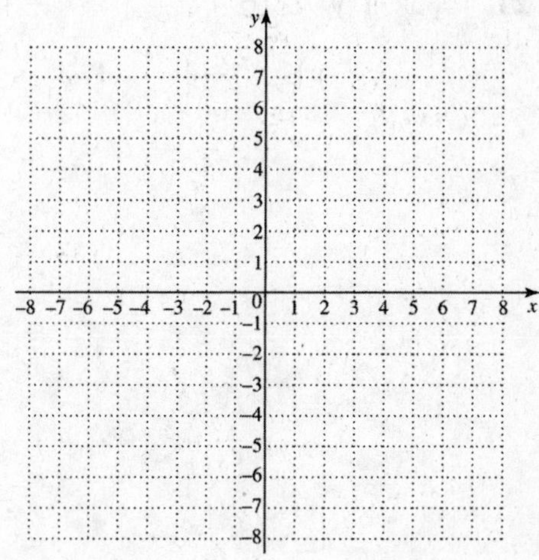

19. $x^2 + y^2 \leq 25$
$3x - 5y > -15$

19.

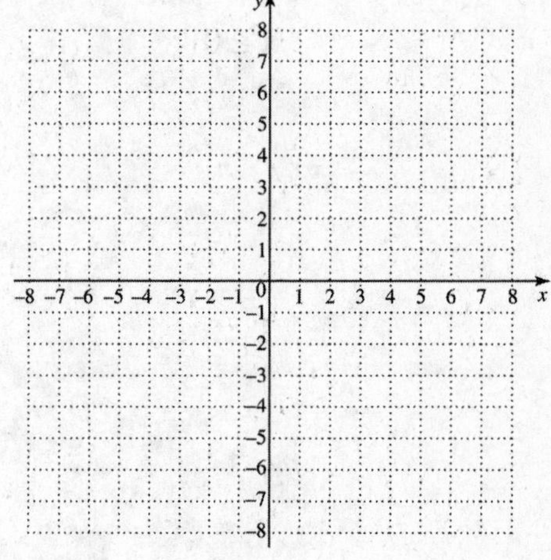

Name: Date:
Instructor: Section:

20. $9x^2 + 16y^2 < 144$
$y^2 - x^2 > 4$

20.

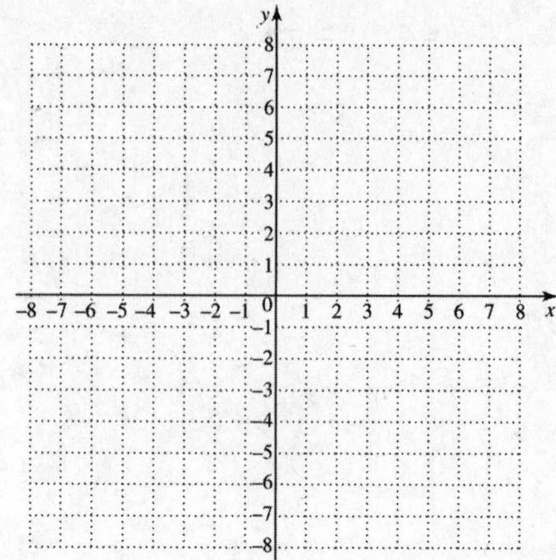

21. $x^2 + y^2 \le 16$
$y \le x^2 - 4$

21.

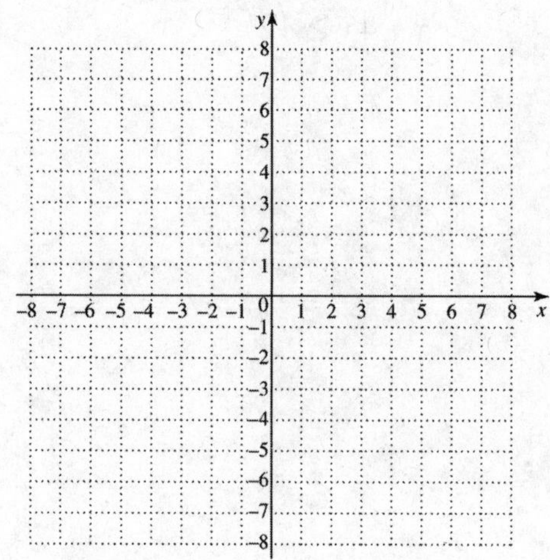

22. $9x^2 + 64y^2 \le 576$
$x \ge 0$

22.

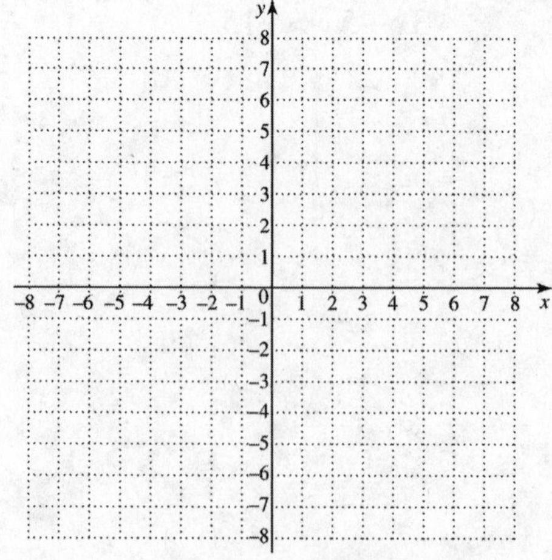

Name: Date:
Instructor: Section:

23. $x^2 > 9 - y^2$
$x \le 0$
$y \ge 0$

23.

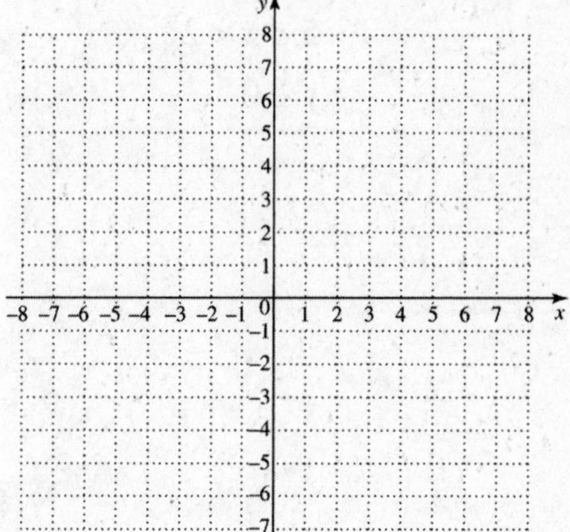

24. $x^2 - y^2 \le 16$
$y \ge 0$

24.

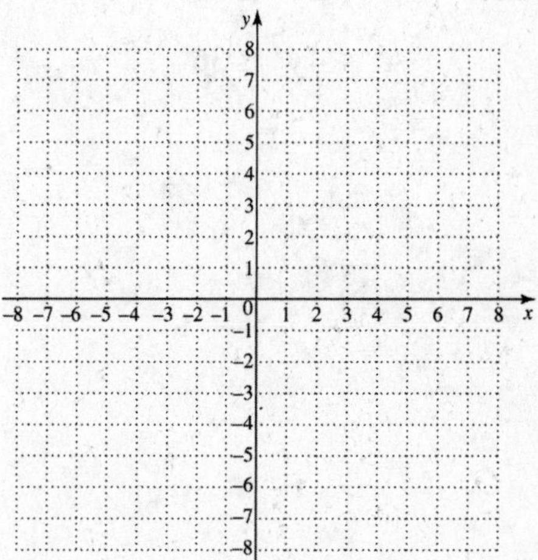

25. $4y + x^2 < 0$
$x \ge 0$

25.

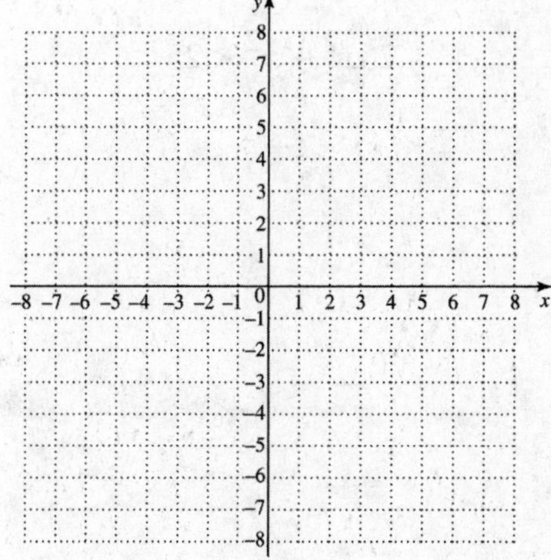

Name:
Instructor:

Date:
Section:

26. $x^2 + 4y^2 \le 36$

$-5 < x < 2$

$y \ge 0$

26.

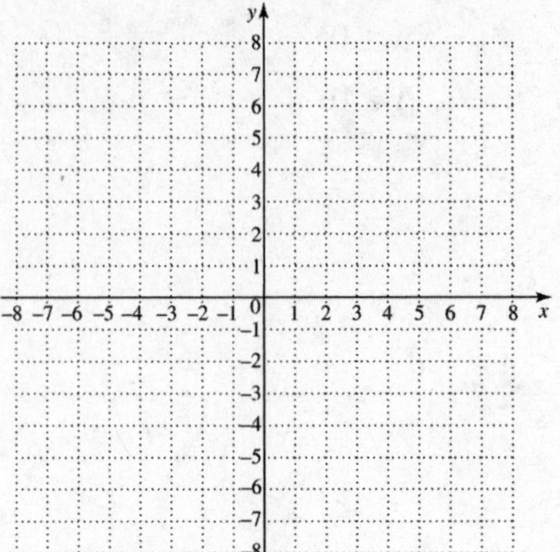

27. $8x^2 + 2y^2 \le 72$

$x^2 + 4y^2 \le 36$

27.

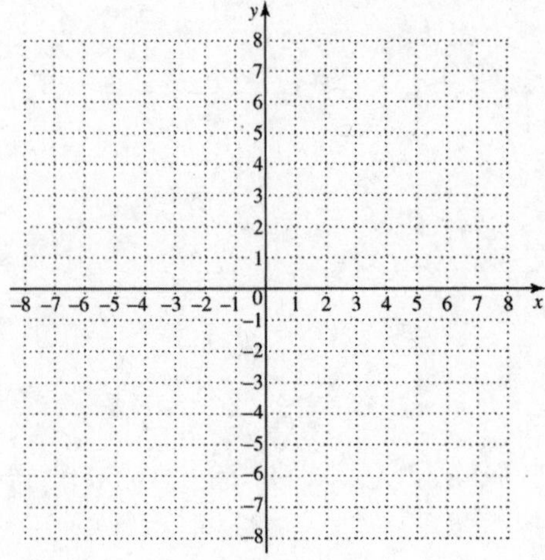

28. $12x^2 + 27y^2 > 108$

$25x^2 + y^2 < 25$

28.

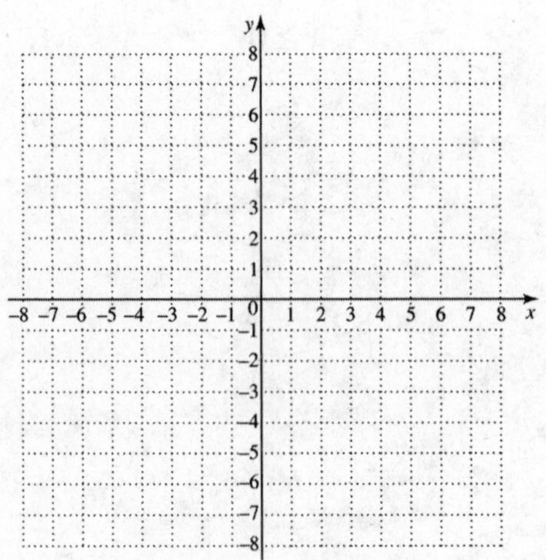

Name: Date:
Instructor: Section:

29. $x^2 + y^2 \geq 9$

$x^2 + y^2 < 36$

29.

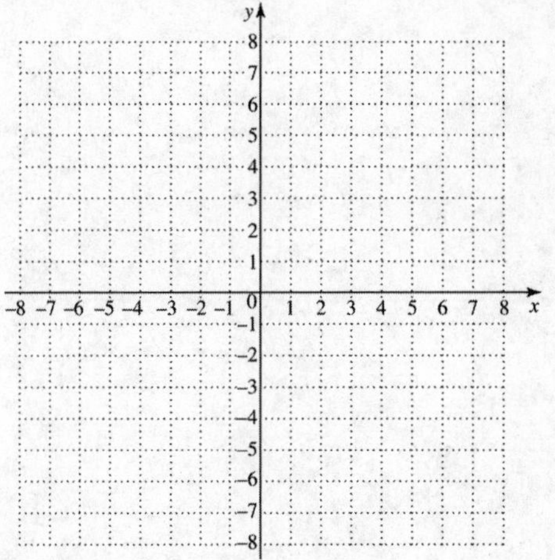

30. $25x^2 + 9y^2 < 225$

$y \leq -x^2 + 4$

$y < -x$

30.

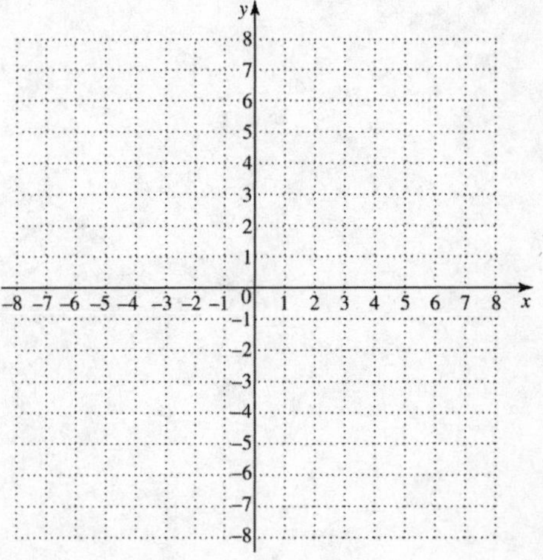

Chapter R PREALGEBRA REVIEW

R.1 Fractions

Key Terms

1. equivalent fractions
2. improper fraction
3. numerator
4. proper fraction
5. denominator
6. composite number
7. prime factorization
8. prime number
9. lowest terms

Objective 1

1. prime
3. neither

Objective 2

5. $2 \cdot 7 \cdot 7$
7. $2 \cdot 2 \cdot 2 \cdot 2 \cdot 2 \cdot 2 \cdot 2 \cdot 2$

Objective 3

9. $\frac{7}{25}$
11. $3\frac{1}{2}$

Objective 4

13. $4\frac{7}{9}$
15. $\frac{122}{9}$

Objective 5

17. $\frac{15}{2}$ or $7\frac{1}{2}$
19. $\frac{45}{4}$ or $11\frac{1}{4}$

Objective 6

21. $\frac{256}{225}$ or $1\frac{31}{225}$
23. $\frac{125}{12}$ or $10\frac{5}{12}$

Objective 7

25. \$220.50
27. $3\frac{1}{4}$ yards
29. $7\frac{5}{16}$ in.^{2}

R.2 Decimals and Percents

Key Terms

1. decimals 2. place value 3. percent

Objective 1

1. $\frac{42}{100}$ 3. $2\frac{54}{10,000}$ 5. $30\frac{5}{10,000}$

Objective 2

7. 15.05 9. 609.168

Objective 3

11. 2.3424 13. 40.32 15. .4292

Objective 4

17. .5625 19. .755

Objective 5

21. .275 23. .004 25. .84%

Objective 6

27. $23.\overline{3}\%$ 29. $36

Chapter 1 THE REAL NUMBER SYSTEM

1.1 Exponents, Order of Operations, and Inequality

Key Terms

1. exponential expression
2. base
3. exponent

Objective 1

1. 27
3. $\frac{16}{81}$
5. 13.824

Objective 2

7. 45
9. $\frac{29}{16}$

Objective 3

11. −8
13. 48
15. 96

Objective 4

17. false
19. false

Objective 5

21. $7 = 13 - 6$
23. $30 - 7 > 20$
25. $20 \geq 2 \cdot 7$

Objective 6

27. $\frac{2}{3} < \frac{3}{4}$
29. $.0002 < .002$

1.2 Variables, Expressions, and Equations

Key Terms

1. equation
2. variable
3. algebraic expression
4. solution

Objective 1

1. 8
3. $-\frac{9}{2}$
5. $\frac{28}{13}$

Objective 2

7. $1+3x$
9. $10x+21$
11. $8x-11$

Objective 3

13. no
15. no
17. yes

Objective 4

19. $5x+2=23$
21. $6(5+x)=19$
23. $61-7x=13+x$

Objective 5

25. expression
27. equation
29. expression

1.3 Real Numbers and the Number Line

Key Terms

1. whole numbers
2. opposite
3. integers
4. natural numbers
5. absolute value
6. number line
7. irrational number
8. coordinate
9. negative number
10. positive number
11. real numbers
12. set-builder notation
13. rational number

Objective 1

1. −75 pounds
3. −396 meters
5. 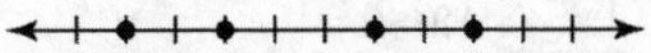

7. 

Objective 2

9. −6.01
11. −4
13. true
15. false

Objective 3

17. 25
19. −22
21. 0
23. $\frac{5}{7}$

Objective 4

25. −10
27. $\frac{5}{6}$
29. 2

1.4 Adding Real Numbers

Key Terms

1. sum
2. addends

Objective 1

1. -40
3. -18
5. $\frac{7}{5}$

Objective 2

7. -5
9. 0
11. $\frac{1}{35}$

Objective 3

13. true
15. false
17. false

Objective 4

19. -16
21. -6
23. $-\frac{7}{8}$

Objective 5

25. $-8+(-4)+(-11); -23$
27. $-10+[20+(-4)];\ 6$
29. \$495

1.5 Subtracting Real Numbers

Key Terms

1. minuend
2. subtrahend
3. difference

Objective 1

1. 3
3. 0
5. −7

Objective 2

7. 7
9. 0
11. 4.4
13. $-\frac{1}{30}$

Objective 3

15. 18
17. 18
19. −2
21. $-\frac{23}{18}$ or $-1\frac{5}{8}$

Objective 4

23. $-4-4;\ -8$
25. $(-4+12)-9;\ -1$
27. −51.2ºC
29. 37ºF

1.6 Multiplying and Dividing Real Numbers

Key Terms

1. quotient 2. reciprocals 3. product

Objective 1

1. −28 3. $-\frac{7}{12}$

Objective 2

5. 40 7. 2.73

Objective 3

9. $-\frac{1}{6}$ 11. 0 13. −2.5

Objective 4

15. 70 17. $\frac{16}{21}$

Objective 5

19. 7 21. 17

Objective 6

23. $(-7)(3)+(-7);\ -28$ 25. $-12+\frac{49}{-7};\ -19$

Objective 7

27. $\frac{2}{3}x=-7$ 29. $\frac{x}{-4}=1$

1.7 Properties of Real Numbers

Key Terms

1. identity element for addition
2. identity element for multiplication

Objective 1

1. 4
3. [10 + (−9)]
5. $\left(\frac{1}{4}\cdot 2\right)$

Objective 2

7. (4)
9. $4a$
11. $[(-r)(-p)]$

Objective 3

13. 4
15. 12
17. $\frac{30}{30}$ or 1

Objective 4

19. 4; inverse
21. 0; identity
23. $-\frac{6}{17}$; inverse

Objective 5

25. $4b+8$
27. $-10y+18z$
29. $-14(x+y)$

1.8 Simplifying Expressions

Key Terms

1. numerical coefficient
2. term
3. like terms

Objective 1

1. $8x+27$
3. $5+s$
5. $35n-9$

Objective 2

7. -2
9. 1
11. $\frac{7}{9}$

Objective 3

13. like
15. unlike
17. unlike

Objective 4

19. $5r-4$
21. $1.7y^2-.5xy$
23. $-1.5y+16$

Objective 5

25. $6x+12+4x=10x+12$
27. $3(9+2x)+4x=10x+27$
29. $4(2x-6x)+6(x+9)=-10x+54$

Chapter 2 EQUATIONS, INEQUALITIES, AND APPLICATIONS

2.1 The Addition Property of Equality

Key Terms

1. equivalent equations
2. linear equation
3. solution set

Objective 1

1. yes
3. no
5. no

Objective 2

7. 20
9. –5
11. $\frac{3}{2}$
13. –5
15. –12.8
17. –10

Objective 3

19. 7
21. –8
23. $\frac{5}{4}$
25. 0
27. $\frac{1}{3}$
29. 7.2

2.2 The Multiplication Property of Equality

Key Terms

1. multiplication property of equality
2. addition property of equality

Objective 1

1. 3
3. 3
5. −14
7. $\frac{8}{7}$
9. $\frac{7}{9}$
11. 3.6
13. −8.2
15. −2.7

Objective 2

17. 25
19. 8
21. −8
23. −1.5
25. 10
27. −24
29. −7

2.3 More on Solving Linear Equations

Key Terms

1. contradiction
2. conditional equation
3. identity

Objective 1

1. $\frac{5}{2}$
3. 2
5. $-\frac{1}{4}$
7. $-\frac{1}{5}$

Objective 2

9. 2
11. $\frac{53}{11}$
13. 30
15. -3

Objective 3

17. none
19. none
21. none
23. infinitely many

Objective 4

25. $\frac{17}{p}$
27. $4x$
29. $x+28$

2.4 An Introduction to Applications of Linear Equations

Key Terms

1. supplementary angles
2. complementary angles
3. right angle
4. straight angle
5. consecutive integers

Objective 1

1. Read the problem; assign a variable to represent the unknown; write an equation; solve the equation; state the answer; check the answer.

Objective 2

3. $4x - 2 = 3 + 6x;\ -\frac{5}{2}$
5. $6(x - 4) = -2x;\ 3$
7. $4x + 7 = 6x - 5;\ 6$

Objective 3

9. $x + (x + 21) = 439;$ 209 votes
11. $x + (x + 5910) = 34{,}730;$ Mt. Rainier: 14,410 ft; Mt. McKinley: 20,320 feet
13. $x + 2x + 8x = 176;$ cranberry juice: 16 oz; orange juice: 32 oz; ginger ale: 128 oz
15. $2x + 36 = 52;$ 8 feet

Objective 4

17. 20°
19. 49°
21. 55°

Objective 5

23. 76, 78
25. 27, 28
27. 13, 15
29. 75, 76, 77

2.5 Formulas and Additional Applications from Geometry

Key Terms

1. vertical angles
2. formula
3. perimeter
4. area

Objective 1

1. $V = 24$
3. $a = 36$
5. $C = 40$
7. $V = 100.48$

Objective 2

9. 8 in.
11. 1.5 years
13. 3052.08 cm^3
15. 31,400 sq ft

Objective 3

17. 35°, 35°
19. 150°, 150°
21. 54°, 126°
23. 148°, 148°

Objective 4

25. $r = -\frac{a-s}{s}$ or $\frac{s-a}{s}$
27. $A = \frac{P}{1-rt}$
29. $n = \frac{S}{180} + 2$ or $n = \frac{S+360}{180}$

2.6 Ratio, Proportion, and Percent

Key Terms

1. proportion
2. ratio
3. terms
4. cross products

Objective 1

1. $\frac{8}{3}$
3. $\frac{3}{4}$
5. 45-count box
7. 44-ounce bottle

Objective 2

9. 15
11. $\frac{20}{19}$
13. $-\frac{7}{4}$

Objective 3

15. 75 minutes
17. 15 inches
19. 14 tanks
21. 10 pounds

Objective 4

23. 87.5
25. 750
27. 4.25%
29. 22%

2.7 Solving Linear Inequalities

Key Terms

1. three-part inequality
2. interval
3. linear inequality
4. inequalities
5. interval notation

Objective 1

1. $(3,\infty)$;

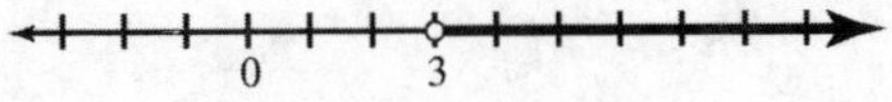

3. $(-1,3)$;

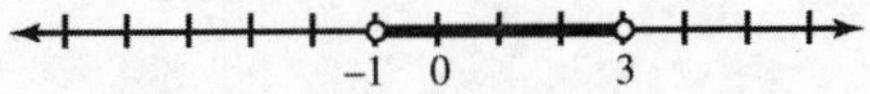

5. $(-3,2]$

Objective 2

7. $[1,\infty)$;

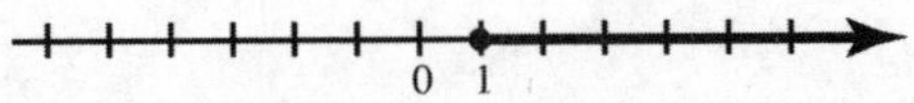

9. $(-\infty,0]$;

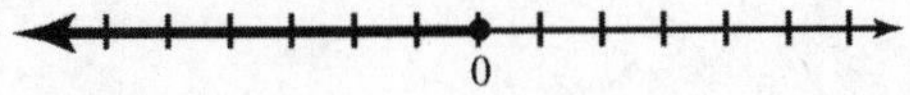

Objective 3

11. $(-2,\infty)$;

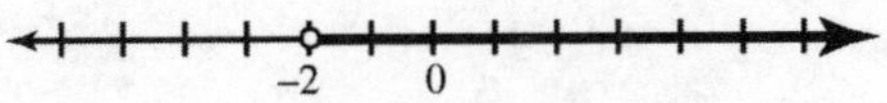

13. $[0,\infty)$;

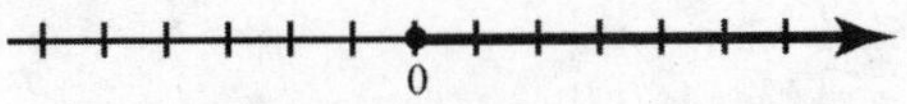

15. $(-\infty,4)$

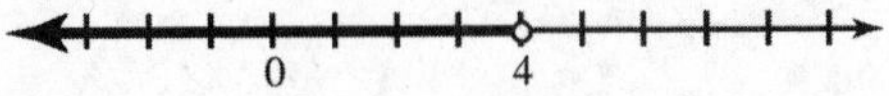

Objective 4

17. $(-\infty,19]$;

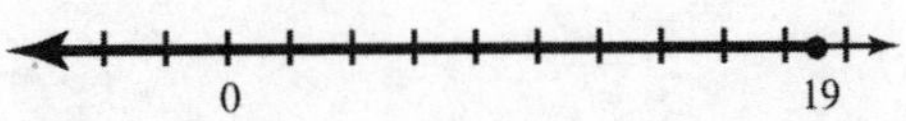

19. $\left[\frac{13}{12},\infty\right)$;

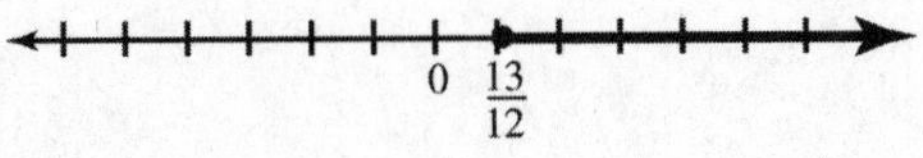

Objective 5

21. $(4,7]$;

23. $(5,6)$;

25. $(-1,5)$

Objective 6

27. $38.84

29. 10 feet

Chapter 3 GRAPHS OF LINEAR EQUATIONS AND INEQUALITIES; FUNCTIONS

3.1 Reading Graphs; Linear Equations in Two Variables

Key Terms

1. line graph
2. linear equation in two variables
3. bar graph
4. coordinates
5. *x*-axis
6. *y*-axis
7. ordered pair
8. rectangular (Cartesian) coordinate system
9. quadrants
10. origin
11. plot
12. scatter diagram
13. table of values
14. plane

Objective 1

1. 2002, 2003, 2005
3. 2003, 2004
5. 325 M.B.A. degrees

Objective 2

7. $\left(\frac{1}{3}, -9\right)$
9. (.2, .3)

Objective 3

11. not a solution
13. a solution
15. a solution

Objective 4

17.(a) (−4, −5); (b) (2, 7); (c) $\left(-\frac{3}{2}, 0\right)$; (d) (−2, −1); (e) (−5, −7)

19.(a) (2, 4); (b) (0, 4); (c) (4, 4); (d) (−4, 4); (e) (.75, 4)

Objective 5

21. $\left(0, \frac{3}{2}\right), (-2, 0), (2, 3)$
23. $(1, -3), (1, 0), (1, 5)$
25. $(0, 4), (3, 0), \left(\frac{15}{4}, -1\right)$

Objective 6

27.–29.

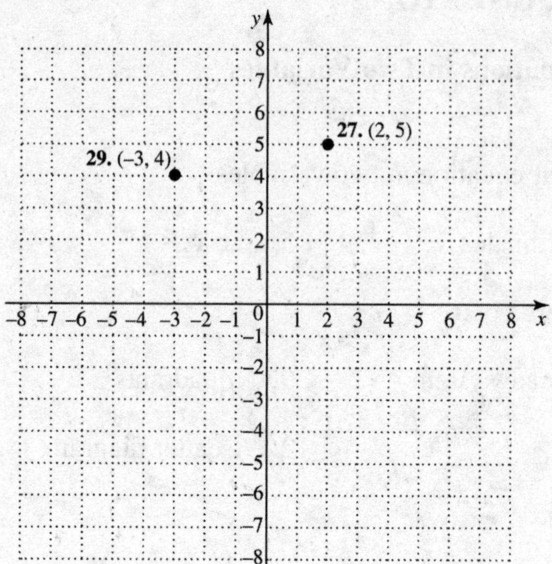

3.2 Graphing Linear Equations in Two Variables

Key Terms

1. *y*-intercept
2. *x*-intercept
3. graphing
4. graph

Objective 1

1. (0, 3), (3, 0), (2, 1)

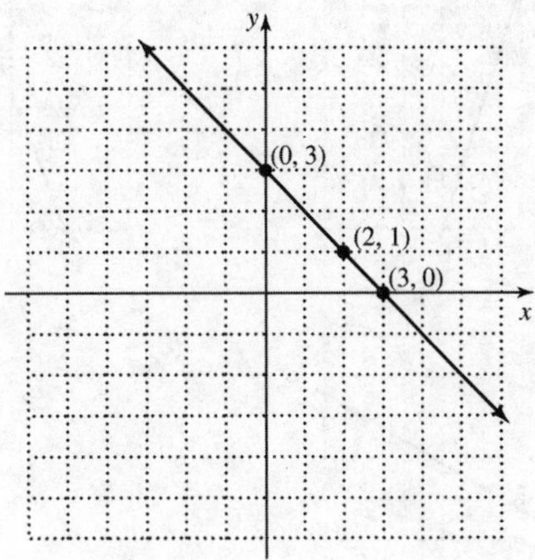

3. (0, −4), (4, 0), (−2, −6)

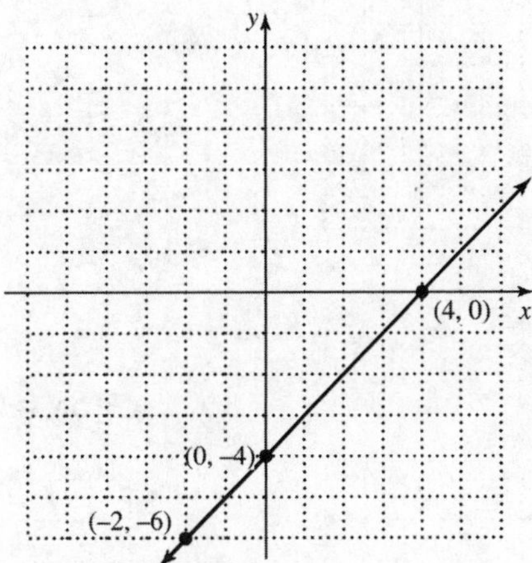

5. (0, 2), (3, 0), (−3, 4)

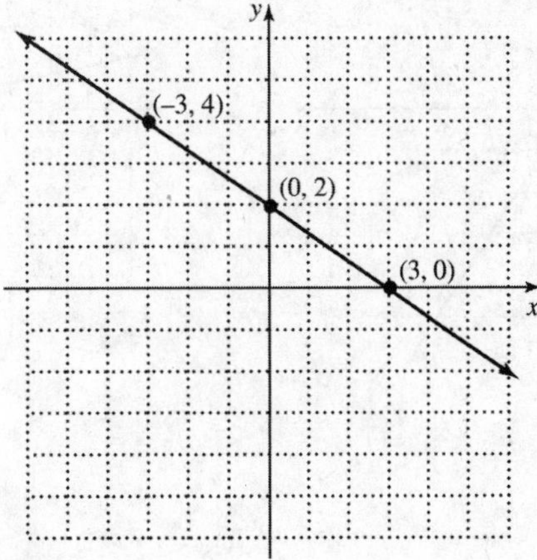

Objective 2

7.

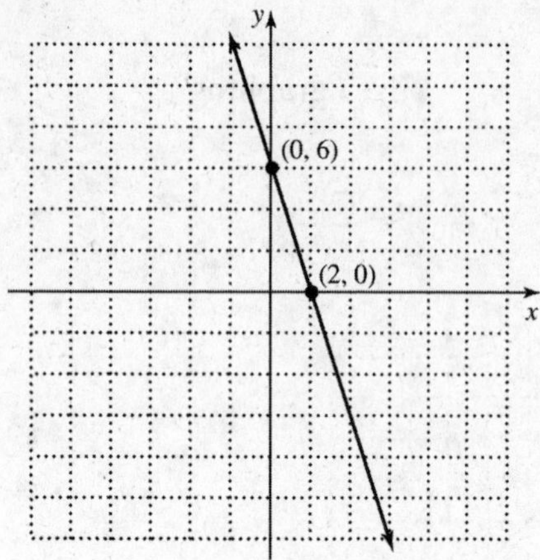

9.

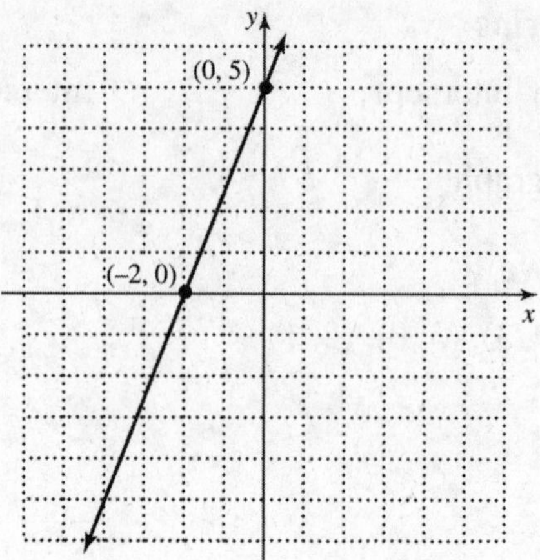

11.

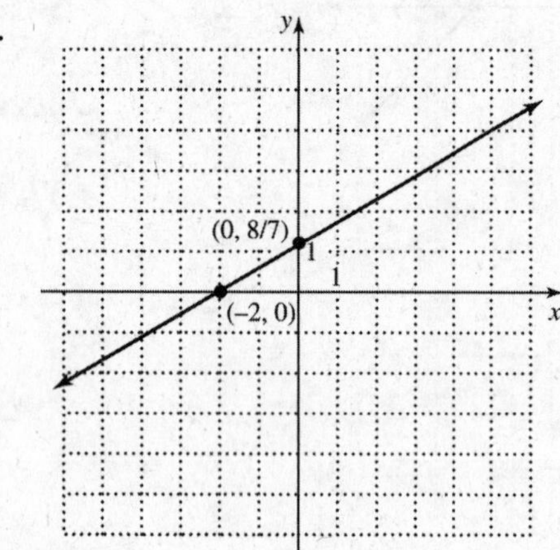

Objective 3

13. $-3x-2y=0$

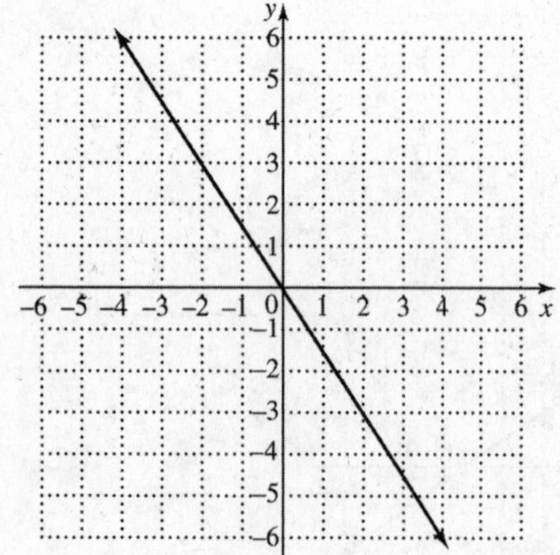

15. $x+5y=0$

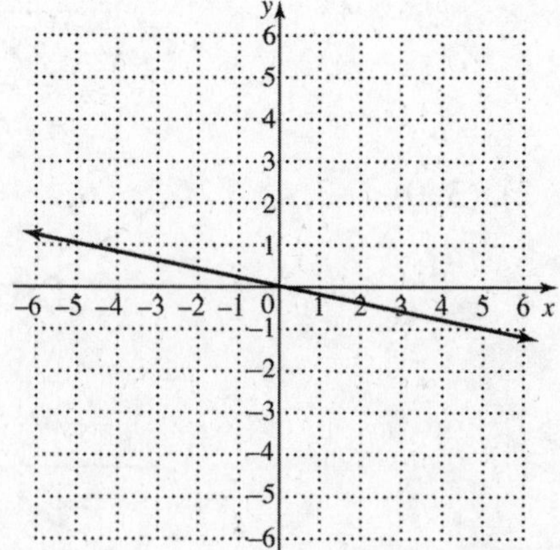

17. $4x = 3y$

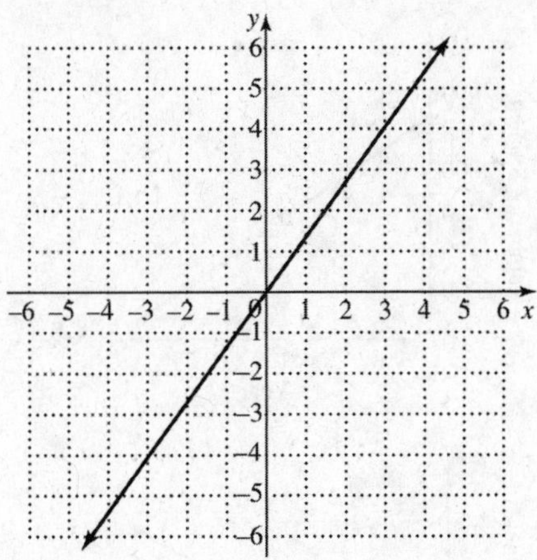

Objective 4

19. $y = -2$

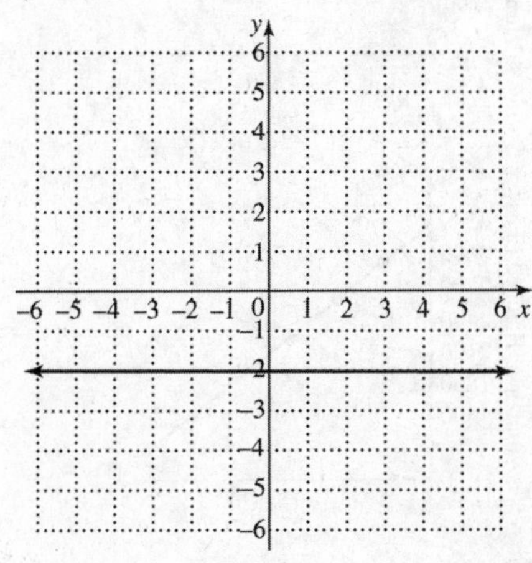

21. $x - 1 = 0$

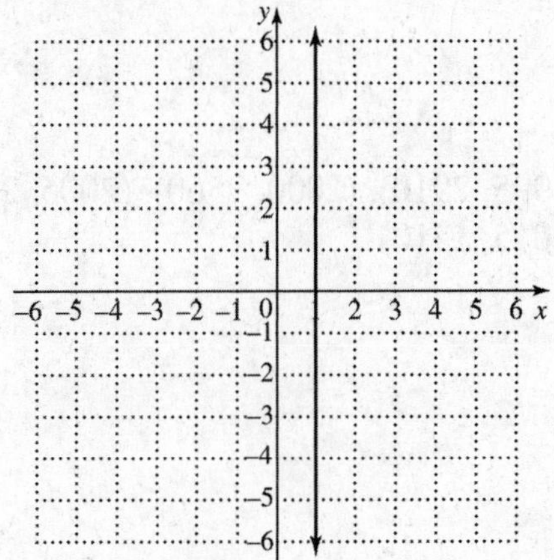

23. $y + 3 = 0$

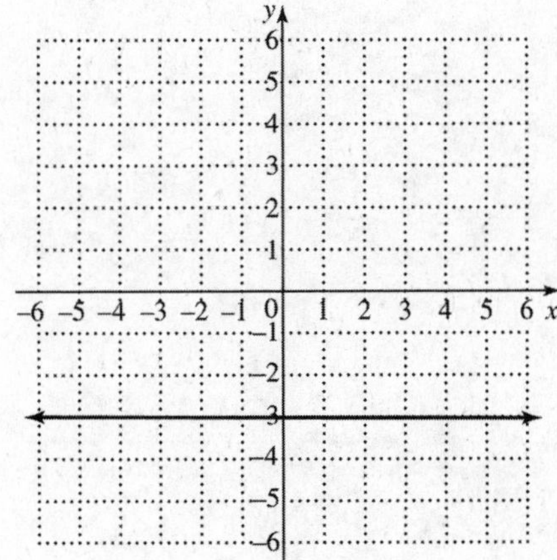

Objective 5

25. (2000, 2435), (2001, 2350), (2002, 2265), (2003, 2180), (2004, 2095), (2005, 2010)

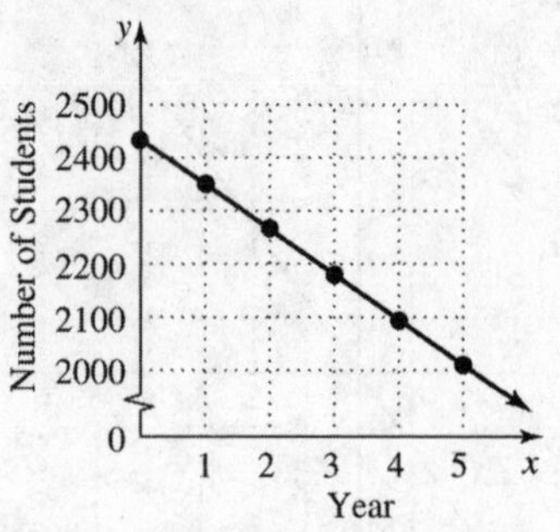

27. (2003, 325), (2004, 367), (2005, 409), (2006, 451)

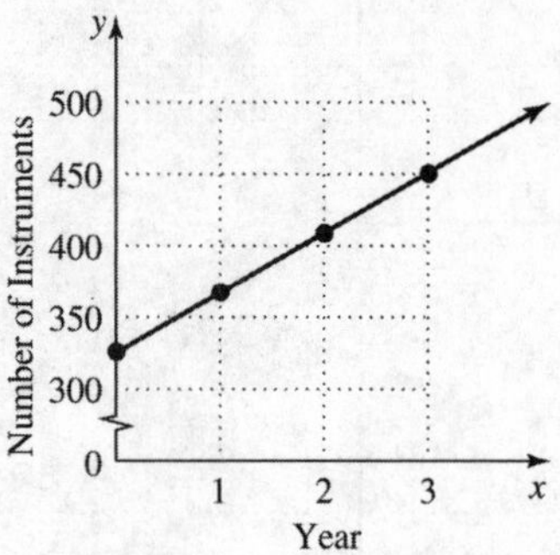

29. (1995, 2910), (2000, 2560), (2005, 2210), (2015, 1510)

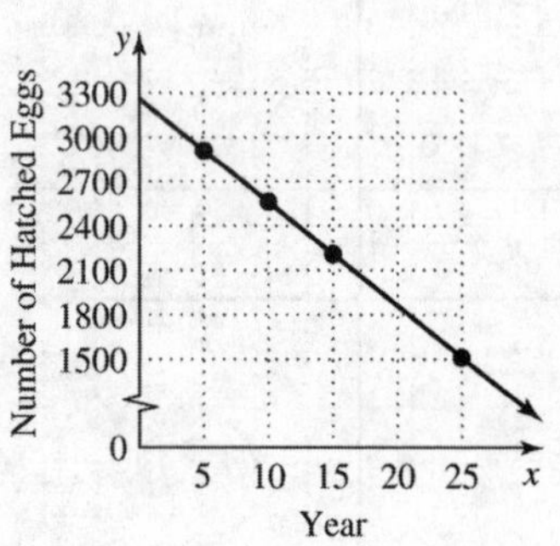

3.3 Slope of a Line

Key Terms

1. perpendicular lines
2. slope
3. rise
4. parallel lines
5. run

Objective 1

1. -2
3. $-\frac{1}{14}$
5. $\frac{5}{3}$
7. 4

Objective 2

9. $\frac{1}{2}$
11. $\frac{4}{7}$
13. 0
15. $-\frac{2}{5}$

Objective 3

17. -5; 5; neither
19. -2; $-\frac{1}{4}$; neither
21. -2; $-\frac{5}{3}$; neither
23. -4; -1; neither

Objective 4

25. $1250/yr
27. 113.5 ft/min
29. an increase of 5.4 employees/yr

3.4 Equations of Lines

Key Terms

1. point-slope form 2. standard form 3. slope-intercept form

Objective 1

1. $y=\frac{3}{2}x-\frac{2}{3}$ 3. $y=-4x$ 5. $y=-3x+3$

Objective 2

7.

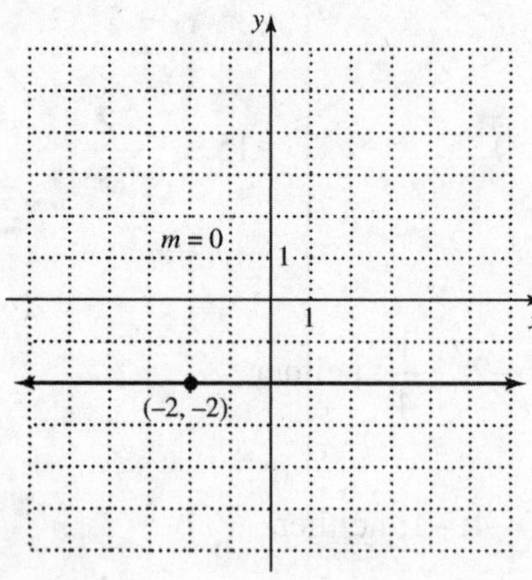

9.

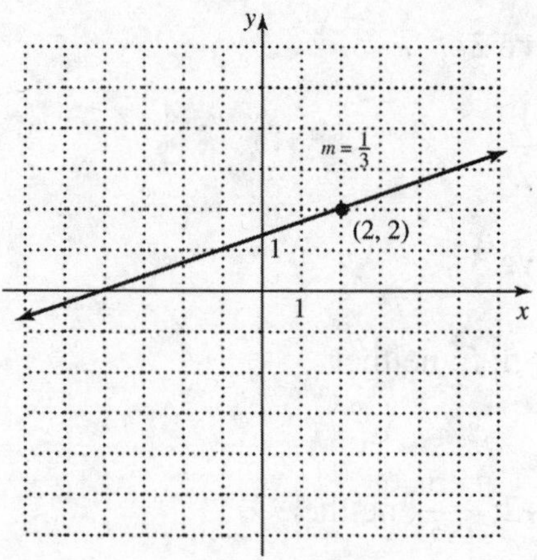

Objective 3

11. $x=-4$ 13. $y=\frac{4}{3}x-\frac{5}{3}$ 15. $y=\frac{2}{3}x+\frac{8}{3}$

Objective 4

17. $5x-8y=-5$ 19. $11x+y=29$

Objective 5

21. $2x+3y=9$ 23. $x=2$ 25. $3x+y=-28$

Objective 6

27. (a) $y=8x+15$; (b) (0, 15), (5, 55), (10, 95)

29. (a) $y=\frac{28}{25}x+10.1$; (b) 43.7% (The year 2015 corresponds to $x=30$)

3.5 Graphing Linear Inequalities in Two Variables

Key Terms

1. boundary line
2. linear inequality in two variables

Objective 1

1.

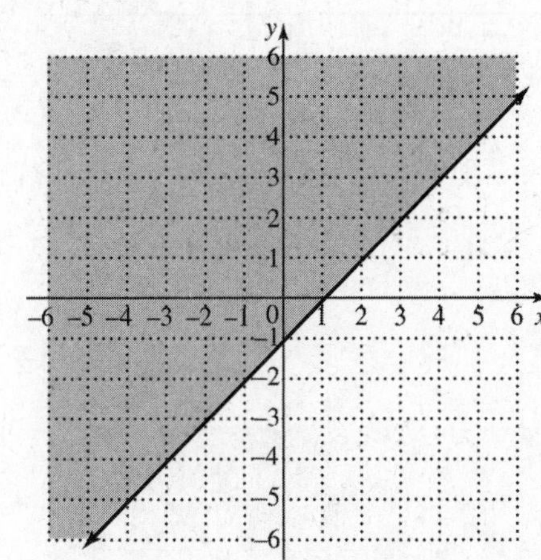

3.

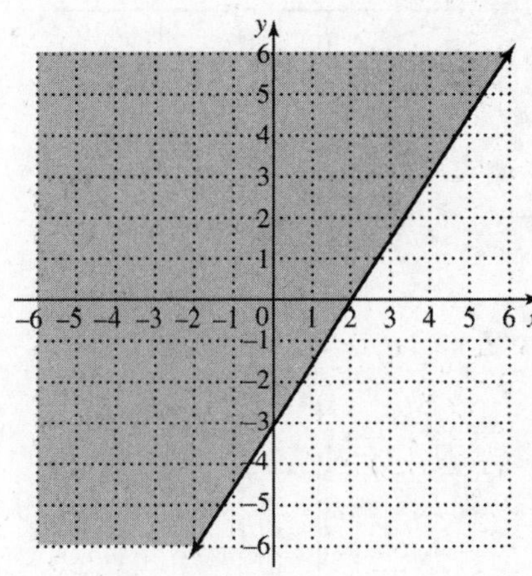

5.

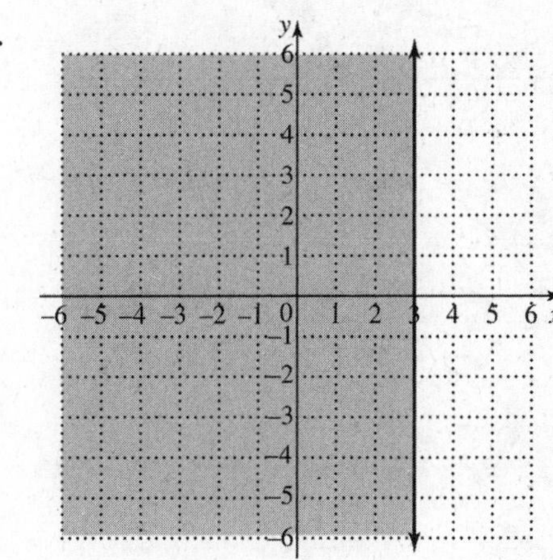

7.

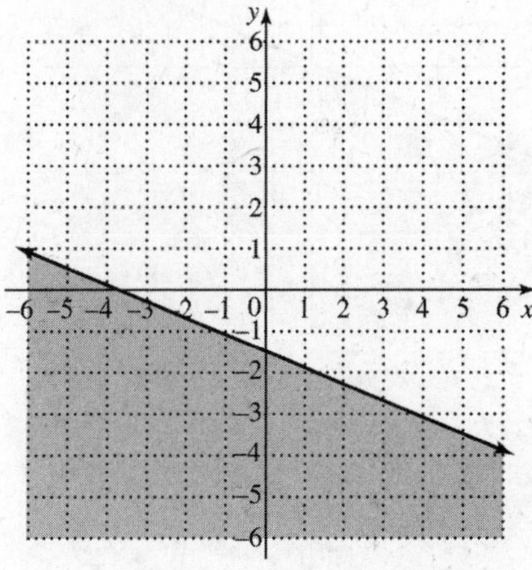

9.

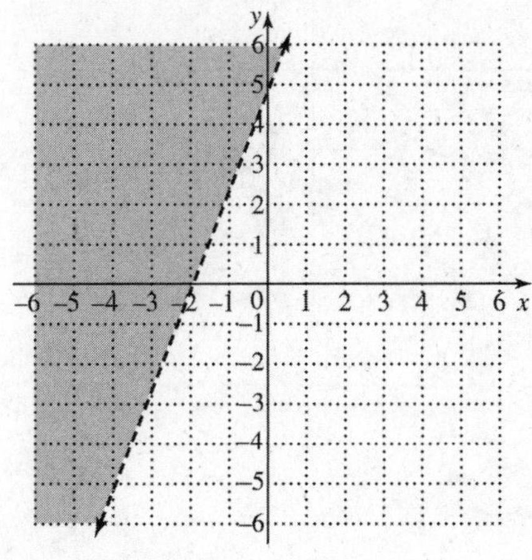

11.

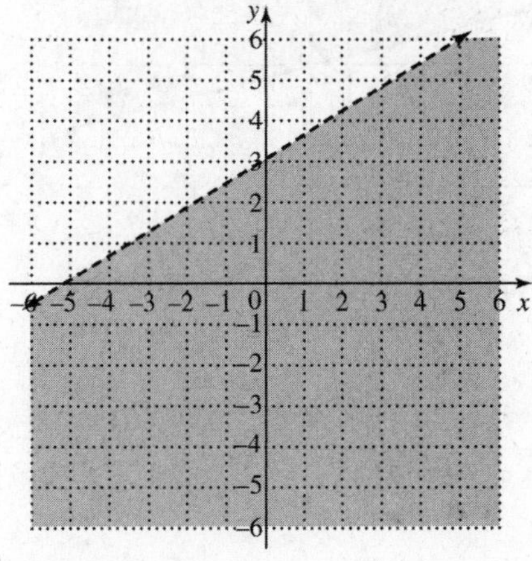

13.

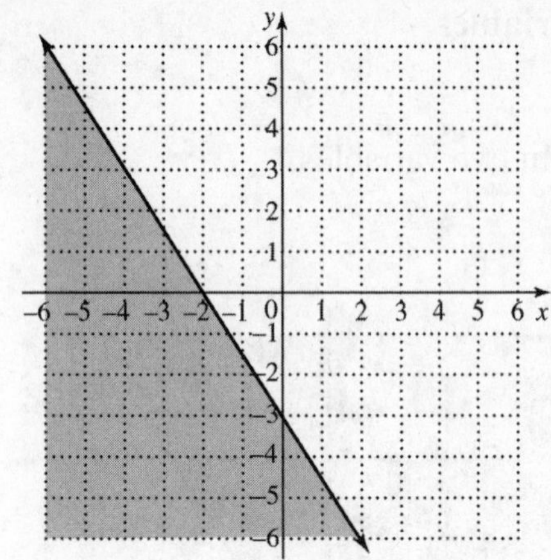

15.

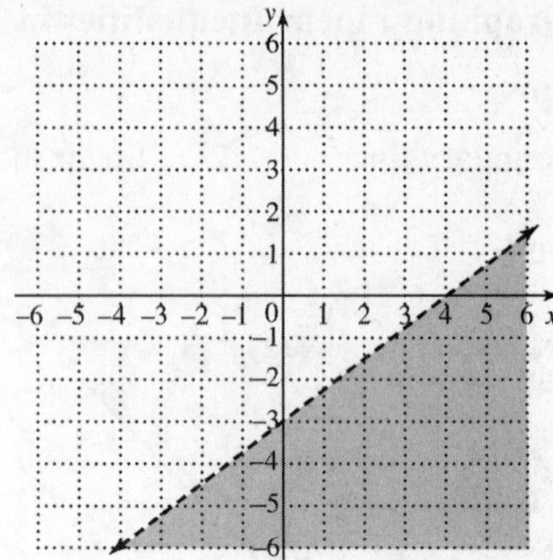

Objective 2

17.

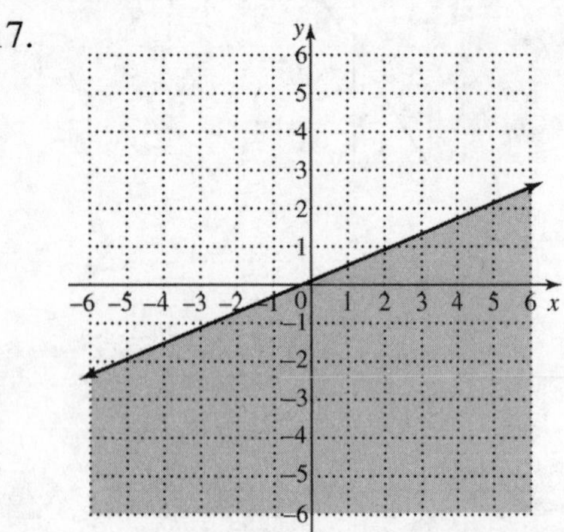

19.

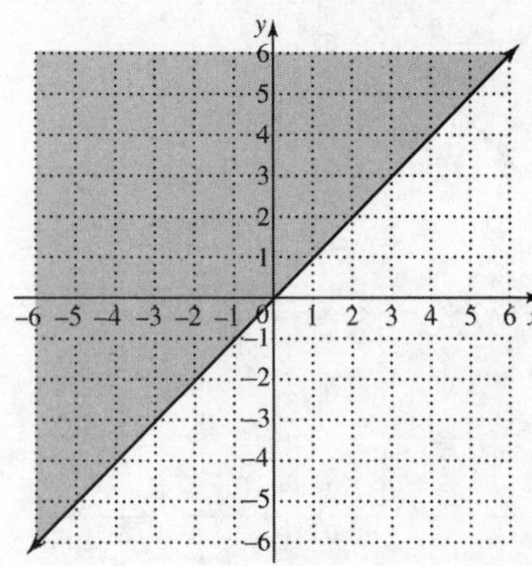

21.

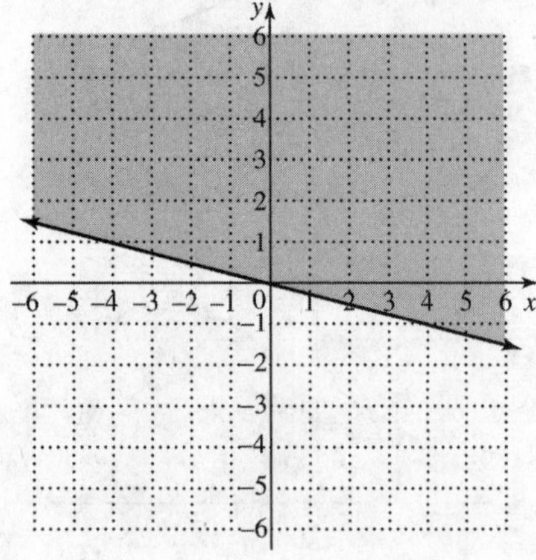

23.

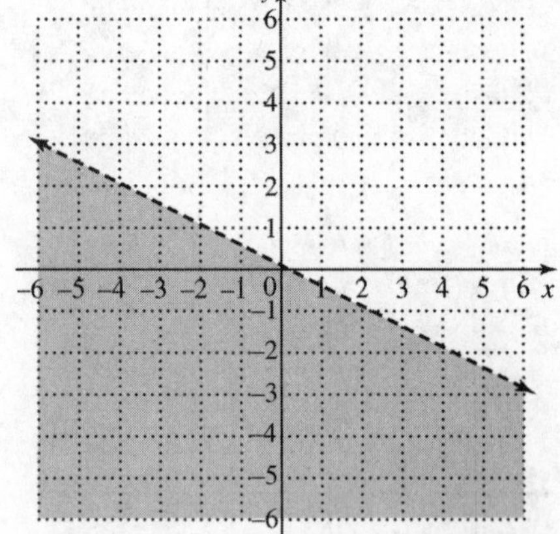

25.

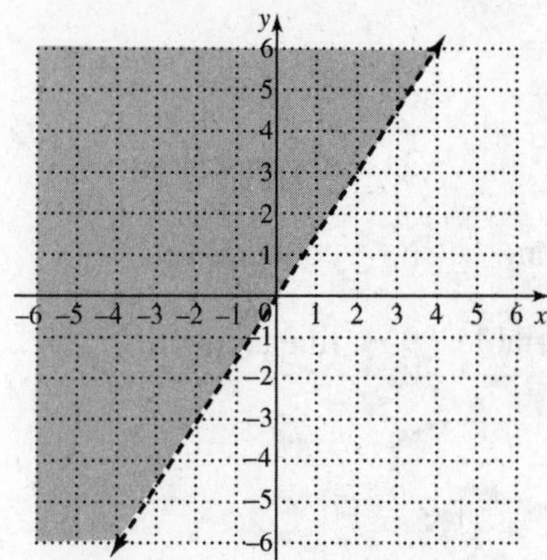

27.

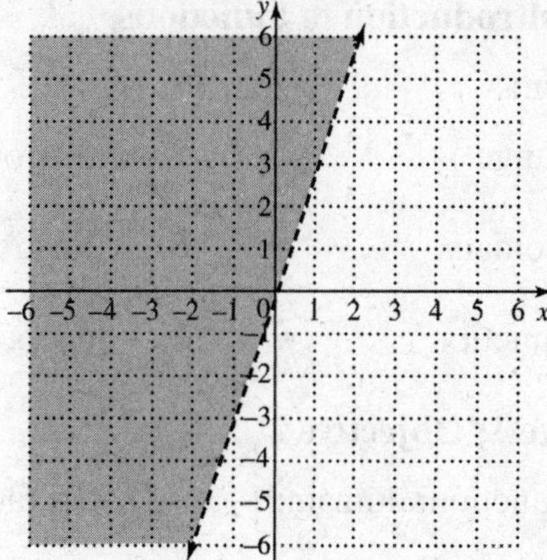

29.

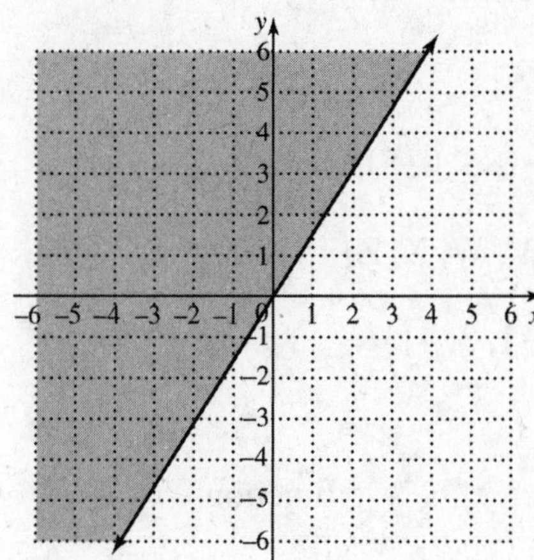

3.6 Introduction to Functions

Key Terms

1. range
2. relation
3. dependent variable
4. domain
5. function notation
6. constant function
7. function
8. independent variable
9. linear function

Objective 1; Objective 2

1. function; domain: {–3, 0, 2, 5}; range: {–8, –4, –1, 2, 7}
3. not a function; domain: {–7, –3, 0, 2}; range: {1, 4, 6, 7, 9}
5. function; domain: {–5, –2, 1, 3, 7}; range: {4}
7. function; domain: {–3, –2, –1, 0, 1}; range: {–5, 0, 5}
9. function; domain: {A, B, C, D, E}; range: {V, W, X, Z}
11. not a function; domain: {–1, 1, 2}; range: {–1, 3, 4}

Objective 3

13. not a function
15. not a function
17. function

Objective 4

19. (a) –13; (b) –7; (c) $-3x - 7$
21. (a) 1; (b) –5; (c) $2x^2 - x - 5$
23. (a) –12; (b) 4; (c) $-x^3 - 2x^2 + 4$

Objective 5

25.

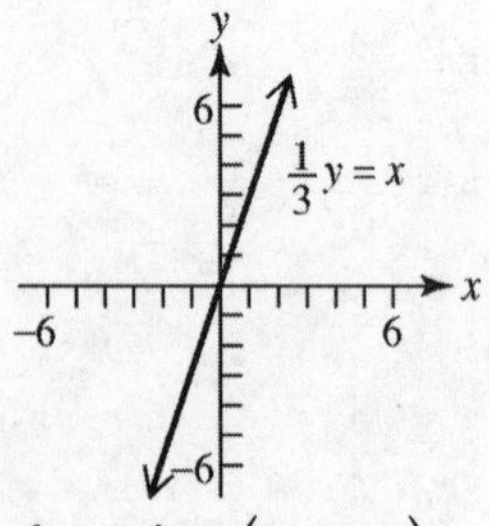

domain: $(-\infty, \infty)$

range: $(-\infty, \infty)$

27.

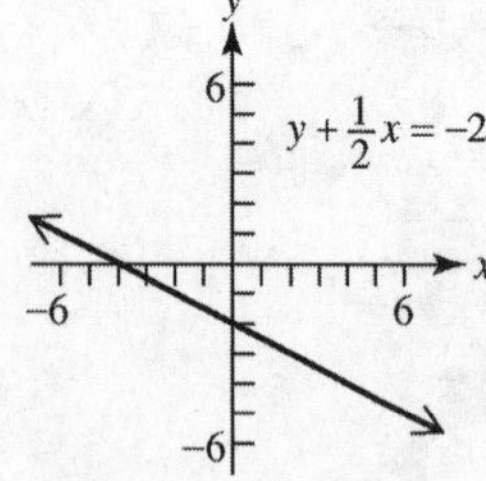

domain: $(-\infty, \infty)$

range: $(-\infty, \infty)$

29.

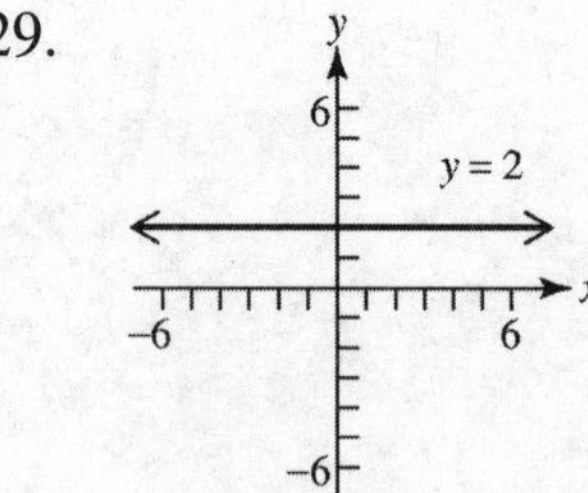

domain: $(-\infty, \infty)$

range: 2

Chapter 4 SYSTEMS OF LINEAR EQUATIONS AND INEQUALITIES

4.1 Solving Systems of Linear Equations by Graphing

Key Terms

1. independent equations
2. consistent system
3. solution set of the system
4. solution of a system
5. dependent equations
6. inconsistent system
7. system of linear equations

Objective 1

1. no
3. yes
5. yes
7. yes

Objective 2

9.

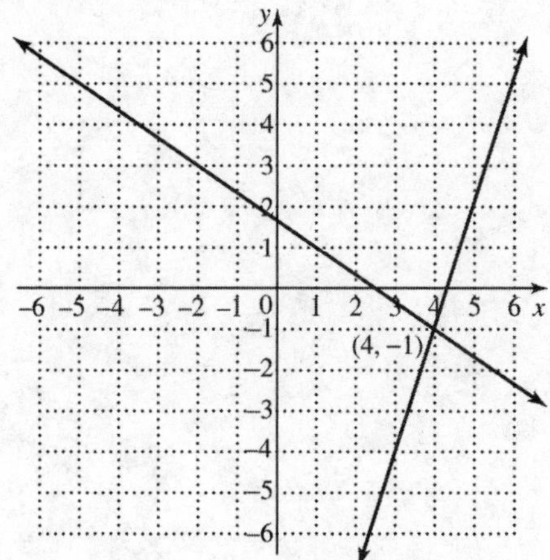

11.

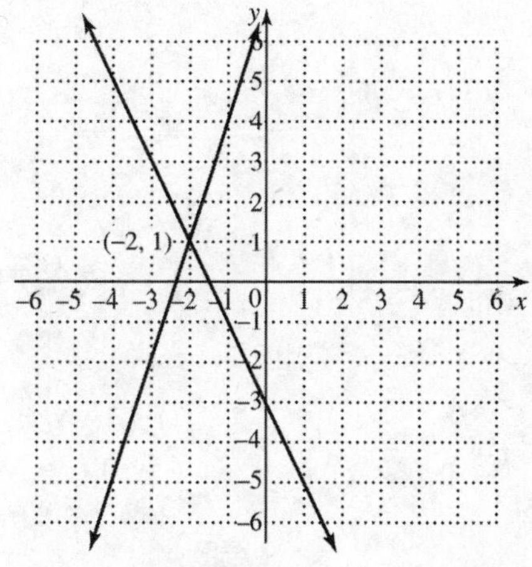

13.

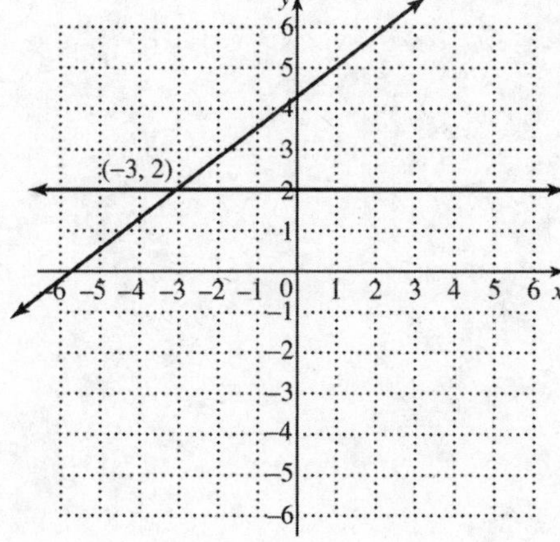

15.

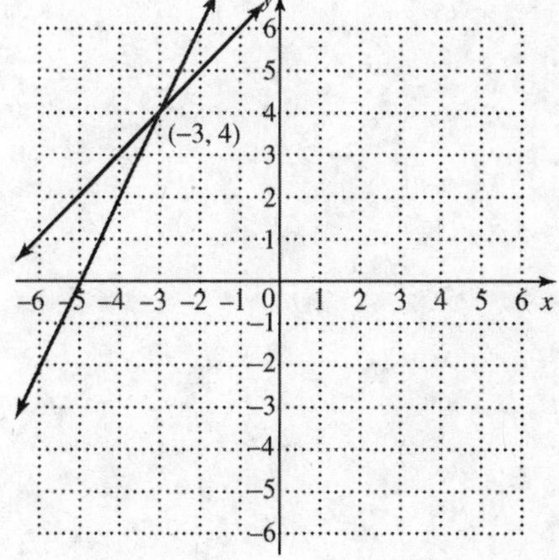

Objective 3

17. infinite number of solutions
19. infinite number of solutions
21. no solution

Objective 4

23. (a) neither (b) intersecting lines (c) one solution
25. (a) dependent (b) one line (c) infinitely many solutions
27. (a) neither (b) intersecting lines (c) one solution
29. (a) neither (b) intersecting lines (c) one solution

4.2 Solving Systems of Linear Equations by Substitution

Key Terms

1. ordered pair
2. substitution
3. dependent system
4. inconsistent system

Objective 1

1. (1, 6)
3. (2, 7)
5. (–2, –1)
7. (–1, –7)
9. $\left(3, -\frac{2}{3}\right)$
11. (–2, –3)

Objective 2

13. ∅
15. ∅
17. ∅

Objective 3

19. (6, 2)
21. (2, 3)
23. (6, 2)
25. (7, –2)
27. (4, –9)
29. $\{(x, y) \mid 0.3x + 0.4y = 0.5\}$

4.3 Solving Systems of Linear Equations by Elimination

Key Terms

1. elimination method 2. addition property of equality

3. substitution

Objective 1

1. (1, 4) 3. (8, 3) 5. (2, –4)

7. (5, 0)

Objective 2

9. $\left(\frac{1}{2}, 1\right)$ 11. (–4, 4) 13. (–3, 2)

15. $\left(\frac{1}{2}, -\frac{3}{2}\right)$

Objective 3

17. (–3, –2) 19. (–4, 1) 21. (3, –2)

23. (–2, 2)

Objective 4

25. $\varnothing$ 27. $\{(x, y) \mid 2x - 4y = 1\}$

29. $\{(x, y) \mid 48x - 56y = 32\}$ or $\{(x, y) \mid -18x + 21y = -12\}$

4.4 Applications of Linear Systems

Key Terms

1. $d = rt$
2. system of linear equations

Objective 1

1. 12, 8
3. 5647 people, 3398 people
5. 56 cm, 26 cm
7. 32 cm, 32 cm, 52 cm

Objective 2

9. 30 $5 bills; 60 $10 bills
11. $6000 at 7%; $4000 at 4%
13. $4000 at 7%; $8000 at 9%
15. 8 $14 ties; 2 $25 ties

Objective 3

17. water: 6 liters; 25% solution: 24 liters
19. $90 coffee: 40 bags; $75 coffee: 10 bags
21. water: 9 oz; 80% solution: 3 oz

Objective 4

23. Bill: 642 kph; Hillary: 582 kph
25. Enid: 44 mph; Jerry: 16 mph
27. John: 54 mph; Mike: 52 mph
29. plane speed: 265 mph; wind speed: 35 mph

4.5 Graphing Linear Inequalities in Two Variables

Key Terms

1. solution set of a system of linear inequalities

2. system of linear inequalities

Objective 1

1. 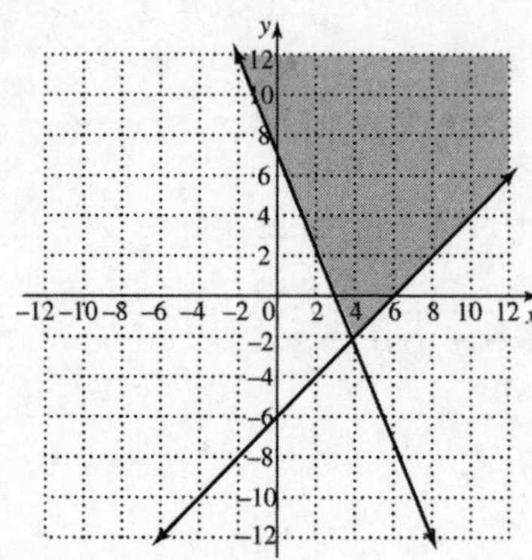

3. 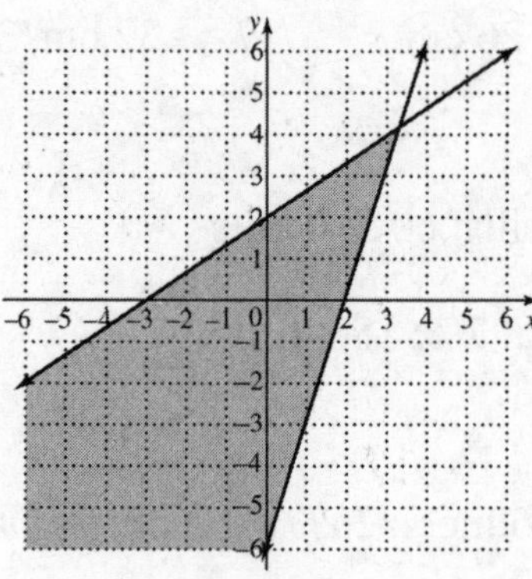

5. 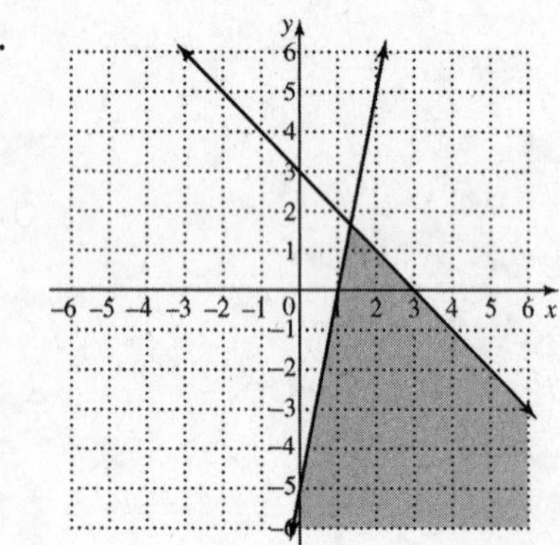

7. 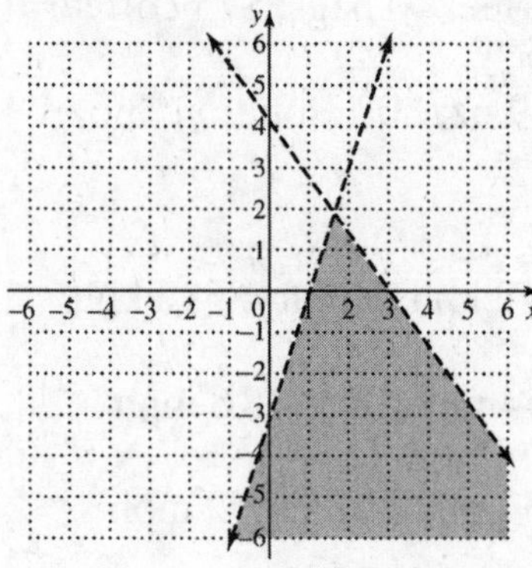

9. 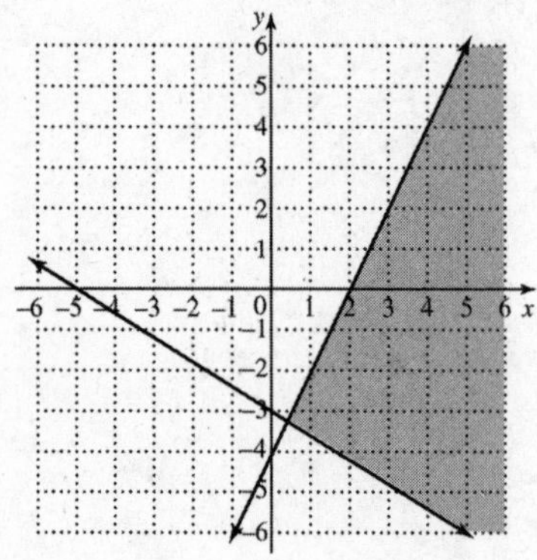

11. 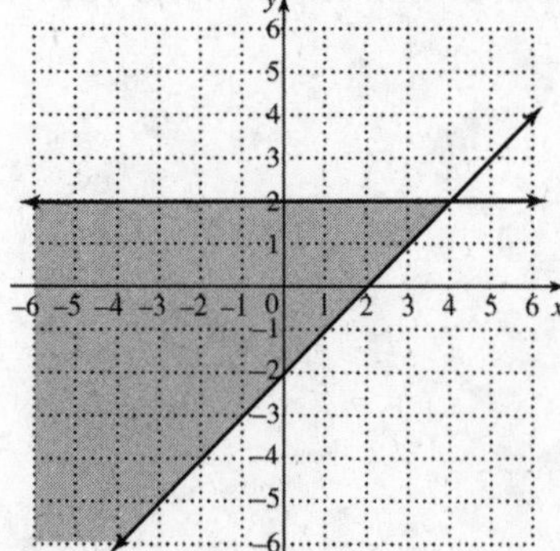

13.

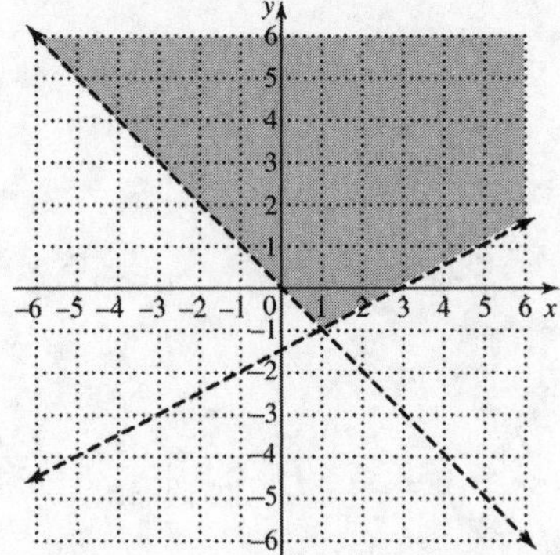

15.

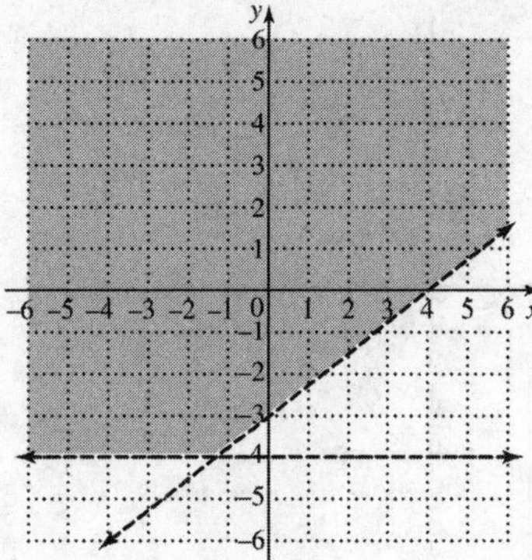

17.

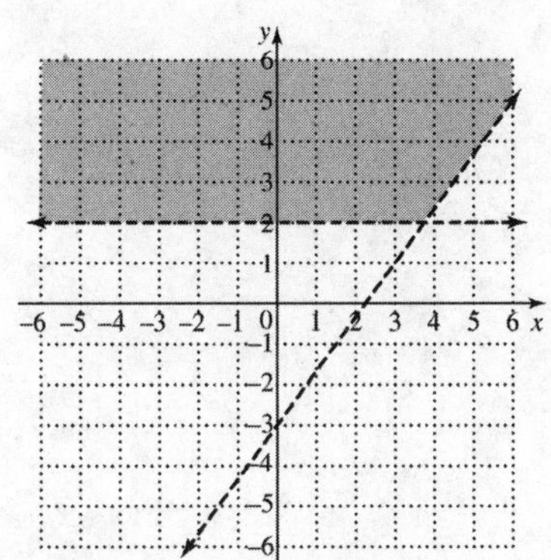

19.

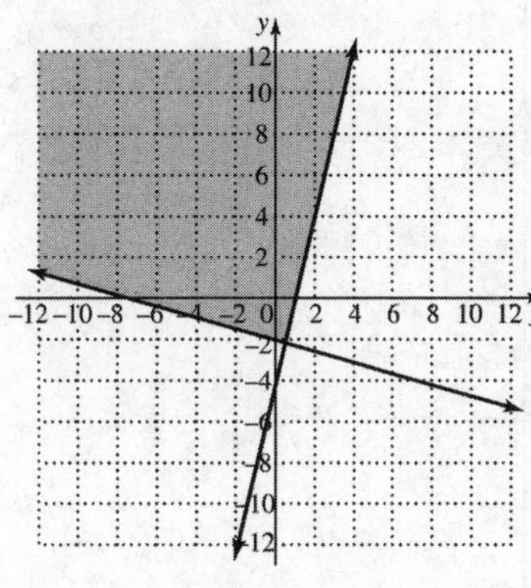

21.

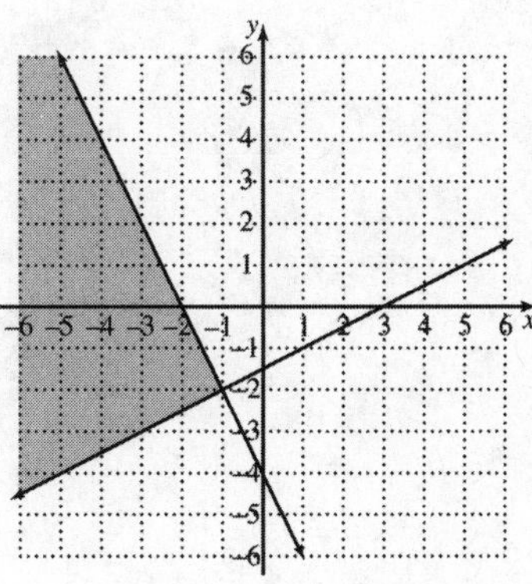

23.

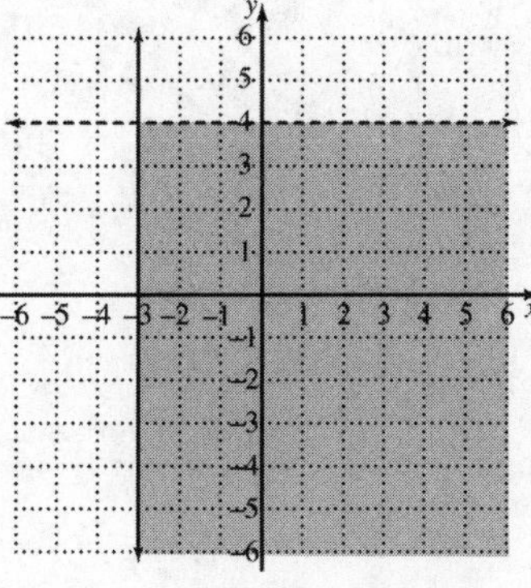

25.

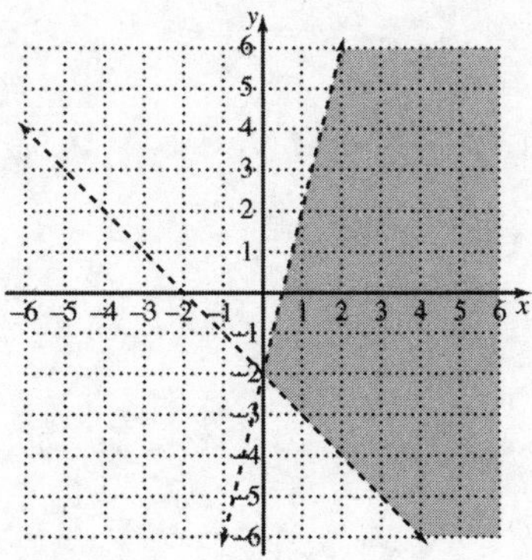

27.

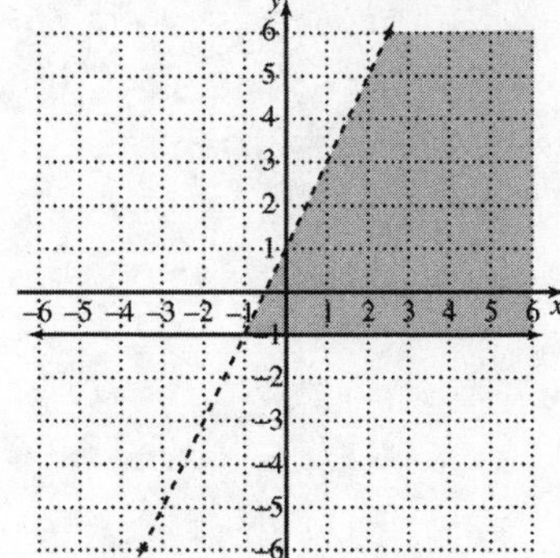

29.

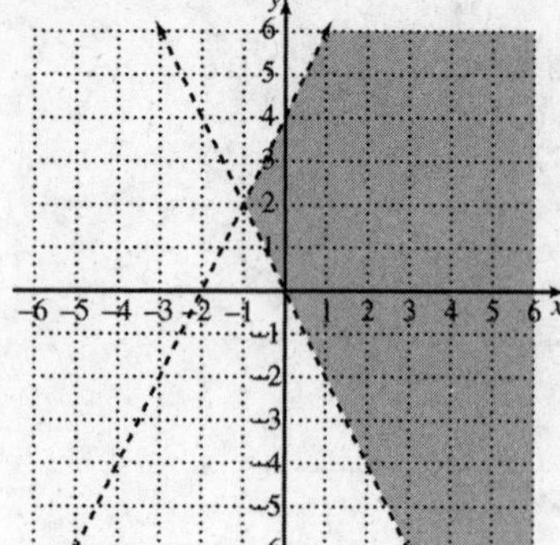

Chapter 5 EXPONENTS AND POLYNOMIALS

5.1 Adding and Subtracting Polynomials

Key Terms

1. degree of a term
2. descending powers
3. term
4. trinomial
5. polynomial
6. numerical coefficient
7. monomial
8. degree of a polynomial
9. binomial
10. like terms

Objective 1

1. $-3z^3$
3. $-\frac{1}{4}r^3$
5. $8c^3 - 8c^2 - 6c + 6$

Objective 2

7. $\frac{1}{2}x^2 - \frac{1}{2}x$; degree 2; binomial

9. $-5.7d^8 - 1.1d^5 + 3.2d^3 - d^2$; degree 8; none of these

Objective 3

11.a. −15; b. 115
13.a. 71; b. −19
15. a. −28; b. 352

Objective 4

17. $5m^3 - 2m^2 - 4m + 4$
19. $3r^3 + 7r^2 - 5r - 2$

Objective 5

21. $-8w^3 + 21w^2 - 15$
23. $8b^4 - 4b^3 - 2b^2 - b - 2$
25. $7d^4 - 3d^2 - 19d + 12$

Objective 6

27. $-8m^2n - 8m - 4n$
29. $-7x^2y + 3xy + 5xy^2$

5.2 The Product Rule and Power Rules for Exponents

Key Terms

1. power
2. exponential expression
3. base

Objective 1

1. $\frac{1}{243}$

3. 256; base: −4; exponent: 4

Objective 2

5. 7^7

7. $8c^{15}$

Objective 3

9. 7^{12}

11. $\left(\frac{1}{3}\right)^{15}$

Objective 4

13. $\frac{1}{9}x^8$

15. $-.008a^{12}b^3$

Objective 5

17. $-\frac{8x^3}{125}$

19. $-\frac{128a^7}{b^{14}}$

Objective 6

21. x^{26}

23. $5^{11}x^{18}y^{37}$

25. $7^3a^{10}b^{18}c^{19}$

Objective 7

27. $36x^5$

29. $3p^8q^6$

5.3 Multiplying Polynomials

Key Terms

1. inner product
2. FOIL
3. outer product

Objective 1

1. $16y^7$
3. $6m+14m^3+6m^4$
5. $-6z^6-18z^4-24z^2-12z$
7. $-35b^4+7b^2-28b^3$
9. $32m^3n+16m^2n^2+56mn^3$

Objective 2

11. $6x^2+23x+20$
13. x^3+27
15. $2r^3+3r^2-4r+15$
17. $6m^5+4m^4-5m^3+2m^2-4m$
19. $y^4-4y^3-2y^2+12y+9$

Objective 3

21. $6x^2-5xy-6y^2$
23. $3+10a+8a^2$
25. $-6m^2-mn+12n^2$
27. $x^2-.2x-.15$
29. $x^2-\frac{5x}{3}+\frac{4}{9}$

5.4 Special Products

Key Terms

1. binomial
2. conjugate

Objective 1

1. $25y^2 - 30y + 9$
3. $49 + 14x + x^2$
5. $4p^2 + 12pq + 9q^2$
7. $16y^2 - 5.6y + .49$
9. $9x^2 + 2xy + \frac{y^2}{9}$

Objective 2

11. $144 - x^2$
13. $49x^2 - 9y^2$
15. $x^2 - .04$
17. $49m^2 - \frac{9}{16}$
19. $16x^2 - \frac{49}{36}$

Objective 3

21. $a^3 - 9a^2 + 27a - 27$
23. $64x^3 + 48x^2y + 12xy^2 + y^3$
25. $\frac{1}{8}t^3 + \frac{3}{2}t^2u + 6tu^2 + 8u^3$
27. $x^4 + 8x^3y + 24x^2y^2 + 32xy^3 + 16y^4$
29. $16x^4 - 32x^3y + 24x^2y^2 - 8xy^3 + y^4$

5.5 Integer Exponents and the Quotient Rule

Key Terms

1. power rule for exponents
2. base; exponent
3. product rule for exponents

Objective 1

1. -1
3. 2
5. -2
7. 0

Objective 2

9. $2r^7$
11. $-\frac{1}{64}$
13. $-\frac{3}{x^2}$
15. 4

Objective 3

17. $\frac{1}{2x^6}$
19. $\frac{x^6}{12^3y^2}$ or $\frac{x^6}{1728y^2}$
21. $\frac{p^8}{3^5m^3}$ or $\frac{p^8}{243m^3}$

Objective 4

23. $\frac{9}{16y^2}$
25. $\frac{1}{9xy}$
27. $a^{16}b^{22}$
29. $\frac{2^4 \cdot 3^8 x^{20}}{y^8}$ or $\frac{104{,}976x^{20}}{y^8}$

5.6 Dividing a Polynomial by a Monomial

Key Terms

1. dividend
2. quotient
3. divisor

Objective 1

1. $2a^3 - 3a$
3. $5a^2 - \frac{9}{4}$
5. $1 + 3p^3$
7. $2 + 3y^3 - 7y^5$
9. $8p^2 - 7p - \frac{3}{p}$
11. $-\frac{r}{3s} - \frac{2}{3} + \frac{s}{r}$
13. $-4a^3 + \frac{3a^2b}{4} - 3ab^2$
15. $-4p^2 - 3 - \frac{5}{p} + \frac{1}{7p^2}$
17. $2z^4 + 9z^2 - 4 + \frac{10}{3z}$
19. $\frac{m}{2} + \frac{7}{2} - \frac{21}{m}$
21. $2y^6 + 8y^3 - 41 - \frac{12}{y^3}$
23. $12 + 16x^3 + \frac{x^7}{2}$
25. $-3 + \frac{2}{y} - \frac{6}{y^2}$
27. $10d^2 + 4d - 7 - \frac{4}{d^2}$
29. $4y^2 + 5y - \frac{2}{3y}$

5.7 Dividing a Polynomial by a Polynomial

Key Terms

1. divisor
2. quotient
3. dividend

Objective 1

1. $6a - 5$
3. $p + 8$
5. $9a - 1$
7. $a - 7 + \frac{37}{2a + 3}$
9. $3m - 2 + \frac{8}{3m - 4}$
11. $5b - 3 + \frac{24}{b + 7}$
13. $3y^2 - 5y + 6$
15. $2m^2 + m - 5 + \frac{26}{3m + 2}$
17. $3x - 2 + \frac{34x - 35}{x^2 - 3x - 5}$
19. $3y^3 - 2y^2 + 2y - 1 + \frac{6y - 8}{4y^2 - 3}$
21. $y^2 - y + 1$
23. $3x^2 + 2x + 1$
25. $16x^4 + 24x^3 + 36x^2 + 54x + 81$

Objective 2

27. $2r^2 - r + 5$ square units
29. $4y^2 + 24y + 100$ units

5.8 An Application of Exponents: Scientific Notation

Key Terms

1. scientific notation
2. power rule
3. quotient rule

Objective 1

1. 3.25×10^2
3. 2.3651×10^4
5. 9.54×10^6
7. 5.03×10^{-2}
9. -2.2208×10^{-4}

Objective 2

11. 72,000,000
13. 23,000
15. 0.0064
17. 0.04007
19. −4.02

Objective 3

21. 2.53×10^2
23. 2×10^2
25. 2.1×10^0 or 2.1
27. 4.86×10^{19} atoms
29. 9.46×10^{12} km

Chapter 6 FACTORING AND APPLICATIONS

6.1 Factors; The Greatest Common Factor

Key Terms

1. factoring
2. factored form
3. greatest common factor
4. factor

Objective 1

1. 15
3. 3
5. 28
7. 6

Objective 2

9. $2ab^2$
11. $w^2x^5y^4$
13. $9xy^2$

Objective 3

15. $-8a(a-3b-5c)$
17. $10x(2x+4xy-7y^2)$
19. $-13x^8(-2+x^4-4x^2)$
21. $8xy^2(7xy^2-3y+4)$

Objective 4

23. $(1+p)(1-q)$
25. $(4x-y)(2x+3y)$
27. $(4x-y^2)(3x^2-y)$
29. $(r^2+s^2)(3r-2s)$

6.2 Factoring Trinomials

Key Terms

1. factoring
2. greatest common factor
3. prime polynomial

Objective 1

1. prime
3. $(s-8)(s+4)$
5. $(x+9)(x+2)$
7. $(x-2)(x+1)$
9. $(x-7)(x+5)$
11. $(x-7y)(x-8y)$
13. $(q-6)(q+2)$
15. $(a-8b)(a-2b)$

Objective 2

17. $2m(m-2)(m+1)$
19. prime
21. $3p^4(p+4)(p+2)$
23. $10k^4(k+5)(k+2)$
25. $x^3(x-2)(x-1)$
27. $2y^2(x-3y)(x+2y)$
29. $qr(r-7q)(r+3q)$

6.3 Factoring Trinomials by Grouping

Key Terms

1. coefficient
2. trinomial

Objective 1

1. $x+3$
3. $4x-2$
5. $(4b+3)(2b+3)$
7. $(5a+2)(3a+2)$
9. $(3b+2)(b+2)$
11. $p(3p+2)(p+2)$
13. $b(7a+4)(a+2)$
15. $(3c+6d)(3c+2d)$
17. $(5c-7t)(2c-3t)$
19. $(2x+7y)(6x-5y)$
21. $m(6m+8n)(m-n)$
23. $(6f-g)(3f+5g)$
25. $2a(4a-5)(5a-4)$
27. $4(2x+y)(x-y)$
29. $(2c+3d)(5c+12d)$

6.4 Factoring Trinomials Using FOIL

Key Terms

1. inner product
2. FOIL
3. outer product

Objective 1

1. $(2x+3)(5x+2)$
3. $(2a+1)(a+6)$
5. $(2m-3)(4m+1)$
7. $(3q-4)(5q+6)$
9. $(3w+2z)(3w+2z)$
11. $(3x-4y)(2x+3y)$
13. $(3y+5)(4y-3)$
15. $(2p+1)(p+5)$
17. $(9y+2)(y-2)$
19. $(3r-1)(3r+5)$
21. $2(2c-d)(c+4d)$
23. $(9r-4t)(3r+2t)$
25. $(7c-3d)(4c+5d)$
27. $(3n-7s)(2n+9s)$
29. $2a(3a+2b)(2a+3b)$

6.5 Special Factoring Techniques

Key Terms

1. difference of squares
2. perfect square trinomial

Objective 1

1. $(5a-6)(5a+6)$
3. $\left(3j-\frac{4}{7}\right)\left(3j+\frac{4}{7}\right)$
5. $\left(4y^2+9\right)(2y-3)(2y+3)$
7. $m^2(mn-1)(mn+1)$

Objective 2

9. $(2x+3)^2$
11. $(4q-5)^2$
13. $\left(10p-\frac{5}{8}r\right)^2$
15. $\left[(p-q)-10\right]^2$

Objective 3

17. $(x-y)(x^2+xy+y^2)$
19. $(6m-5p^2)(36m^2+30mp^2+25p^4)$
21. $8(3x-y)(9x^2+3xy+y^2)$
23. $3x^2-3x+1$

Objective 4

25. $(3r+2s)(9r^2-6rs+4s^2)$
27. $(5p+q)(25p^2-5pq+q^2)$
29. $2x(x^2+3y^2)$

6.6 A General Approach to Factoring

Key Terms

1. factoring by grouping
2. FOIL

Objective 1

1. $-6x(2x+1)$
3. $5rt(r-2+t)$
5. $3a[a+2(x-y)]$
7. $(m-n)^2$

Objective 2

9. $2xy^2(xy+6)(xy-6)$
11. $[(r+s)+2][(r+s)^2-2(r+s)+4]$ or $[(r+s)+2][r^2+2rs+s^2-2r-2s+4]$
13. $(y^2+1)(y^4-y^2+1)$
15. $[(3a-1)+y^3][(3a-1)-y^3]$

Objective 3

17. $(a-6)(2a-5)$
19. $(5x-2y)(5x+y)$
21. $(4m+5)(3m-1)$
23. $(4b+1)(b+2)(b-1)$

Objective 4

25. $(x-y)(a+b)$
27. $(x-3)(x^2+7)$
29. $(a-3b+5)(a-3b-5)$

6.7 Solving Quadratic Equations by Factoring

Key Terms

1. standard form
2. quadratic equation

Objective 1

1. $\{-2, -5\}$
3. $\{-7, 7\}$
5. $\{-7, 9\}$
7. $\left\{-\frac{2}{3}, 3\right\}$
9. $\left\{-\frac{2}{3}, -\frac{2}{3}\right\}$
11. $\left\{-\frac{4}{3}, \frac{4}{3}\right\}$
13. $\left\{-\frac{2}{7}, \frac{3}{2}\right\}$
15. $\left\{-3, \frac{1}{2}\right\}$

Objective 2

17. $\left\{-\frac{3}{2}, 0, 5\right\}$
19. $\{-7, 0, 7\}$
21. $\{-4, 0, 2\}$
23. $\left\{-\frac{3}{2}, \frac{3}{2}, 2\right\}$
25. $\left\{-\frac{4}{3}, 0, 1\right\}$
27. $\{-6, 2, 3, 6\}$
29. $\left\{-5, \frac{3}{2}, 7\right\}$

6.8 Applications of Quadratic Equations

Key Terms

1. legs
2. hypotenuse

Objective 1

1. width: 8 in., length: 24 in.

3. rectangle 1: width: 4 m, length: 12 m; rectangle 2: width: 6 m; length: 8 m

5. base: 12 cm; height: 7 cm

7. width: 3m, length: 5 m

Objective 2

9. −4, −3 or 3, 4

11. 12, 14

13. 8, 10

15. 6 in., 8 in., 10 in.

Objective 3

17. car: 60 mi; train: 80 mi

19. 45 mi

21. 20 mi

Objective 4

23. a. $\frac{1}{2}$ sec; b. 1 sec; c. 2 sec

25. 110 items

27. 40 items or 110 items

29. 404 ft

Chapter 7 RATIONAL EXPRESSIONS AND FUNCTIONS

7.1 Rational Expressions and Functions; Multiplying and Dividing

Key Terms

1. rational function
2. rational expression

Objective 1; Objective 2

1. $\frac{7}{4}; \left\{a \mid a \neq \frac{7}{4}\right\}$
3. none; $(-\infty, \infty)$
5. 1, 2; $\{q \mid q \neq 1, 2\}$

Objective 3

7. $-\frac{4+x}{2}$
9. $\frac{9(x+3)}{2}$
11. -1

Objective 4

13. $\frac{x+4}{2(x-4)}$
15. $-\frac{x-3}{2x+3}$
17. $\frac{m-2}{m-4}$

Objective 5

19. $\frac{s-8}{8-s}$ or -1
21. $\frac{3x-6}{x^2+4}$ or $\frac{3(x-2)}{x^2+4}$
23. no reciprocal for 0

Objective 6

25. $-\frac{2(m-5)}{m+3}$
27. $\frac{4(y-1)}{3(y-3)}$
29. $\frac{2(z+1)}{3z(z-2)}$

7.2 Adding and Subtracting Rational Expressions

Key Terms

1. equivalent expressions
2. least common denominator (LCD)

Objective 1

1. $z+y$
3. $\frac{4n-7}{m+3}$
5. $\frac{1}{r-s}$
7. $\frac{1}{x-5}$
9. $\frac{1}{q-7}$

Objective 2

11. $5a(a+2)$
13. $3(x+2)(x+3)$
15. $a(a-2)(2a+5)$
17. $n(3+n)(3-n)$
19. $(p-4)(p+4)^2$

Objective 3

21. $\frac{3x-2}{2(x-2)(x+2)}$
23. $-\frac{2(m-5)}{3(m+4)(3m-1)}$
25. $\frac{9b+14}{(3b+4)(3b-4)(b+2)}$
27. $\frac{-6n^2+27n+12}{(n+4)^2(n-4)}$
29. $\frac{6z^2+19z-4}{(z+2)(z+3)(z+4)}$

7.3 Complex Fractions

Key Terms

1. complex fraction
2. LCD

Objective 1

1. $\frac{1}{rs}$

3. $\frac{5r-2rs}{3rs-4s}$ or $\frac{r(5-2s)}{s(3r-4)}$

5. $(a+2)^2$

7. $\frac{q(q^2+q+1)}{(q+1)^2(q-1)}$

Objective 2

9. $\frac{x+3}{x}$

11. $\frac{r^2+3}{5+r^2t}$

13. $\frac{9}{10}$

Objective 3

15. $\frac{1-r}{r(1+r)}$

17. $\frac{7(5k-m)}{4}$ or $\frac{35k-7m}{4}$

19. $\frac{3(8-x)}{2(15+x)}$

21. $\frac{2(s-9)}{5}$

Objective 4

23. $\frac{1}{5x+1}$

25. $\frac{2z^3+xy^2z^3}{x}$

27. $\frac{2y^3}{x^2y^3+3x^2}$

29. $\frac{1}{xy-1}$

7.4 Equations with Rational Expressions and Graphs

Key Terms

1. vertical asymptote
2. domain of the variable in a rational equation
3. discontinuous
4. horizontal asymptote

Objective 1

1. a. 0, –1; b. $\{x \mid x \neq 0, -1\}$
3. a. 7, –8; b. $\{x \mid x \neq 7, -8\}$
5. a. 0, 2; b. $\{x \mid x \neq 0, 2\}$

Objective 2

7. $\left\{-\frac{18}{7}, 1\right\}$
9. $\{4\}$
11. $\left\{\frac{1}{3}\right\}$
13. $\left\{-\frac{4}{3}, 1\right\}$
15. $\varnothing$
17. $\left\{\frac{4}{3}\right\}$
19. $\{2\}$
21. $\{-24, 1\}$
23. $\left\{-\frac{15}{4}\right\}$

Objective 3

25.

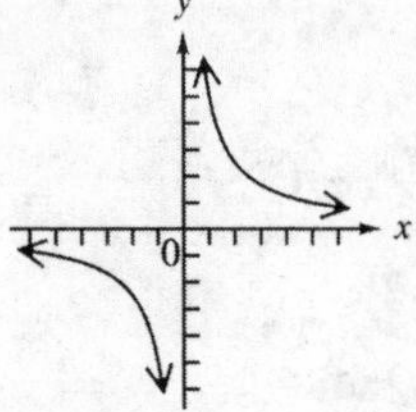

Vertical asymptote: $x = 0$
Horizontal asymptote: $y = 0$

27.

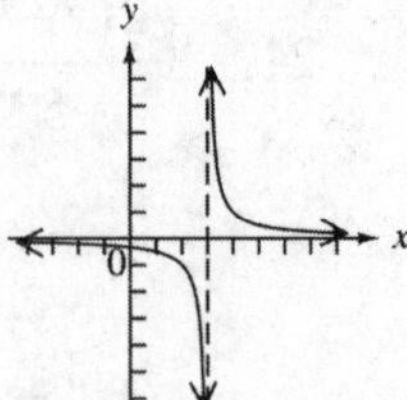

Vertical asymptote: $x = 3$
Horizontal asymptote: $y = 0$

29.

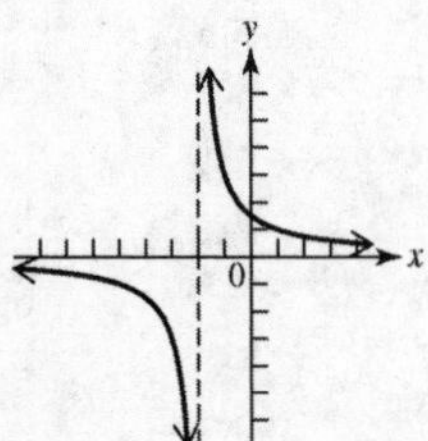

Vertical asymptote: $x = -2$
Horizontal asymptote: $y = 0$

7.5 Applications of Rational Expressions

Key Terms

1. ratio
2. proportion

Objective 1

1. $m = \frac{75}{8}$
3. $d_i = \frac{50}{3}$
5. $L = 20$

Objective 2

7. $T_2 = \frac{T_1 V_2 P_2}{V_1 P_1}$
9. $F = \frac{9C}{5} + 32$
11. $v_s = \frac{Fv - fv - fv_0}{F}$

Objective 3

13. 5 hr
15. \$3.90
17. 50,000 people

Objective 4

19. 16 mi
21. 120 mi
23. 350 mi

Objective 5

25. $2\frac{2}{5}$ hr
27. $\frac{2}{5}$ hr
29. 2 hr

7.6 Variation

Key Terms

1. constant of variation
2. varies inversely
3. varies directly

Objective 1

1. $k = 3$; $y = 3x$
3. $k = .25$; $y = .25x$
5. $k = \frac{14}{9}$; $y = \frac{14}{9}x$

Objective 2

7. 72
9. 69.08 cm
11. 100 newtons

Objective 3

13. $\frac{5}{2}$ or $2\frac{1}{2}$
15. 100 amps
17. 40 lb

Objective 4

19. 96
21. 96
23. 1470 joules

Objective 5

25. $\frac{1}{12}$
27. 120
29. 9 hr

Chapter 8 EQUATIONS, INEQUALITIES AND SYSTEMS REVISITED

8.1 Review of Solving Equations and Linear Inequalities

Key Terms

1. equivalent equations
2. linear (first-degree) equation in one variable
3. solution set
4. contradiction
5. conditional equation
6. identity
7. solution
8. linear inequality in one variable

Objective 1

1. $\left\{-\frac{2}{7}\right\}$
3. $\left\{\frac{5}{9}\right\}$
5. $\{4\}$
7. $\{-5\}$
9. $\{3\}$
11. contradiction; $\varnothing$

Objective 2

13. $[3,\infty)$;
15. $(2,\infty)$; 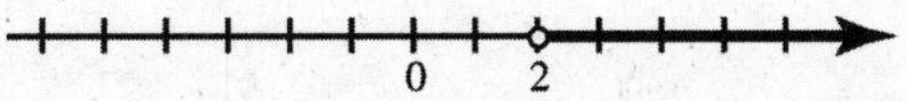
17. $(-\infty,-2)$; 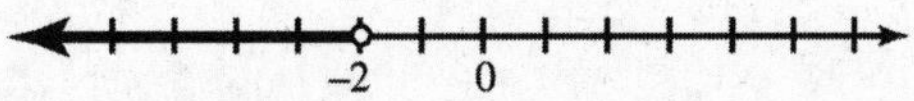
19. $(-2,\infty)$; 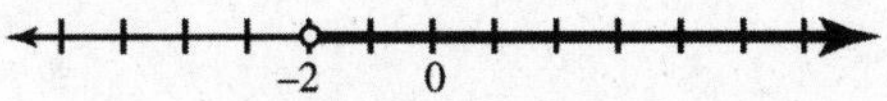
21. $[0,\infty)$;
23. $(-\infty,4)$

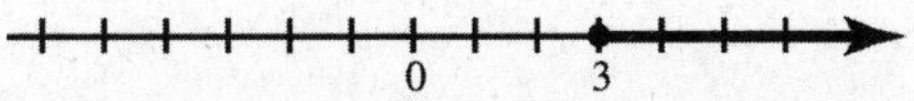

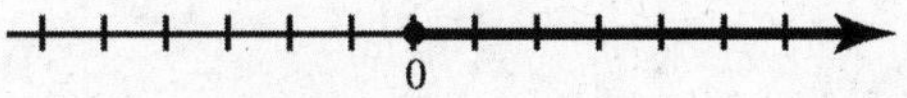

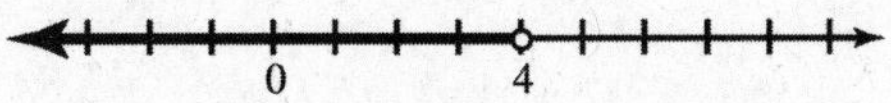

Objective 3

25. $(4,7]$;

27. $(5,6)$;

29. $(-1,5)$;

8.2 Set Operations and Compound Inequalities

Key Terms

1. union 2. compound inequality 3. intersection

Objective 1

1. $\{2,4\}$ 3. $\varnothing$ 5. $\{0\}$ 7. $\varnothing$

Objective 2

9. $[5,9]$

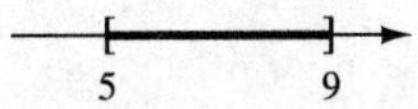

11. $\varnothing$

13. $(0,1]$

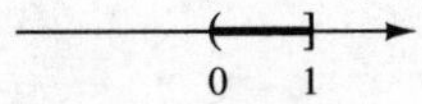

15. $\left(-\infty, -\frac{2}{7}\right]$

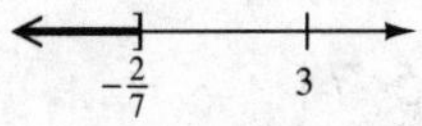

Objective 3

17. $\{0,1,2,3,4,5,7,9\}$ 19. $\{1,2,3,4,5,6,7,8,9,10\}$

21. $\{0,1,2,3,4,5\}$

Objective 4

23. $(-\infty, -1) \cup (5, \infty)$

25. $(-\infty, -4] \cup (4, \infty)$

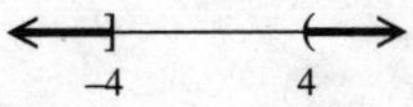

27. $(-\infty, -2) \cup (3, \infty)$

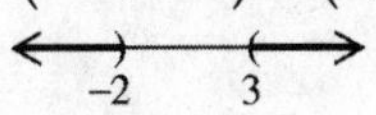

29. $(-\infty, \infty)$

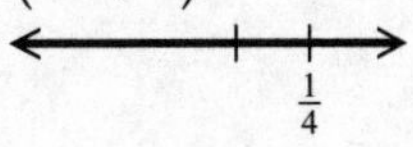

8.3 Absolute Value Equations and Inequalities

Key Terms

1. absolute value equation
2. absolute value inequality

Objective 1

1. 

3. 

5. 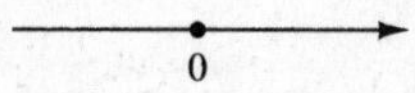

Objective 2

7. $\left\{-\dfrac{5}{3}, \dfrac{7}{3}\right\}$

9. $\{-2, 14\}$

Objective 3

11. $(-\infty, -6) \cup (10, \infty)$

13. $\left[-\dfrac{1}{4}, \dfrac{3}{4}\right]$

15. $\left(-4, \dfrac{16}{5}\right)$
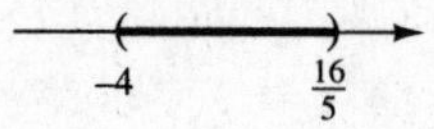

Objective 4

17. $\varnothing$

19. $\{-42, 50\}$

Objective 5

21. $\left\{-\dfrac{15}{4}, \dfrac{1}{8}\right\}$

23. $\left\{-3, \dfrac{11}{3}\right\}$

25. $\left\{-\dfrac{1}{2}, \dfrac{5}{2}\right\}$

Objective 6

27. {all real numbers}

29. $\varnothing$

8.4 Review of Systems of Linear Equations in Two Variables

Key Terms

1. independent equations
2. consistent system
3. system of equations
4. solution set of a system
5. dependent equations
6. inconsistent system
7. linear system

Objective 1

1.

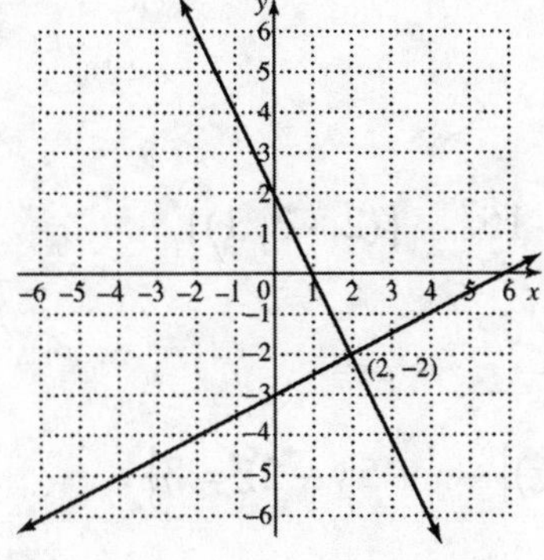

3.

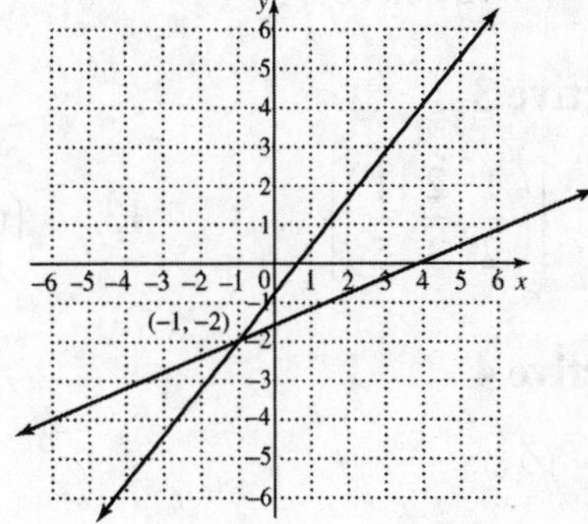

5.

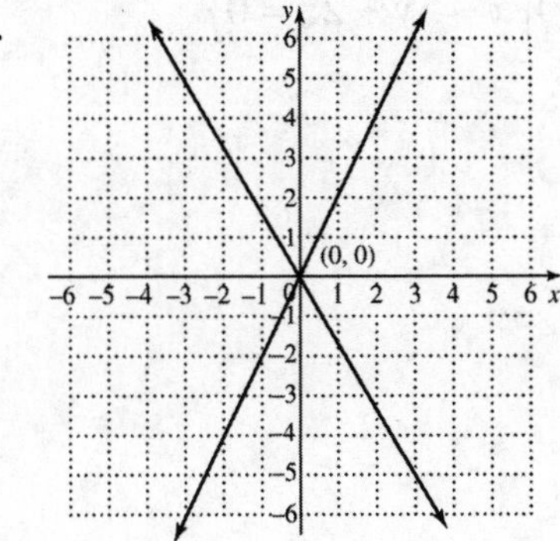

7. {(−4, −8)}
9. {(−3, −4)}
11. {(3, 5)}
13. {(1, 3)}
15. {(−3, −1)}
17. {(6, 8)}
19. $\left\{\left(\frac{49}{13}, \frac{15}{13}\right)\right\}$
21. $\{(-3,-1)\}$
23. {(−5, −4)}

Objective 2

25. ∅
27. ∅
29. $\{(x, y) \mid -3x + 2y = 6\}$

8.5 Systems of Linear Equations in Three Variables; Applications

Key Terms

1. ordered triple 2. dependent system 3. inconsistent system

Objective 1

1. The planes intersect in one point.

Objective 2

3. (–2, 1, 1)} 5. {(0, –2, 5)} 7. {(4, –4, 1)}

9. {(–15, 0, 16)}

Objective 3

11. $\left\{\left(\frac{1}{2}, \frac{2}{3}, \frac{1}{5}\right)\right\}$ 13. {(0, –2, 5)} 15. {(2, –3, 1)}

Objective 4

17. $\varnothing$ 19. $\{(x, y, z) | -x+5y-2z=3\}$

21. $\varnothing$ 23. $\{(x, y, z) | x-5y+2z=0\}$

Objective 5

25. 9, 10, 12

27. \$5 bills: 7, \$10 bills: 12, \$20 bills: 32

29. 60 nickels, 20 dimes, 5 quarters

8.6 Solving Systems of Linear Equations by Matrix Methods

Key Terms

1. augmented matrix
2. matrix
3. square matrix
4. row echelon form
5. elements of the matrix

Objective 1

1. 3×2
3. 3×4
5. 6, 0, 1

Objective 2

7. $\left[\begin{array}{cc|c} 3 & -4 & 7 \\ 2 & 1 & 12 \end{array}\right]$

9. $\left[\begin{array}{cc|c} \frac{1}{2} & \frac{1}{2} & -16 \\ -3 & 1 & 2 \end{array}\right]$

11. $\left[\begin{array}{ccc|c} -2 & 3 & -5 & 7 \\ 6 & 2 & -4 & 12 \\ 5 & -2 & 1 & -1 \end{array}\right]$

Objective 3

13. {(2, –3)}
15. {(3, –2)}
17. {(–5, –3)}

Objective 4

19. {(2, –1, –3)}
21. {(1, 3, –2)}
23. {(1, 2, 4)}

Objective 5

25. ∅
27. $\{(x, y) | 2x + y = 10\}$
29. {(1, 0, 0)}

Chapter 9 ROOTS, RADICALS, AND ROOT FUNCTIONS

9.1 Radical Expressions and Graphs

Key Terms

1. radicand
2. perfect square
3. index (order)
4. square root
5. radical expression
6. principal square root
7. irrational number
8. radical
9. cube root

Objective 1

1. 25, –25
3. $\frac{11}{14}, -\frac{11}{14}$
5. $-\frac{50}{80}$ or $-\frac{5}{8}$

Objective 2

7. not a real number
9. irrational

Objective 3

11. –4
13. –3
15. 4

Objective 4

17.

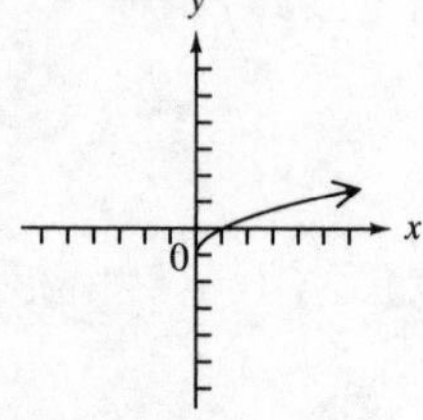

Domain: $[0, \infty)$
Range: $[-1, \infty)$

19.

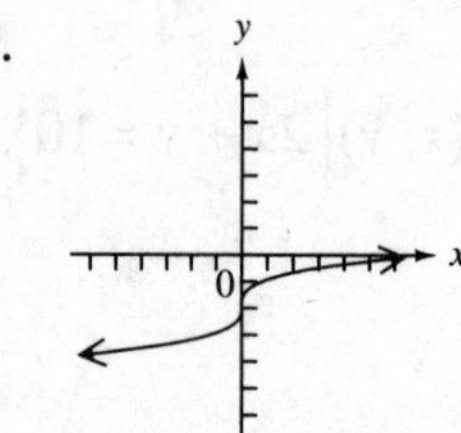

Domain: $(-\infty, \infty)$
Range: $(-\infty, \infty)$

Objective 5

21. x^5
23. x^2
25. $-x^4$

Objective 6

27. 2.546
29. 2.418

9.2 Rational Exponents

Key Terms

1. power rule for exponents
2. product rule for exponents
3. quotient rule for exponents

Objective 1

1. -2
3. -3
5. -4
7. 15

Objective 2

9. $\frac{1}{125}$
11. $-\frac{1}{25}$
13. 7776

Objective 3

15. $\sqrt[3]{2t^2}$
17. $\sqrt[8]{y^7}$
19. $\sqrt[6]{k^5}$
21. $\sqrt[15]{x}$

Objective 4

23. y
25. 1
27. 8
29. $\frac{y^{10/3}}{x^{14/5}}$

9.3 Simplifying Radical Expressions

Key Terms

1. hypotenuse
2. index; radicand
3. legs

Objective 1

1. $\sqrt[4]{4x}$
3. $\sqrt[4]{120}$
5. $\sqrt[7]{48a^3t^6}$

Objective 2

7. $\frac{\sqrt{15}}{13}$
9. $-\frac{a^2}{5}$

Objective 3

11. $\sqrt[3]{x^2}$
13. $\sqrt[4]{9x^2y^3}$
15. $3bc^2\sqrt[3]{10bc^2}$

Objective 4

17. $\frac{1}{\sqrt[4]{2}}$
19. $\sqrt[6]{5400}$

Objective 5

21. 26
23. $6\sqrt{2}$
25. $3\sqrt{5}$

Objective 6

27. $\sqrt{34}$
29. $\sqrt{x^2+4y^2}$

9.4 Adding and Subtracting Radical Expressions

Key Terms

1. unlike radicals
2. like radicals

Objective 1

1. $22\sqrt[3]{3}$
3. $5\sqrt[4]{2}$
5. $9\sqrt[3]{3}$
7. $12\sqrt{x}$
9. $-\sqrt{6}$
11. $24\sqrt[3]{5}$
13. $13\sqrt{2z}$
15. $28z\sqrt[3]{2}$
17. $17\sqrt{5x}$
19. $\dfrac{17}{w^2}$
21. $\dfrac{26\sqrt{2}}{r}$
23. $\dfrac{(2t-10)\sqrt{3y}}{t^2}$
25. $11\sqrt{5}$ cm
27. $35\sqrt{15}$ cm^2
29. width: $3\sqrt{3}$ cm; area: 135 cm^2

9.5 Multiplying and Dividing Radical Expressions

Key Terms

1. conjugate
2. rationalizing the denominator

Objective 1

1. $-55-25\sqrt{6}$
3. $2\sqrt{15}-\sqrt{110}+3\sqrt{2}-\sqrt{33}$
5. $6x-13\sqrt{x}+6$
7. $4-\sqrt[3]{25}$

Objective 2

9. $\dfrac{ab\sqrt{30b}}{6}$
11. $\dfrac{m^2\sqrt{k}}{k^2}$
13. $\dfrac{\sqrt[3]{35x^2}}{7x}$

Objective 3

15. $-4\sqrt{3}+8$
17. $\dfrac{4\left(\sqrt{5}-\sqrt{2}\right)}{3}$
19. $20\sqrt{5}+25\sqrt{3}$
21. $-\sqrt{15}+2\sqrt{5}+2\sqrt{3}-4$

Objective 4

23. $\dfrac{1-\sqrt{2}}{2}$
25. $\dfrac{25+2\sqrt{5x}}{5}$
27. $1+\sqrt{2x}$
29. $\dfrac{\sqrt{5}-1}{7}$

9.6 Solving Equations with Radicals

Key Terms

1. extraneous solution 2. radical equation

Objective 1

1. {10} 3. {11} 5. {7}

7. ∅ 9. ∅

Objective 2

11. $\left\{\frac{3}{2}\right\}$ 13. {−1} 15. {0}

17. {1} 19. {7}

Objective 3

21. {−31} 23. {−4} 25. {−3}

27. {1} 29. {21}

9.7 Complex Numbers

Key Terms

1. complex number	2. complex conjugate	3. imaginary part
4. real part	5. standard form	6. pure imaginary number

Objective 1

1. $2i\sqrt{15}$
3. $6i\sqrt{30}$
5. $-i\sqrt{105}$
7. $\frac{i\sqrt{70}}{5}$

Objective 2

9. imaginary

Objective 3

11. $-4 + i$
13. $2 - 3i$
15. $7 + 5i$

Objective 4

17. $-8 + 6i$
19. $-1 - 2i\sqrt{6}$

Objective 5

21. $\frac{4}{5} - \frac{7}{5}i$
23. $\frac{15}{13} + \frac{16}{13}i$
25. $\frac{1}{5} + \frac{1}{5}i$

Objective 6

27. -1
29. i

Chapter 10 QUADRATIC EQUATIONS, INEQUALITIES, AND FUNCTIONS

10.1 Solving Quadratic Equations by the Square Root Property

Key Terms

1. quadratic equation 2. zero-factor property

Objective 1

1. $\left\{-\frac{7}{3}, 4\right\}$

3. $\{3\}$

Objective 2

5. $\{-30, 30\}$

7. $\varnothing$

9. $\left\{-\frac{2\sqrt{6}}{11}, \frac{2\sqrt{6}}{11}\right\}$

11. $\left\{-7\sqrt{2}, 7\sqrt{2}\right\}$

Objective 3

13. $\{-6, 2\}$

15. $\left\{4-\sqrt{7}, 4+\sqrt{7}\right\}$

17. $\left\{\frac{1}{5}, \frac{4}{5}\right\}$

19. $\{-20, 36\}$

21. $\left\{9-15\sqrt{2}, 9+15\sqrt{2}\right\}$

Objective 4

23. $\{-1-6i, -1+6i\}$

25. $\left\{-\frac{5}{4}-\frac{\sqrt{3}}{2}i, -\frac{5}{4}+\frac{\sqrt{3}}{2}i\right\}$

27. $\{2-2i, 2+2i\}$

29. $\left\{-2-\frac{7}{8}i, -2+\frac{7}{8}i\right\}$

10.2 Solving Quadratic Equations by Completing the Square

Key Terms

1. perfect square trinomial
2. square root property
3. completing the square

Objective 1

1. {–4, 1}
3. $\{-2-\sqrt{6}, -2+\sqrt{6}\}$
5. {–9, 7}
7. $\left\{\frac{1-\sqrt{11}}{2}, \frac{1+\sqrt{11}}{2}\right\}$
9. {1, 8}

Objective 2

11. {–2, 7}
13. $\left\{\frac{-3-\sqrt{11}}{2}, \frac{-3+\sqrt{11}}{2}\right\}$
15. ∅
17. $\left\{-1, \frac{2}{3}\right\}$
19. $\left\{\frac{-7-3\sqrt{17}}{4}, \frac{-7+3\sqrt{17}}{4}\right\}$

Objective 3

21. {2}
23. ∅
25. {–8, 2}
27. $\left\{\frac{-1-3\sqrt{5}}{2}, \frac{-1+3\sqrt{5}}{2}\right\}$
29. $\{6-\sqrt{33}, 6+\sqrt{33}\}$

10.3 The Quadratic Formula

Key Terms

1. discriminant
2. quadratic formula

Objective 1

1. $\{2, 4\}$

3. $\left\{\frac{-1-\sqrt{2}}{2}, \frac{-1+\sqrt{2}}{2}\right\}$

5. $\left\{\frac{1-\sqrt{13}}{2}, \frac{1+\sqrt{13}}{2}\right\}$

7. $\left\{\frac{-2-\sqrt{14}}{5}, \frac{-2+\sqrt{14}}{5}\right\}$

9. $\left\{\frac{7}{6}\right\}$

11. $\left\{-\frac{5}{14}-\frac{\sqrt{59}}{14}i, -\frac{5}{14}+\frac{\sqrt{59}}{14}i\right\}$

13. $\left\{1-i\sqrt{2}, 1+i\sqrt{2}\right\}$

15. $\{-4-i, -4+i\}$

17. $\left\{1-\frac{\sqrt{6}}{2}i, 1+\frac{\sqrt{6}}{2}i\right\}$

19. $\left\{-\frac{1}{2}-\frac{\sqrt{11}}{2}i, -\frac{1}{2}+\frac{\sqrt{11}}{2}i\right\}$

Objective 2

21. A
23. D
25. C
27. B
29. A

10.4 Equations Quadratic in Form

Key Terms

1. standard form   2. quadratic in form

Objective 1

1. $\left\{-\frac{5}{3}, 3\right\}$ 3. $\left\{-7, \frac{5}{4}\right\}$ 5. $\left\{-\frac{35}{4}, -3\right\}$

7. $\left\{-7, -\frac{7}{2}\right\}$

Objective 2

9. 15 hr; 30 hr   11. 550 mph

13. bike: 12 mph; hike: 2 mph

15. 18.6 hr

Objective 3

17. $\{2, 5\}$ 19. $\{3, 5\}$ 21. $\left\{\frac{1}{4}\right\}$

23. $\left\{\frac{1}{16}, \frac{1}{9}\right\}$

Objective 4

25. $\left\{-\frac{1}{2}, \frac{1}{2}, -\sqrt{5}, \sqrt{5}\right\}$

27. $\{9\}$ 29. $\{-2, -1, 4, 5\}$

10.5 Formulas and Further Applications

Key Terms

1. quadratic function
2. Pythagorean theorem

Objective 1

1. $d = \dfrac{k^2 l^2}{F^2}$

3. $k = \dfrac{p^2 g}{l}$

5. $c = \dfrac{(a-1)^2}{b}$

7. $a = \dfrac{-c \pm c\sqrt{2}}{b}$

Objective 2

9. east: 54 mi; south: 72 mi

11. 34 cm

13. 24 in.

Objective 3

15. 8 in. × 5 in. × 4 in.

17. 9 in. × 14 in.

19. 3 ft

21. 1.5 ft

Objective 4

23. 17.1 hr

25. 9.1 sec

27. 27 items

29. 4.1 sec

10.6 Graphs of Quadratic Functions

Key Terms

1. axis
2. vertex
3. quadratic function
4. parabola

Objective 1; Objective 2

1. 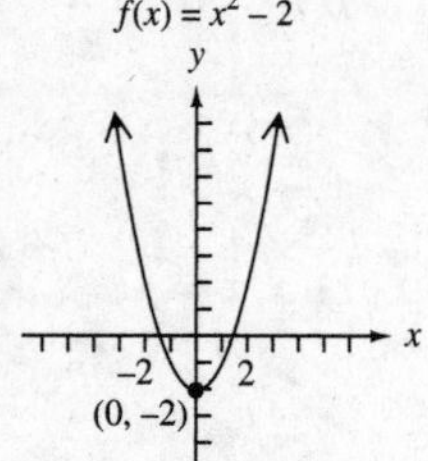

Vertex: (0, −2)
Axis: $x = 0$
Domain: $(-\infty, \infty)$
Range: $[-2, \infty)$

3. 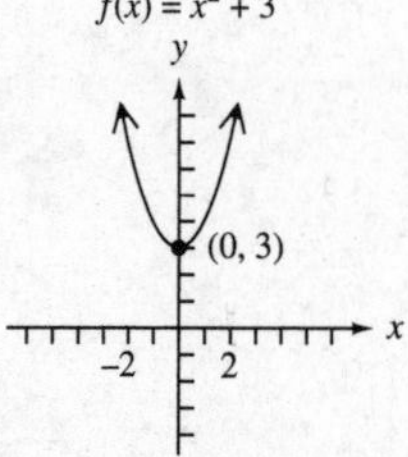

Vertex: (0, 3)
Axis: $x = 0$
Domain: $(-\infty, \infty)$
Range: $[3, \infty)$

5. 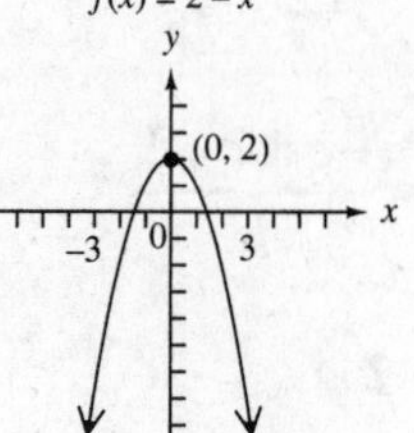

Vertex: (2, 0)
Axis: $x = 0$
Domain: $(-\infty, \infty)$
Range: $(-\infty, 2]$

7. 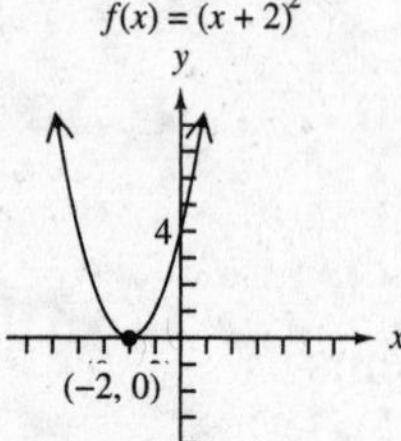

Vertex: (−2, 0)
Axis: $x = -2$
Domain: $(-\infty, \infty)$
Range: $[0, \infty)$

9. 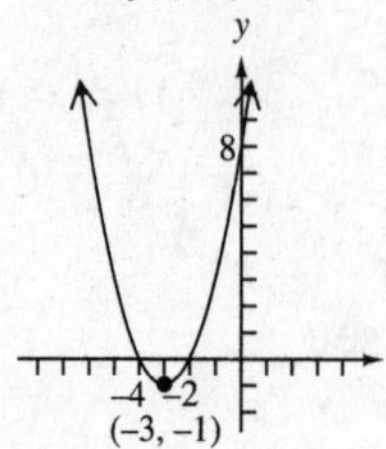

Vertex: (−3, −1)
Axis: $x = -3$
Domain: $(-\infty, \infty)$
Range: $[-1, \infty)$

11. 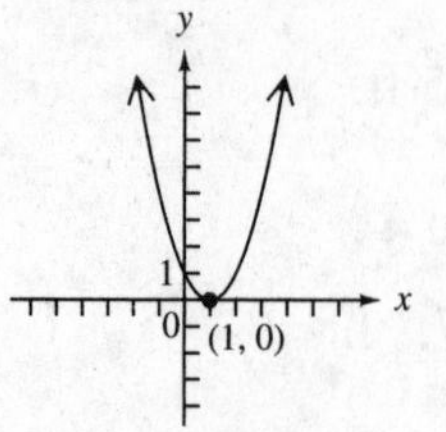

Vertex: (1, 0)
Axis: $x = 1$
Domain: $(-\infty, \infty)$
Range: $[0, \infty)$

13. 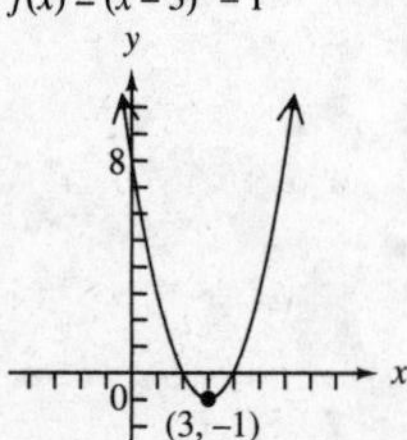

Vertex: (3, −1)
Axis: $x = 3$
Domain: $(-\infty, \infty)$
Range: $[-1, \infty)$

15. 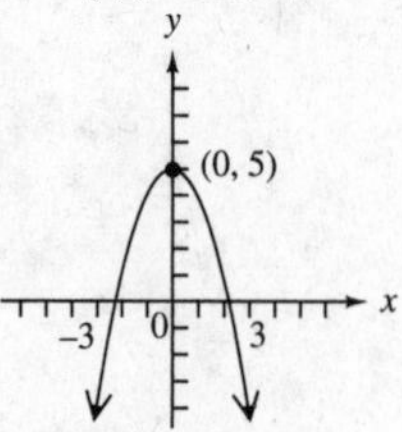

Vertex: (0, 5)
Axis: $x = 0$
Domain: $(-\infty, \infty)$
Range: $(-\infty, 5]$

Objective 3

17. down; narrower; vertex: (0, 0); domain: $(-\infty, \infty)$; range: $(-\infty, 0]$

19. up; wider; vertex: (0, 5); domain: $(-\infty, \infty)$; range: $[-5, \infty)$

21. down; narrower; vertex: (−1, 0); domain: $(-\infty, \infty)$; range: $[0, \infty)$

23. up; narrower; vertex: (1, 7); domain: $(-\infty, \infty)$; range: $[7, \infty)$

Objective 4

25. quadratic; negative 27. quadratic; positive

29. (a)

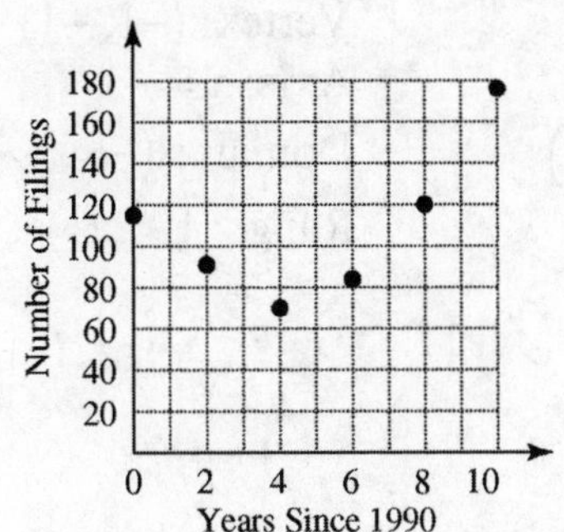

(b) quadratic; positive

(c) $y = 2.969x^2 - 23.125x + 115$ (d) 265

10.7 More about Parabolas and Their Applications

Key Terms

1. discriminant
2. vertex

Objective 1; Objective 2

1. 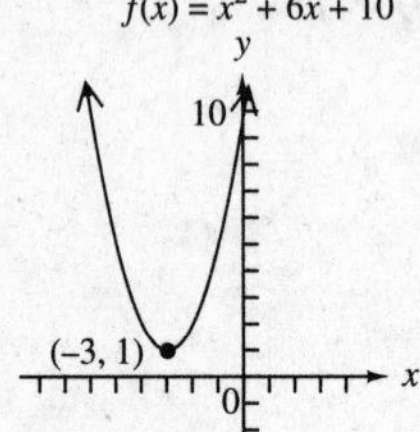

Vertex: (−3, 1)
Axis: $x = -3$
Domain: $(-\infty, \infty)$
Range: $[3, \infty)$

3. $f(x) = -x^2 + 8x - 10$

y
(4, 6)
0
x

Vertex: (4, 6)
Axis: $x = 4$
Domain: $(-\infty, \infty)$
Range: $(-\infty, 6]$

5. 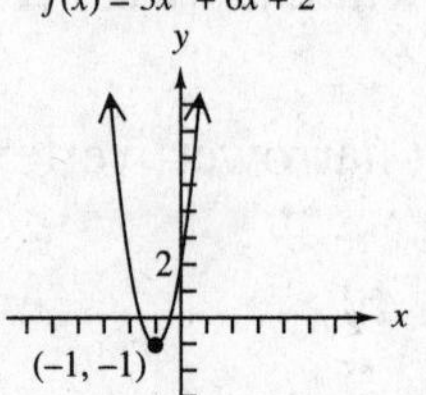

Vertex: (−1, −1)
Axis: $x = -1$
Domain: $(-\infty, \infty)$
Range: $[-1, \infty)$

7. 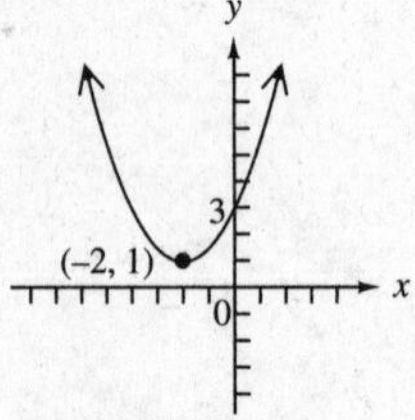

Vertex: (−2, 1)
Axis: $x = -2$
Domain: $(-\infty, \infty)$
Range: $[1, \infty)$

9. 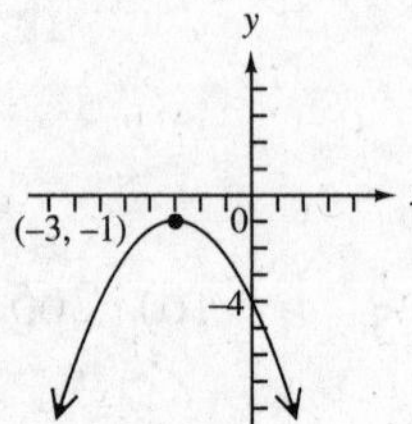

Vertex: (−3, −1)
Axis: $x = -3$
Domain: $(-\infty, \infty)$
Range: $(-\infty, -1]$

11. 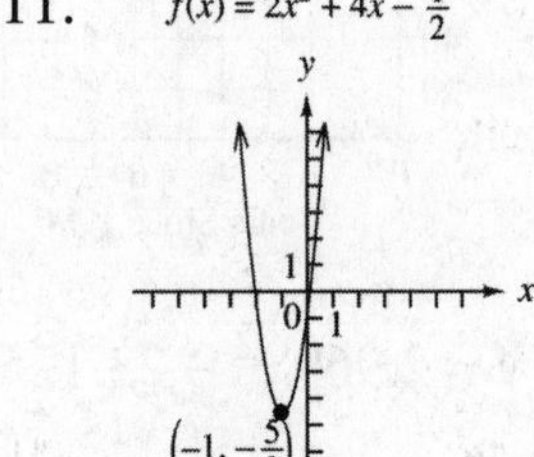
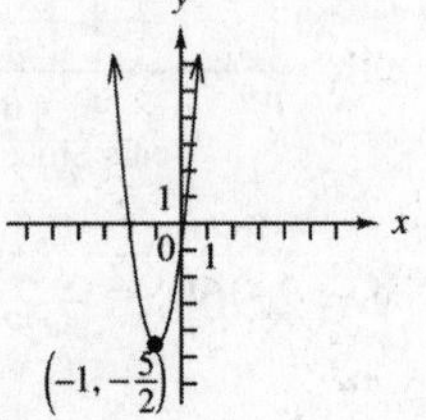

Vertex: $\left(-1, -\frac{5}{2}\right)$
Axis: $x = -1$
Domain: $(-\infty, \infty)$
Range: $\left[-\frac{5}{2}, \infty\right)$

Objective 3

13. 0
15. 1
17. 1

Objective 4

19. 25 units; \$3650
21. 50 pots; \$200
23. 256 ft; $\frac{5}{2}$ sec

Objective 5

25. 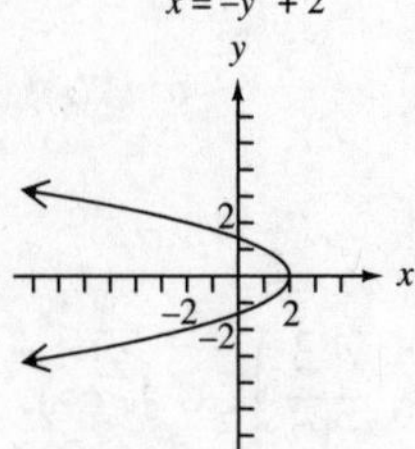

Vertex: (2, 0)
Axis: $y = 0$
Domain: $(-\infty, 2]$
Range: $(-\infty, \infty)$

27. 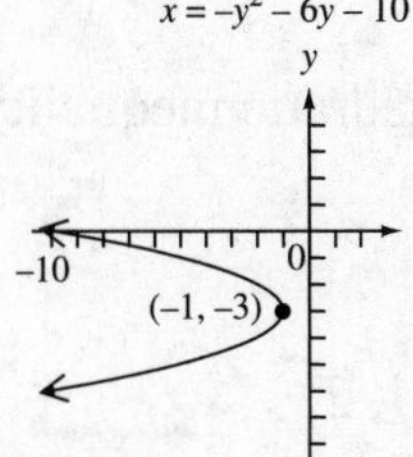

Vertex: (−1, −3)
Axis: $y = -1$
Domain: $(-\infty, -3]$
Range: $(-\infty, \infty)$

29. 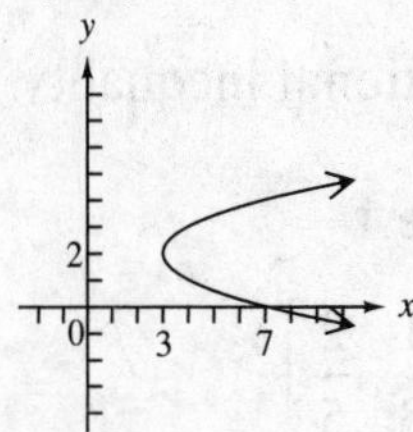

Vertex: (3, 2))
Axis: $y = 2$
Domain: $[3, \infty)$
Range: $(-\infty, \infty)$

10.8 Polynomial and Rational Inequalities

Key Terms

1. rational inequality
2. quadratic inequality

Objective 1

1. $\left[\frac{1}{3}, \frac{2}{5}\right]$

3. $[-1, 2]$

5. $\left(-\infty, -\frac{3}{2}\right) \cup (4, \infty)$

7. $\left(-\infty, -\frac{3}{2}\right) \cup \left(\frac{1}{4}, \infty\right)$

9. $(-\infty, \infty)$

Objective 2

11. $(-\infty, -4] \cup [-1, 2]$

13. $(-\infty, -5] \cup [-3, 1]$

15. $(-\infty, -3] \cup [-1, 4]$

17. $\left(-\infty, -\frac{3}{2}\right] \cup \left[-\frac{1}{3}, \frac{1}{2}\right]$

19. $\left[\frac{3}{4}, \frac{10}{3}\right] \cup \left[\frac{7}{2}, \infty\right)$

Objective 3

21. $(-\infty, 1) \cup [8, \infty)$

23. $\left(-\infty, -\frac{19}{9}\right] \cup \left(-\frac{5}{3}, \infty\right)$

25. $\left[0, \frac{3}{4}\right)$

27. $(-\infty, -2] \cup \left(-\frac{1}{3}, \infty\right)$

29. $[-2, 3)$

Chapter 11 INVERSE, EXPONENTIAL, AND LOGARITHMIC FUNCTIONS

11.1 Inverse Functions

Key Terms

1. one-to-one function
2. inverse of a function f

Objective 1

1. $\{(-1, -3), (0, -2), (1, -1), (2, 0)\}$
3. not one-to-one
5. $\{(0, 0), (1, 1), (-1, -1), (2, 2), (-2, -2)\}$
7. not one-to-one

Objective 2

9. not one-to-one
11. not one-to-one
13. one-to-one

Objective 3

15. $f^{-1}(x) = \dfrac{x+5}{3}$
17. $f^{-1}(x) = -\dfrac{\sqrt{2-2x}}{2}, \; x \le 1$
19. $f^{-1}(x) = \dfrac{x^2}{12}, \; x \ge 0$
21. not one-to-one

Objective 4

23. 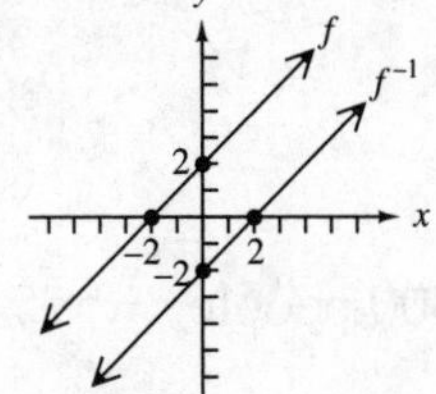

25. not one-to-one

27. 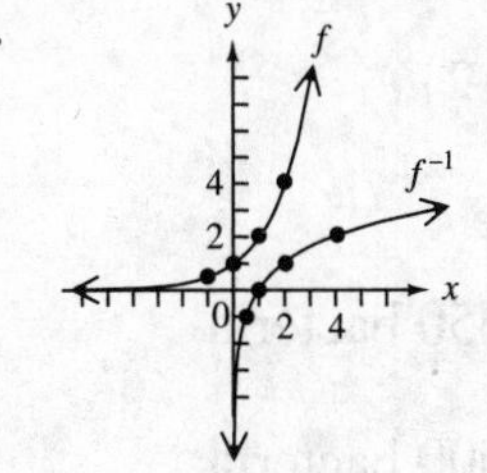

29. 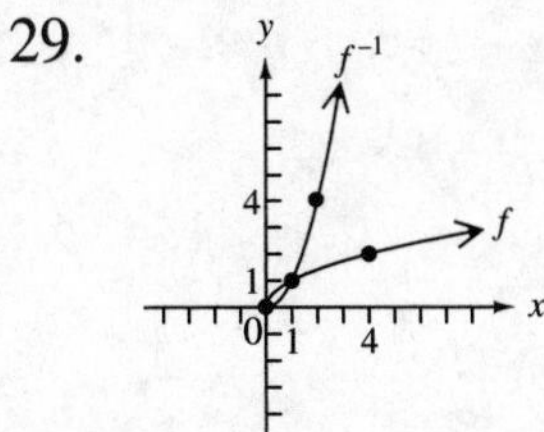

11.2 Exponential Functions

Key Terms

1. inverse
2. exponential equation

Objective 1

1. exponential function
3. not an exponential function
5. not an exponential function

Objective 2

7. 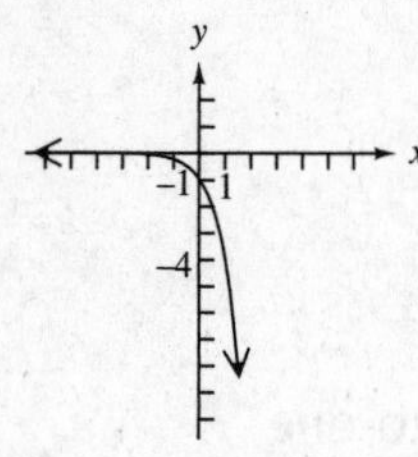

9. 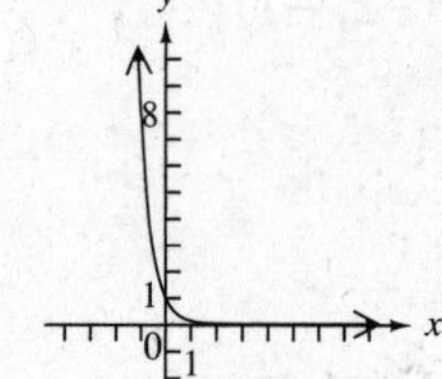

11. 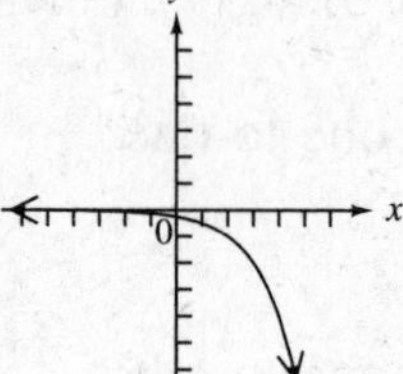

13. 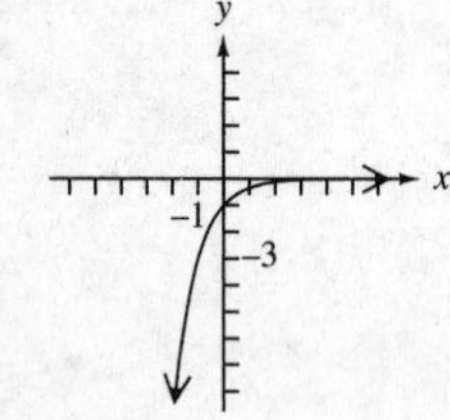

Objective 3

15. $\left\{\frac{3}{4}\right\}$
17. $\left\{\frac{3}{4}\right\}$
19. $\left\{-\frac{1}{2}\right\}$
21. $\{-2\}$

Objective 4

23. 3650 bacteria
25. \$27,048.14
27. 35,000 people
29. 1000 bacteria

11.3 Logarithmic Functions

Key Terms

1. logarithm
2. logarithmic equation

Objective 1

1. 2
3. −4
5. $\frac{1}{2}$

Objective 2

7. $10^{-3} = 0.001$
9. $4^{-2} = \frac{1}{16}$
11. $\log_3 9 = 2$

Objective 3

13. {6}
15. {81}
17. {16}

Objective 4

19.

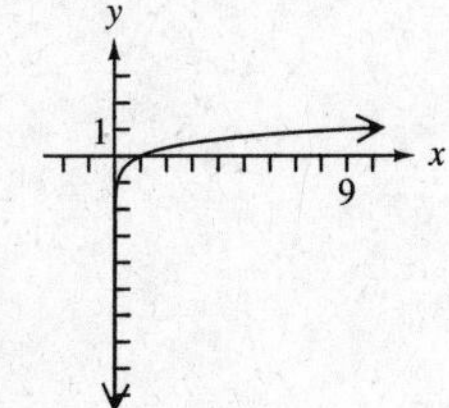

21.

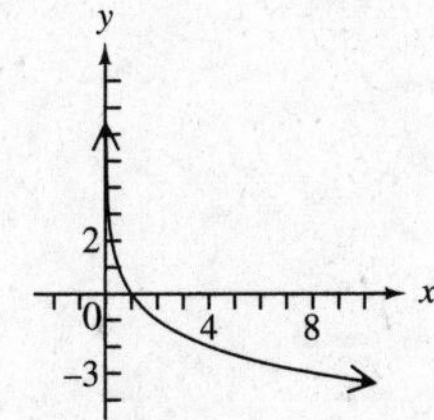

23.

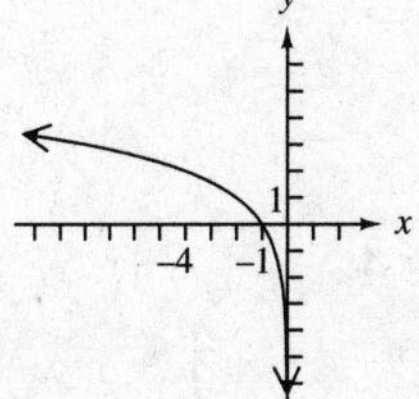

Objective 5

25. $100,000
27. 239 foxes
29. 335,000 items

11.4 Properties of Logarithms

Key Terms

1. logarithmic function
2. base; exponent

Objective 1

1. $\log_3 6 + \log_3 5$
3. $\log_2 6 + \log_2 x + \log_2 y$
5. $\log 12$
7. $\log_9 120r^5$

Objective 2

9. $\log_6 k - \log_6 3$
11. $\log_2 \frac{7q^2}{5}$
13. $\log_7 \frac{3}{5r^4}$

Objective 3

15. $3\log_3 4$
17. $\frac{1}{2}\log_b 5$
19. 2
21. 1

Objective 4

23. $\frac{1}{4}\log_5 3 + \frac{1}{4}\log_5 p$
25. $\log_4 3 + \log_4 m - \log_4 (m+2)$
27. $\log_5 \frac{1}{m^4}$ or $-\log_5 m^4$
29. 0

11.5 Common and Natural Logarithms

Key Terms

1. natural logarithm
2. common logarithm

Objective 1

1. 1.7576
3. −1.0390
5. 4.9401
7. 0.6022

Objective 2

9. 8.8
11. 6.3×10^{-11}
13. 5.0×10^{-2}
15. 85 dB

Objective 3

17. 4.3347
19. 6.7731
21. 3.9120

Objective 4

23. 23 yr
25. 11 yr
27. 2018 yr
29. $t = 4.6$ hr; time of death: 7:24 pm

11.6 Exponential and Logarithmic Equations; Further Applications

Key Terms

1. quotient rule for logarithms
2. product rule for logarithms
3. power rule for logarithms

Objective 1

1. {−0.631}
3. {1}
5. {1}

Objective 2

7. {1000}
9. {3}
11. $\left\{\frac{9}{8}\right\}$

Objective 3

13. 7.3 yr
15. 7.7 yr
17. $68,098.05

Objective 4

19. 8.25 g
21. 264 g
23. 11.3 mg

Objective 5

25. 1.5
27. −6.2288
29. −3.9694

Chapter 12 NONLINEAR FUNCTIONS, CONIC SECTIONS, AND NONLINEAR SYSTEMS

12.1 Additional Graphs of Functions; Operations and Composition

Key Terms

1. step function
2. composition; composite function
3. greatest integer function
4. asymptotes

Objective 1

1. $f(x) = |x-2| + 3$

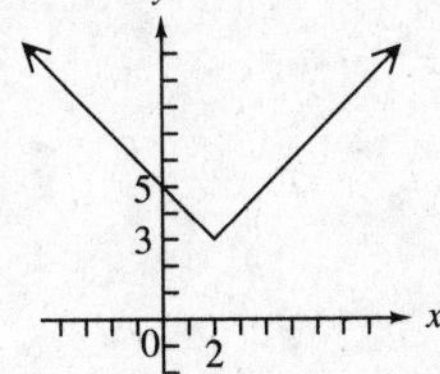

Domain: $(-\infty, \infty)$
Range: $[3, \infty)$

3. $f(x) = \frac{1}{x-1}$

Domain:
$(-\infty, 1) \cup (1, \infty)$
Range:
$(-\infty, 0) \cup (0, \infty)$

5. $f(x) = \frac{1}{x} + 3$

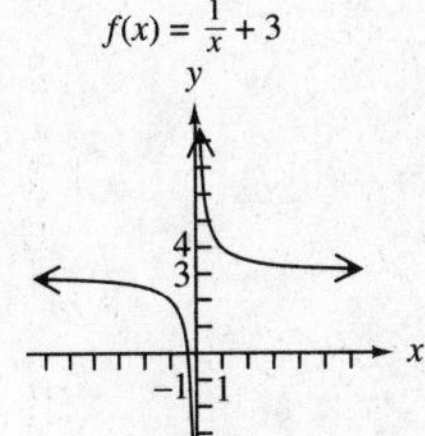

Domain: $(-\infty, 0) \cup (0, \infty)$
Range: $(-\infty, 3) \cup (3, \infty)$

7. $f(x) = \frac{1}{x-3}$

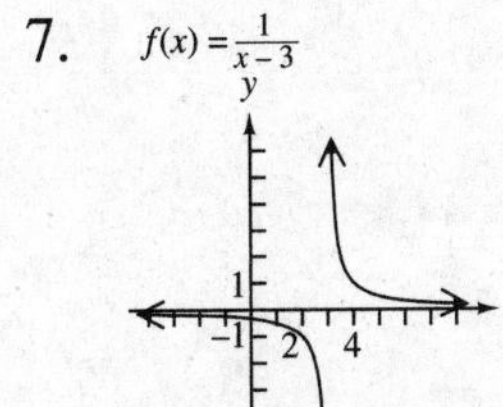

Domain: $(-\infty, 3) \cup (3, \infty)$
Range: $(-\infty, 0) \cup (0, \infty)$

Objective 2

9.

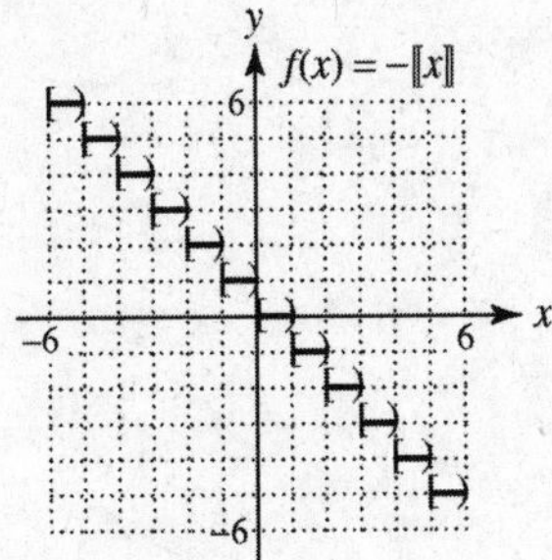

11.

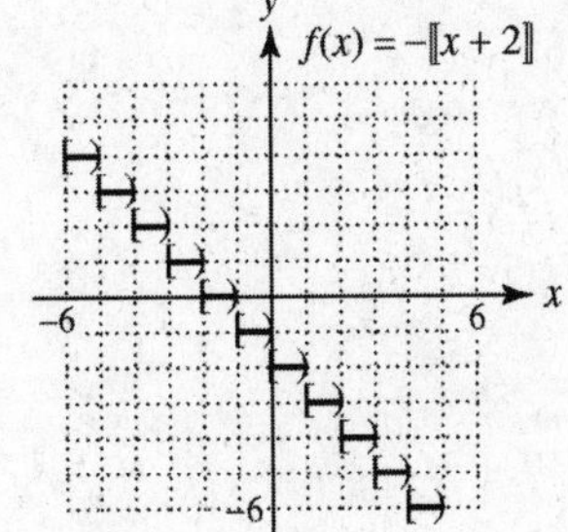

13.

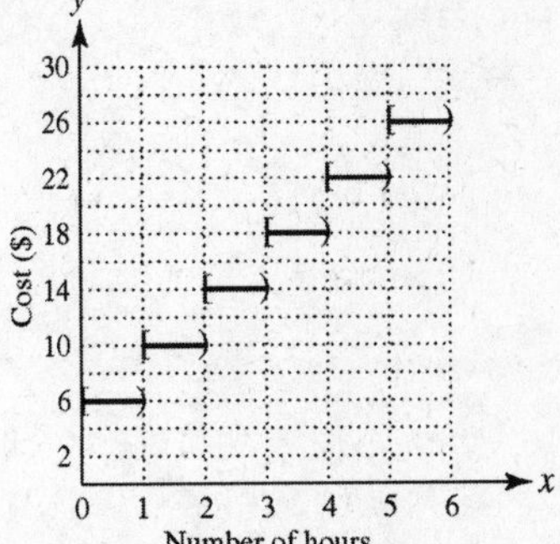

Objective 3

15. (a) $3x^2 - 6x + 21$; (b) $9x^2 - 8x + 3$

17. -5

19. 56

21. undefined

Objective 4

23. 38

25. $(3x+2)^2 + 3$ or $9x^2 + 12x + 7$

27. $(x+4)^2 + 3$ or $x^2 + 8x + 19$

29. 7

12.2 The Circle and the Ellipse

Key Terms

1. circle
2. ellipse
3. conic sections
4. center
5. radius

Objective 1

1. $(x+3)^2+(y-2)^2=25$
3. $(x-3)^2+(y+4)^2=25$
5. $x^2+(y-3)^2=2$
7. $(x-7)^2+(y-1)^2=4$

Objective 2

9. center: (5, –6); radius: 3
11. center: (3, –2); radius: $\frac{3}{2}$
13. center: (–4, –2); radius: 7
15. 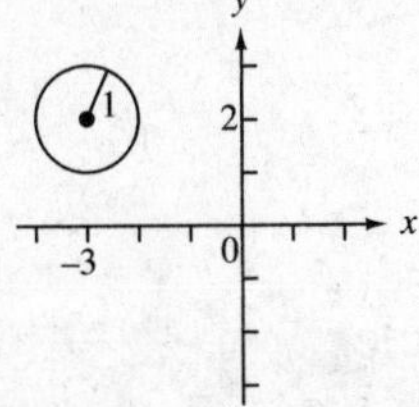

center: (–3, 2)
radius: 1

Objective 3; Objective 4

17. 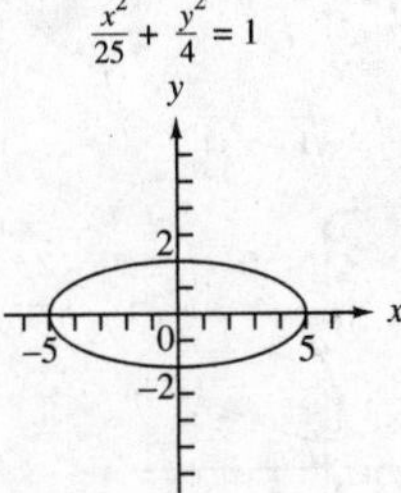

19. 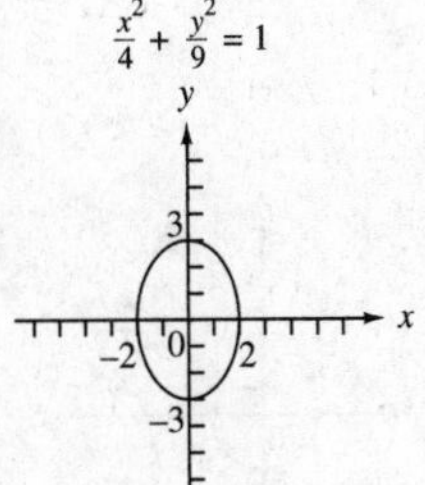

21. 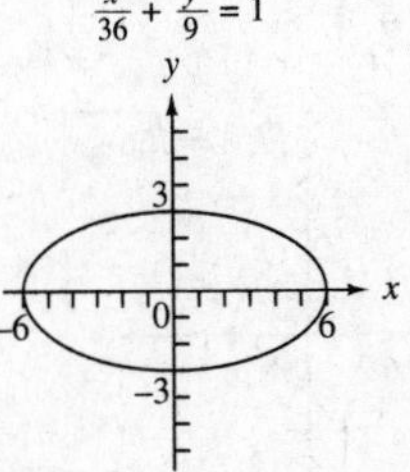

23. 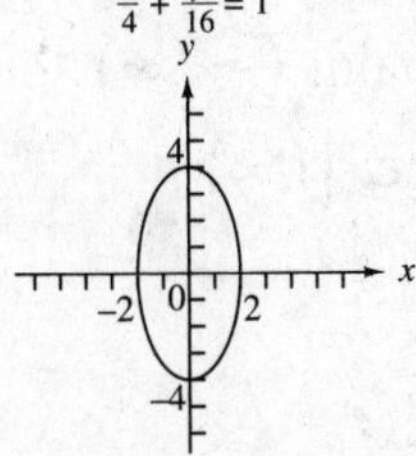

25. 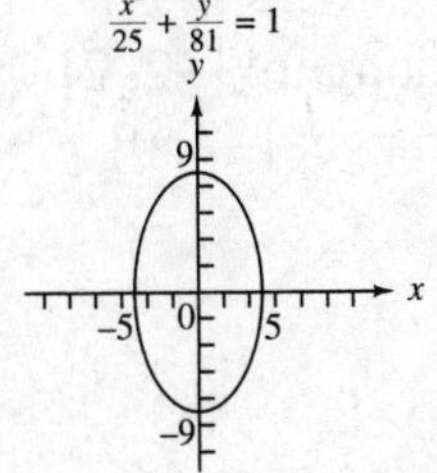

27. (a) $184\text{ m}\times 140\text{ m}$; (b) yes
29. 7.8 ft

12.3 The Hyperbola and Other Functions Defined by Radicals

Key Terms

1. hyperbola
2. asymptotes of a hyperbola
3. fundamental rectangle

Objective 1; Objective 2

1. 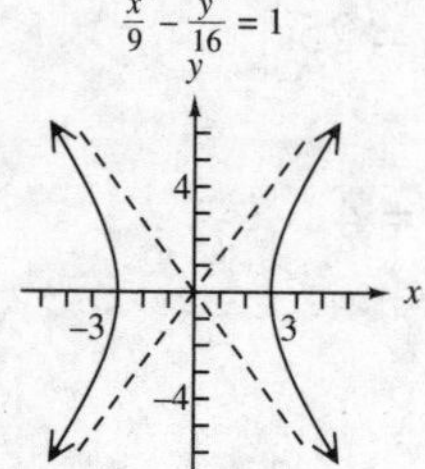

3. 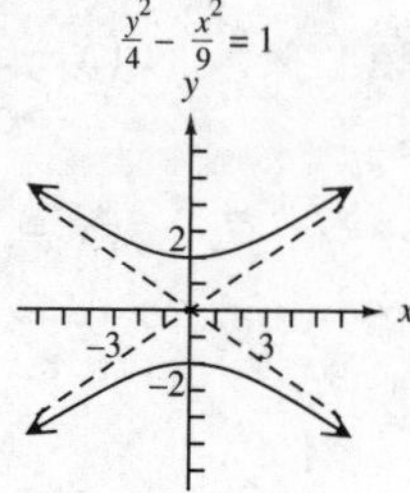

5. 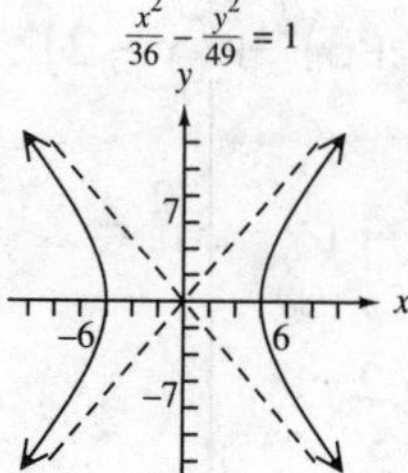

7. 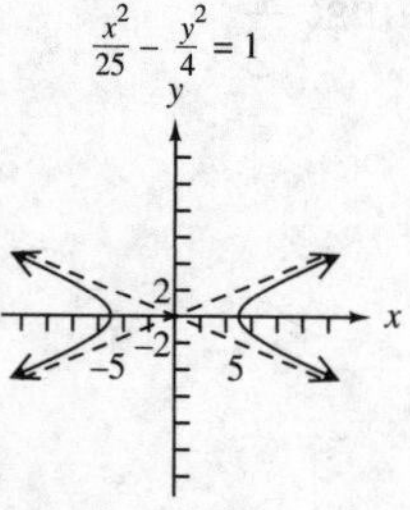

Objective 3

9. parabola
11. hyperbola
13. circle
15. circle
17. parabola
19. circle
21. hyperbola

Objective 4

23. 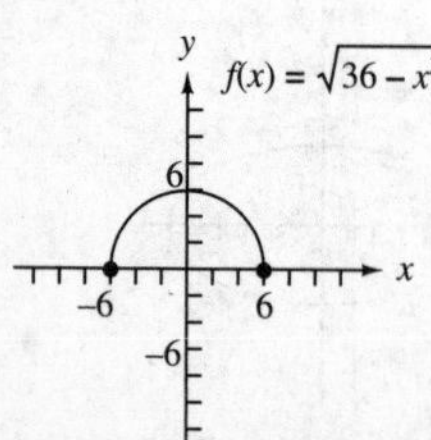

domain: [−6, 6]
range: [0, 6]

25. 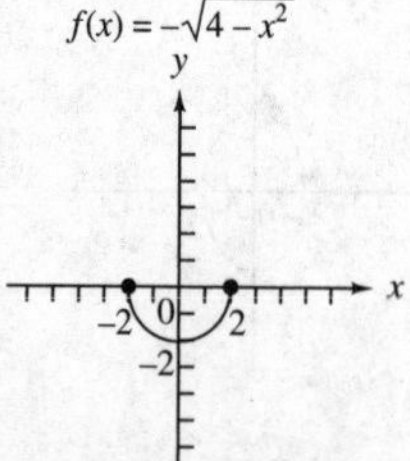

domain: [−2, 2]
range: [−2, 0]

27. 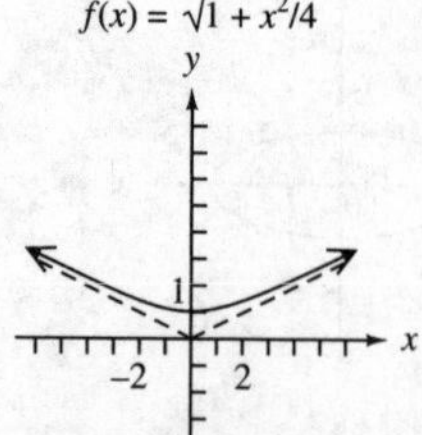

domain: $(-\infty, \infty)$
range: $[1, \infty)$

29. 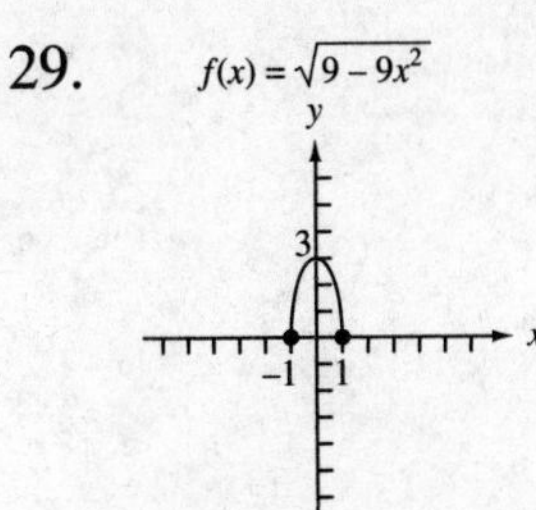

domain: [−1, 1]
range: [0, 3]

12.4 Nonlinear Systems of Equations

Key Terms

1. nonlinear equation
2. nonlinear system of equations

Objective 1

1. $\left\{(4, -1), \left(\frac{16}{5}, -\frac{13}{5}\right)\right\}$
3. $\left\{\left(\frac{1}{4}, \frac{3}{2}\right), (-1, 1)\right\}$
5. $\{(5, 2), (-1, -4)\}$
7. $\{(3, -2), (-2, 3)\}$
9. $\left\{\left(\frac{5}{2}, -4\right), (2, -5)\right\}$

Objective 2

11. $\{(1, 3), (1, -3), (-1, 3), (-1, -3)\}$
13. $\{(2, 3), (2, -3), (-2, 3), (-2, -3)\}$
15. $\{(3, 1), (3, -1), (-3, 1), (-3, -1)\}$
17. $\{(1, 1), (1, -1), (-1, 1), (-1, -1)\}$
19. $\{(2, 1), (2, -1), (-2, 1), (-2, -1)\}$

Objective 3

21. $\{(6, 1), (1, 6), (-6, -1), (-1, -6)\}$
23. $\left\{(1, 1), (-1, -1), \left(\sqrt{3}, \frac{\sqrt{3}}{3}\right), \left(-\sqrt{3}, -\frac{\sqrt{3}}{3}\right)\right\}$
25. $\{(2, -3), (-2, 3), (3, -2), (-3, 2)\}$
27. $\left\{(3, -2), (-3, 2), \left(\frac{2\sqrt{6}}{3}, -\frac{3\sqrt{6}}{2}\right), \left(-\frac{2\sqrt{6}}{3}, \frac{3\sqrt{6}}{2}\right)\right\}$
29. $\{(2, 2), (-2, -2), (2i, -2i), (-2i, 2i)\}$

12.5 Second-Degree Inequalities and Systems of Inequalities

Key Terms

1. system of inequalities. 2. second-degree inequality

Objective 1

1.

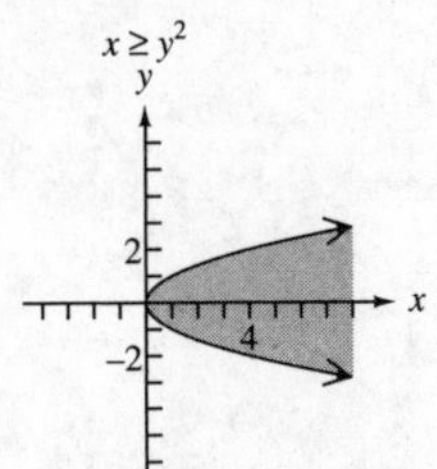

3.

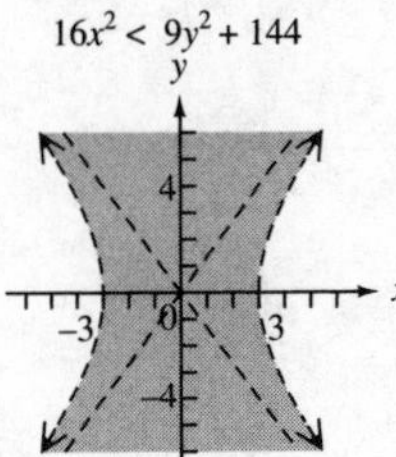

5.

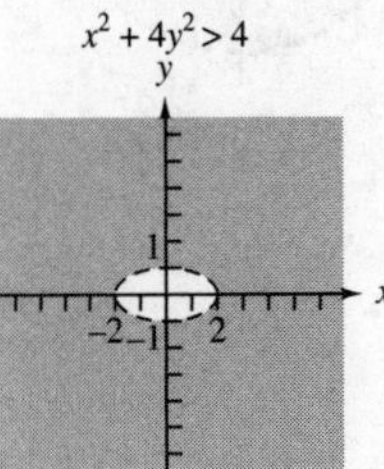

7.

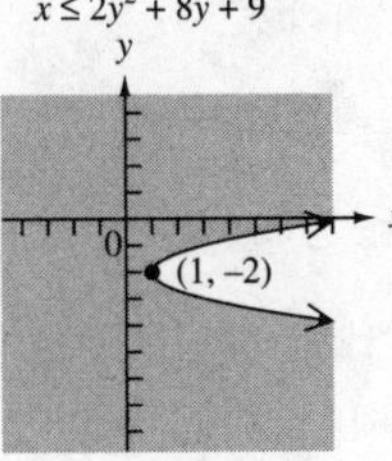

9.

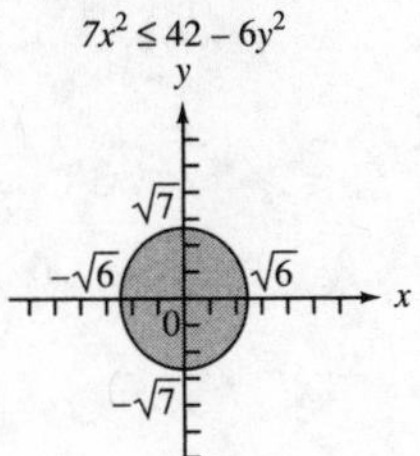

11.

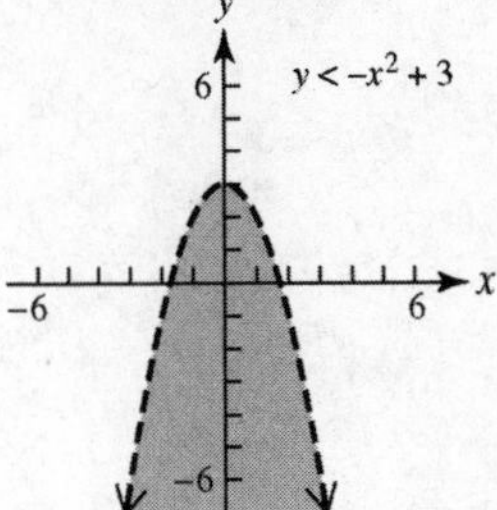

13.

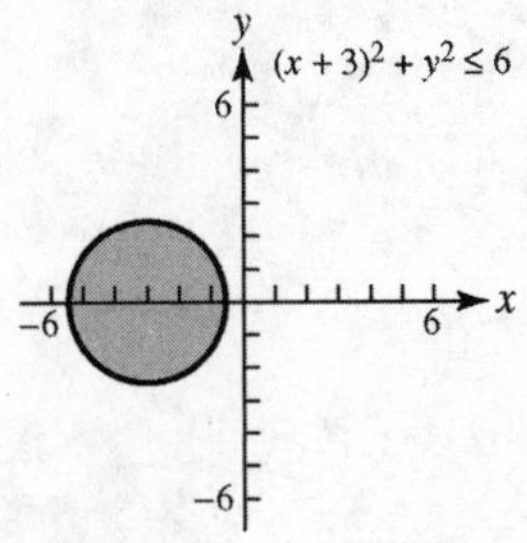

15.

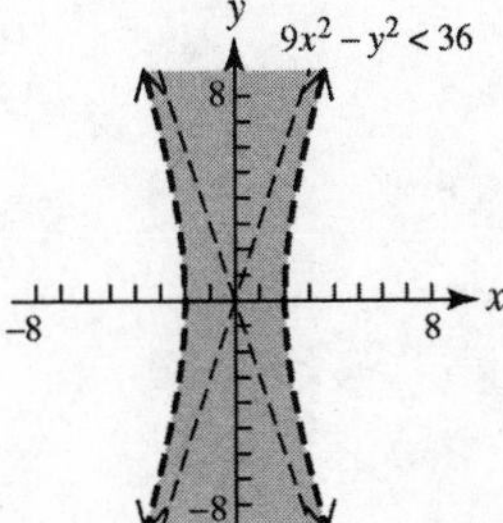

Objective 2

17.

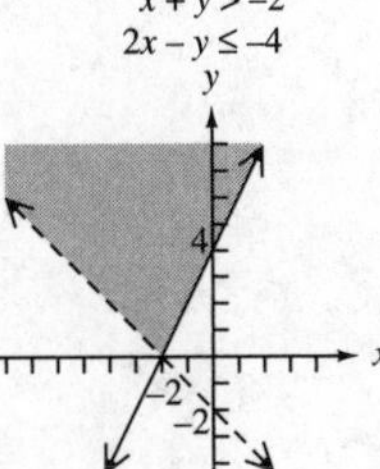

19.

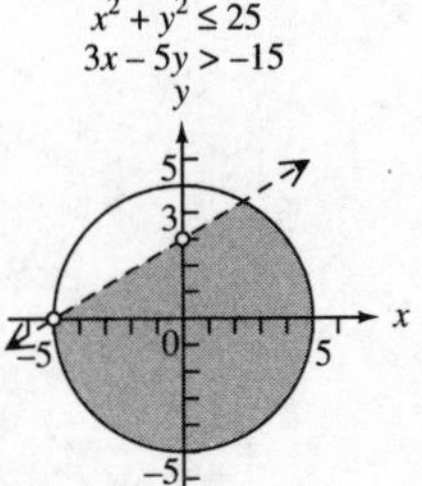

21.

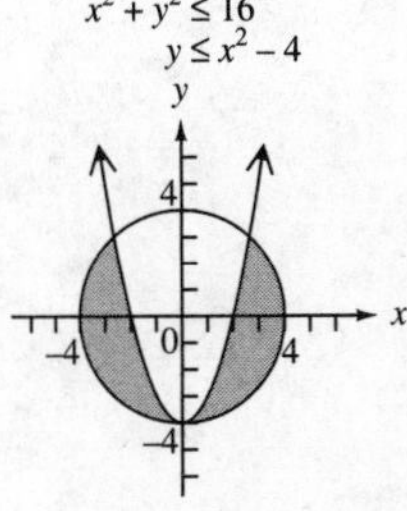

23.

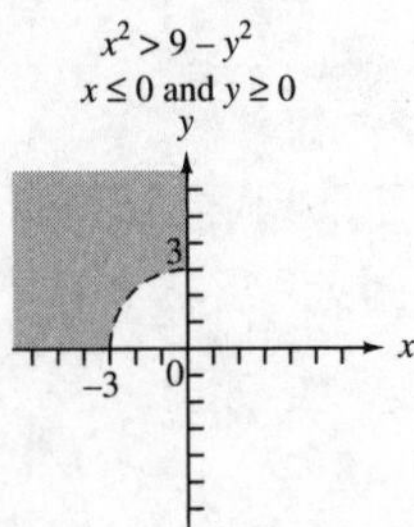

25.

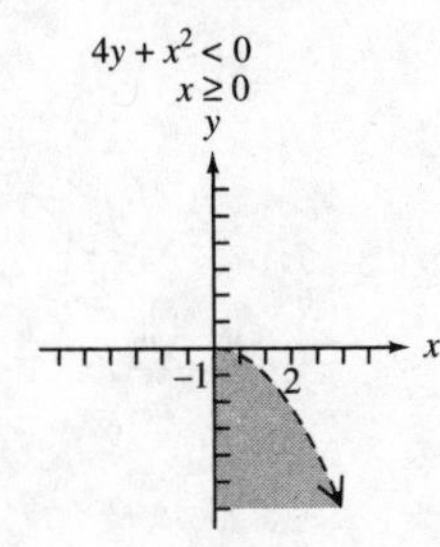

27.

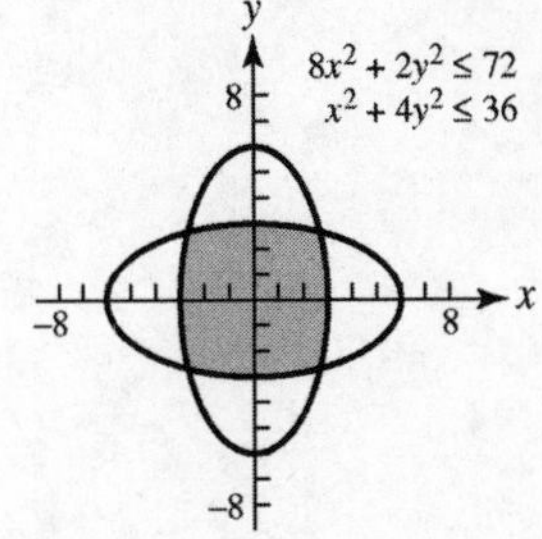

29. 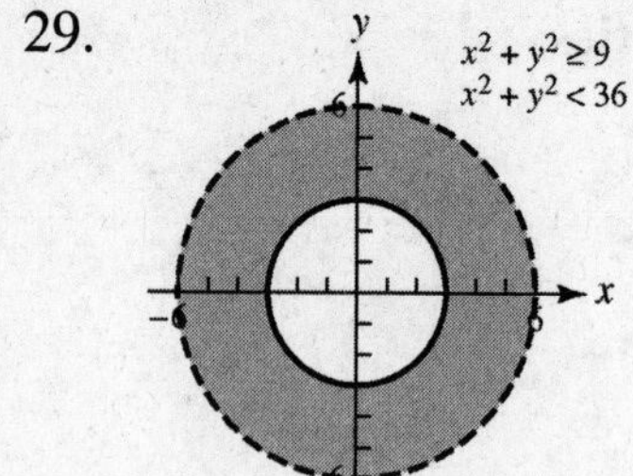